D0138218

כתבי האקדמיה הלאומית הישראלית למדעים

PUBLICATIONS OF THE ISRAEL ACADEMY
OF SCIENCES AND HUMANITIES

SECTION OF SCIENCES

THE GENUS TAMARIX

THE GENUS TAMARIX

BY

BERNARD R. BAUM

JERUSALEM 1978

THE ISRAEL ACADEMY OF SCIENCES AND HUMANITIES

Printed in Israel
Set at Jerusalem Academic Press
Plates by Emil Pikovsky Ltd.
Printed at Central Press, Jerusalem

"... en effet, l'auteur systématique ayant le choix parmi tous les caractères des plantes,... serait bien maladroit s'il ne choisissait des organes tres apparents et faciles à voir pour base de sa classification, tandis que l'auteur d'une méthode naturelle n'a pas la liberté du choix; il est conduit par des principes rigoureux à observer tous les organes, et à donner à chacun une importance relative, non à la facilité que nous avons de la voir, mais au rôle que cet organe joue dans la vie des êtres: or, ces organes les plus importants peuvent être, et sont souvent en effet, les plus difficiles à voir."

DC., Théorie Elémentaire de la Botanique, 52–53, 1819

PREFACE

TAXONOMICALLY *Tamarix* is one of the most difficult genera among the Angiosperms, partly because its members show few distinctive external features. The floral morphology does include many characteristics which can be used for interspecific delimitation, but they can only be discerned by meticulous examination under a powerful lens. Awkward dissections of the tiny flowers are often indispensable; for dissection part of the raceme or single flower must be soaked in hot, almost boiling, water, and sometimes clarification with lactophenol (Strasburger, 1923, p. 783) is also necessary, especially for study of the androecium.

The present study was based partly on living material observed in France, Turkey, Iran and Israel, but most of the examinations were made on dry herbarium material. The names of the herbaria mentioned in the text are abbreviated according to the Lanjouw and Stafleu Index Herbariorum.

Authors' names are written mainly according to Kirpicznikov (1962).

For citation of periodicals and journals Schwarten & Rickett (1958 and 1961) have been consulted.

For publication dates Steenis-Kruseman & Stearn (1954), Rickett & Stafleu (1959, 1960, 1961), Stearn (unpublished personal communications) and other sources, including Stearn and Sherborn, have been consulted.

For morphological terminology the author has followed the suggestions advanced in *Descriptive Biological Terminology* published in *Taxon*, 9:245–257, 1960; 11: 145–156, 1962.

I was only able to detect and confirm the nomenclatural types after consulting such texts as: de Candolle (1880), Clokie (1964), Deleuze (1823), The History of Collections Contained in the Natural History Department of the British Museum (1904), Urban (1917) and others, and the special accounts of Buchenau (1868) on Herb. Roth, Boulger (1897) on the Roxburgh collection, Gomes (1868) on the Loureiro specimens, Juel (1918) on Hasselquist's collection, Lambert (1811) on the Pallas Herbarium, Stansfield (1953) on Royle's types, Stephen (1888) on the Arnott Herbarium and others. The collections of autographs of the British Museum, Muséum National d'Histoire Naturelle and Naturhistorisches Museum have been especially useful in this task.

ACKNOWLEDGEMENTS

The author is greatly indebted to the directors and curators of the following herbaria for putting their collections at his disposal either by sending him material on loan or by rendering him facilities during his visits to the herbaria:

Aaron Aaronsohn Herbarium, Zikhron-Ya'aqov, Israel (AAR).
Ankara Universitesi Fen Fakültesi Botanik Enstitüsü, Ankara, Turkey (ANK).
Botanisches Museum, Berlin, Germany (B).
Universitetets Botaniske Museum, Bergen, Norway (BG).
British Museum (Natural History), London, Great Britain (BM).
Department of Botany, Cairo University, Egypt (CAI).
Botany School, University of Cambridge, Great Britain (CGE).
Forest Research Institute and Colleges, Dehra Dun, India (DD).
Department of Botany, University of Delhi, India (DEL).
Royal Botanic Garden, Edinburgh, Great Britain (E).
The East African Herbarium, Nairobi, Kenya (EA).
Forest Herbarium, Department of Forestry, Commonwealth Forestry Institute, University of Oxford, Great Britain (FHO).
Herbarium Universitatis Florentinae, Istituto Botanico, Florence, Italy (FI).
Conservatoire et Jardin botaniques, Geneva, Switzerland (G).
Department of Botany, University of Glasgow, Great Britain (GL).
Staatsinstitut für allgemeine Botanik und Botanischer Garten, Hamburg, Germany (HBG).
Department of Botany, Hebrew University of Jerusalem, Israel (HUJ).
The Herbarium and Library, Kew, Richmond, Surrey, Great Britain (K).
Rijksherbarium, Leiden, Netherlands (L).
Herbarium of the Komarov Botanical Institute of the Academy of Sciences of the USSR, Leningrad, USSR (LE).
The Linnean Society of London, Great Britain (LINN).
Herbiers de la Faculté des Sciences de Lyon, France (LY).
Instituto 'Antonio Jose Cavanilles', Jardin Botanico, Madrid, Spain (MA).
Institut de Botanique, Université de Montpellier, France (MPU).
Fielding Herbarium, Druce Herbarium, Department of Botany, Oxford, Great Britain (OXF).
Muséum National d'Histoire Naturelle, Laboratoire de Phanérogamie, Paris, France (P).
Muséum National d'Histoire Naturelle, Laboratoire de Biologie Végétale Appliquée, Paris, France (PCU).
Herbarium of Pomona College, Claremont, California, USA (POM).
Universitatis Carolinae Facultatis Biologicae Scientiae Cathedra (Institutum Botanicum Universitatis Carolinae), Prague, Czechoslovakia (PRC).
Botanical Research Institute, National Herbarium, Pretoria, South Africa (PRE).
Institut Scientifique Chérifien, Laboratoire de Phanérogamie et Laboratoire de Cryptogamie, Rabat, Morocco (RAB).

Rancho Santa Ana Botanic Garden, Claremont, California, USA (RSA).
Botanical Department, Naturhistoriska Riksmuseum, Stockholm, Sweden (S).
Herbarium of Southern Methodist University, Dallas, Texas, USA (SMU).
Botanical Museum–Herbarium, Utrecht, Netherlands (U).
Herbarium of the University of California, Berkeley, California, USA (UC).
Institute of Systematic Botany, University of Uppsala, Sweden (UPS).
U. S. National Museum (Department of Botany), Washington, D. C., USA (US).
Naturhistorisches Museum, Vienna, Austria (W).
Botanisches Institut und Botanischer Garten der Universität Wien, Vienna, Austria (WU).
Herbarium Mouterde, Université St. Josef, Beirut, Lebanon (Herb. Mouterde).
Herbier Pabot, Tehran, Iran (duplicates of it are in HUJ).
Herbarium Esfandiari, Teheran, Plant Pest and Diseases Research Institute (Herb. Esfand.).

This study could not have been carried out without the warm interest and many facilities rendered to me by Prof. M. Zohary of the Hebrew University, to whom I am deeply grateful.
I am especially indebted to the librarians of the Naturhistorisches Museum, Vienna; Conservatoire Botanique, Geneva; British Museum (Natural History), Department of Botany, London; Laboratoire de Panégrogramie Muséum d'Histoire Naturelle, Paris, for their help and most valuable advice. I am also very grateful for the useful suggestions given to me by Dr W. T. Stearn of the British Museum and by Dr P. Jovet and Dr A. Lourteig of the Muséum d'Histoire Naturelle. Thanks are due to Dr H. K. Airy Shaw of the Royal Botanic Garden for his editorial work; to Dr I. Greenberg-Fertig of the Hebrew University of Jerusalem for the Latin translation in the Appendix; to Mr F. Schaefer for the drawings of the species and to my wife Danielle for the distribution maps and some other drawings.
I would also like to thank the Publication Department of the Israel Academy of Sciences and Humanities, especially Norma Schneider, for their help in preparing the manuscript for publication and seeing it through the press

CONTENTS

GENERAL PART

Historical Survey

LINNAEUS altered the name *Tamariscus* of Tournefort (*Institutiones*, p. 661, 1719) to *Tamarix*, the name used in all the editions of Linnaeus' *Genera Plantarum* and *Species Plantarum*. Linnaeus (1753) described two species of *Tamarix: T. gallica* and *T. germanica*.

Willdenow (1816) was the first monographer of this genus. He recorded seven new species in addition to the nine known in 1812.

Desvaux (1824), to whom the monograph of Willdenow was unknown, described many new species, some of which had already been published by Willdenow. Desvaux, however, separated from *Tamarix* the genus *Myricaria*, based on the name *Myrica* which Pliny used for the *Tamarix*. Desvaux' separation of the two genera was based mainly on the morphology of the androecium and of the seed apex.

Although *Myricaria* Desvaux has been accepted until now by most botanists, the delimitation of *Tamarix* and *Myricaria* has remained a matter of dispute for many years.

Ehrenberg (1827) was concerned with only part of the genus. He tried to set a clearer limit between *Tamarix* and *Myricaria*, and in so doing he established a new genus: *Hololachna*, which was based on *Tamarix songarica* Pall. Ehrenberg recognized eight species of *Tamarix* proper. He was very accurate in his observations[1] and especially so in those made on *T. gallica*, where he distinguished nine varieties, most of which were proved by later monographers to be well-discernible species. He was, however, inconsistent in his taxonomic criteria.[2]

A. P. de Candolle (1828) published his *Prodromus* shortly afterwards, and so Ehrenberg's findings were not included; most of the species in the *Prodromus* are based on Desvaux' records and some later publications, together with several species described by de Candolle for the first time.

Ledebour (together with Meyer and Bunge, 1829) cleared up the delimitation of *Tamarix* and *Myricaria*.

1 He was the first to emphasize the taxonomic value of the disk and the insertion of the stamens: 'Es findet ein regelmässiges Verhältniss zwischen den Randzähnen der Drüse und der Zahl und Stellung der Staubfäden statt, welches als Eintheilungsgrund benutzt werden kann'; Ehrenberg, *op. cit.*, pp. 251–252.

2 'Mir scheint es: dass nur die Samen-Kapel ein constantes Merkmal bietet um mehrere dieser, ähnlichen Formen unter einem festem Gesichtspunkt zu fassen . . .'; Ehrenberg, *op. cit.*, pp. 256–257.

Wight and Walker-Arnott (1834) excluded *T. ericoides* Rottler from the genus *Tamarix* and founded the new genus *Trichaurus*.

Decaisne (1835) is worth mentioning because of his clear observations on some local species.[3] He was indeed the first to use the configuration of the disk as a species characteristic.

Webb (1841) continued investigations into the androecium of *Tamarix* and evaluated its diagnostic importance for several species. *T. gallica* L. was, however, erroneously interpreted by him.

Bunge (1852) was the most eminent monographer of the genus *Tamarix*. Though he had relatively little material at hand, his clear observations (as seen from his notes on herbarium specimens) and his very accurate descriptions of the species cast a new light on this genus on a world-wide scale. Almost all the species described up to 1852 were dealt with in his monograph, and most of them were treated critically. The number of species accepted by Bunge was 51, including several varieties, from a total of 178 binomials and trinomials (including misidentifications also put into the list of synonyms). Twenty-one of the species were described by him[4] for the first time. It is unfortunate that, like many of his predecessors, he made extensive use of the position of the racemes (whether *vernales*, produced by old branches, or *aestivales*, produced by the current year's branches), a feature which has recently been shown by Baum (1964) to be diagnostically unreliable.

J. Gay (about 1852–1853) made useful observations on this genus on a world-wide scale. These were never published, but the manuscript containing his comments on *Tamarix* can be found in the library of the Kew Herbarium.

Niedenzu (late 1895) coined new terms for the structure of the floral disk as observed by Bunge; in other respects he faithfully followed Bunge's concepts. In his monograph some new species were added to give a total of 67, as against the 64 species given by Niedenzu himself (early in 1895 in Engler & Prantl's *Natürlichen Pflanzenfamilien*). Later, in his revised account of the genus for the second edition of Engler & Prantl's *Natürlichen Pflanzenfamilien*, Niedenzu increased the number of accepted species to 78.

Arendt (1926), the last monographer of *Tamarix*, reduced the number of accepted species to 64. His classification does not use Bunge's concepts; even the disk character is rejected for classification purposes, and the delimitation of species is based mainly on the relative length of bract and pedicel. His work failed to contribute towards a better understanding of the genus *Tamarix*.

Some local revisions which are worthy of mention are: Vierhapper (1907), Pau (mainly 1927), Sennen (1928–1933), Maire (1931–1940), Gorschkova (1937 and 1949), Corti (1942), Rusanov (1949), and Zohary (1956).

3 '. . . l'espéce que je rapporte au *T. mannifera* Ehrenberg de même que dans le *T. orientalis* Forsk. ce disque est à cinq lobes tronqués, entre les intervalles desquels s'insèrent les filets des etamines. Dans le *T. gallica, africana*, etc. . . . ce sont les lobes mêmes du disque qui vont s'atténuant et forment les filets nectarifères'; Decaisne, *op. cit.*, p. 261.

4 Some of the species had already been published by Bunge.

Morphological Features and their Diagnostic Value

The Root

Tamarix is usually deep-rooted (Thomas, 1921; Zohary & Fahn, 1952). Tap roots may reach 30 m in depth; subsuperficial side roots may reach a length of 50 m and are capable of producing adventitious buds. Adventitious roots are abundantly produced when the plant is buried by shifting dunes, and some species of *Tamarix* are therefore excellent dune binders. In *T. aphylla* large amounts of water are stored in the roots.

Trunk and Bark

Tamarisks are either true trees with a well-developed trunk (e.g., *T. aphylla*) or shrubs (e.g., *T. macrocarpa*). The bark colour varies little with the species, and only in younger branches (not in the trunk or old branches) is it of some diagnostic value in identification. The common colours are black to dark purple, brown, reddish-brown, blackish-brown, and grey. Fasciation of the branches is rather common and was observed by Buchenau as early as 1878.

Indumentum

The young green branchlets or the rachis of the racemes, may be papillose. The papillae may be dense, long and hair-like (up to 300 microns long; e.g., *T. hispida*), or very short (15 microns or less; e.g., *T. aralensis*). Accordingly, the plant is hairy or papillulose (subpapillose, according to Brunner, 1909), respectively. The occurrence of papillae (whether papillae proper or hairs) on various species was first observed by Brunner (1909). In the present study it is used for the first time as a diagnostic characteristic.

The Leaves

The leaves of *Tamarix* are usually herbaceous, small and scale-like. Five main leaf forms can easily be recognized and all are diagnostically valuable:

1. sessile with narrow base (e.g., *T. gallica*);
2. sessile with auriculate base (e.g., *T. hispida*);
3. amplexicaul (e.g., *T. aucheriana*);
4. vaginate (e.g., *T. aphylla*);
5. pseudo-vaginate, i.e., strongly amplexicaul with close pressed margins, adpressed to the branchlets along their major part, and closely resembling vaginate leaves (e.g., *T. bengalensis* or the young leaves of *T. indica*).

For the anatomy of branches, leaves and salt-secreting glands, see Brunner (1909), Brunswick (1920), Chapman (1934), Decker (1961), Fahn (1958), Gupta (1952), Klebahn (1884), Lewin & Reibenbach (1857), Marloth (1887), Maury (1888), Messeri (1938), Möller (1876), Pujiula (1942), Saint-Laurent (1932), Trabut (1926), and Volkens (1887).

Inflorescence

The inflorescence of *Tamarix* is racemose. Simple inflorescences consist of solitary racemes; compound inflorescences have many racemes, are often paniculately branched, and occur on current-year branches, either densely congested or loosely scattered on the common axis. The inflorescence is termed 'vernal' when the racemes come out directly from older or at least previous-year branches. They are 'aestival' when borne on green current-year branches. The character of the 'seasonality' of the inflorescence, whether vernal or aestival, is of no diagnostic value (Baum,[5] 1964). Until recently it was used, however, as a principal key character in *Tamarix.*

Racemes

While the width of the racemes varies somewhat from species to species, the length is more variable within the species. The width of the racemes is thus used diagnostically in the present study. Aestival racemes are usually somewhat narrower than vernal ones, and the flowers are slightly to considerably smaller in aestival racemes than in vernal ones. As to the length of the racemes, there are species with long vernal racemes and shorter aestival ones (e.g., *T. tetragyna*, *T. africana*), and vice versa (e.g., *T. aralensis*, *T. hispida*).

Bracts

The flowers of *Tamarix* are usually subtended by a single bract; in one species, *T. rosea* Bge., however, at least some flowers in each vernal raceme are subtended by 2–3 bracts.

Bracts possess several diagnostic characteristics:

1. Length of bract: In some species (e.g., *T. polystachya*) the bracts are shorter than the pedicels; in others (e.g., *T. canariensis*) they are longer than the pedicels. In certain species (e.g., *T. meyeri*) the bracts exceed the calyx or the whole flower. There are also species (e.g., *T. chinensis*) in which the lengths of the bracts differ according to their position on the raceme or in different kinds of racemes.

2. Structure: Some species have purely herbaceous bracts (e.g., *T. leptostachya*). In several species the upper part of the bract is diaphanous and the lower herbaceous (the vernal racemes of most species), and in others the bracts are altogether diaphanous (e.g., *T. kotschyi*).

3. Shape: Bracts, like leaves, appear sessile, auriculate, amplexicaul or vaginate. However, the shape of the bracts is not correlated with the shape of the leaves. *T. aphylla*, for instance, has vaginate leaves and only slightly clasping bracts.

In some species the bracts are unique in shape, such as the long acuminate bracts of *T. elongata*, which are not found in related species. The spoon-like bracts of *T. polystachya* are also almost unique.

5 To the list of previous evidence given by several authors as to the unreliability of the *vernales-aestivales* character in *Tamarix*, and cited by Baum (1964), we add Sickenberger's 1901 remark on *T. effusa* Ehrenb.: 'Cette espèce fait la transition entre les vernales et les aestivales. On pourrait la prendre pour un *T. tetragyma* pentamére.'

Sepals

The aestivation of sepals is imbricate (Fig. 1). Thus, in pentamerous calyces there are two outer and two inner sepals, and one intermediate, and in tetramerous ones there are two outer and two inner sepals only.

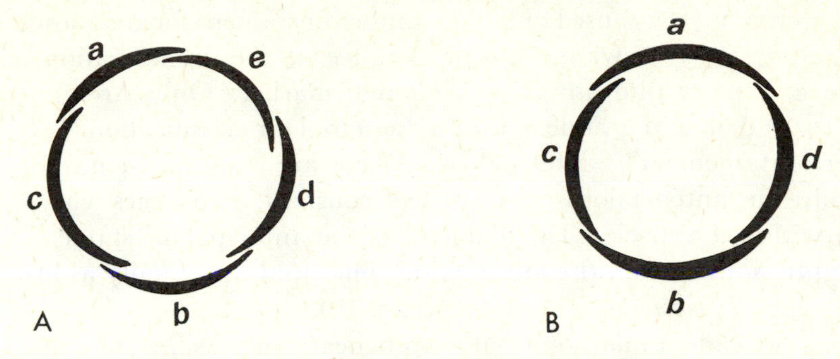

Fig. 1 Aestivation of sepals in pentamerous species (A) and in tetramerous species (B)

In some species (e.g., *T. hampeana*) meiomery of the calyx, in respect of the *e* and *d* sepals, is often found; in others meiomery is rare. The shapes and dimensions of the outer and the inner sepals may be reliably used as characteristics. The margin of the sepals, entire, dentate or incised, is also an important characteristic in identification.

Petals

The aestivation of the petals is contorted (Fig. 2).

The petals provide most useful diagnostic characteristics in the following respect:

1. Persistence: Persistent petals may remain until ripening of the fruit (e.g., *T. smyrnensis*). In subpersistent, only one or two petals may remain until maturity (e.g., *T. tetragyna*). Caducous petals are shed immediately after anthesis (e.g., *T. canariensis*).

Fig. 2 Aestivation of petals in *Tamarix*

2. Shape: The shape of the petals is usually constant. There are a few species with a wide range of variability in the petals (e.g., *T. aucheriana*). Three main types may be distinguished as to shape: (a) ovate; (b) elliptic; (c) obovate.

In several species the petals display a particular form. In *T. smyrnensis*, for instance, the petals are broadly ovate to orbicular and keeled in their lower part.

Androecium

The androecium provides the most important distinguishing characteristics in *Tamarix*. Ehrenberg (1827) used only the number of stamens for diagnostic purposes. From Decaisne (1835) onwards, the importance of the configuration of the disk came more and more into use as a taxonomic marker. Only Arendt (1926) and Rusanov (1949) denied the value of the disk as a tool for classification.

1. Number of stamens: The antesepalous stamens are constant in number in most species while the antepetalous ones are less constant, sometimes varying greatly in number within the species. The filaments of the antesepalous stamens are always at least slightly longer than the antepetalous ones and are usually a little broader at the base (e.g., *T. macrocarpa*; see Plate 51: 7a–b).

2. Insertion: A filament may come out from beneath the disk, sometimes very near its margin (hypodiscal insertion) or from the periphery of the disk (peridiscal insertion). Among the species with hypodiscal insertion of filaments are *T. smyrnensis*, *T. nilotica*, and *T. dioica*. There are, however, species with hypo-peridiscal insertion of filaments, i.e., with one or two filaments inserted hypodiscally and three or four filaments inserted peridiscally (e.g., *T. palaestina*, *T. arabica*, and *T. bengalensis*).

It is important to note that in species with hypodiscal insertion only the antesepalous filaments are hypodiscal. The difference in length between antesepalous and antepetalous stamens, the position of the stamen in relation to perigonial organs (whether epi- or ante-), and the type of filament insertion, have led to the recognition of a two-whorled androecium in the *Tamarix* flower (Zohary & Baum, 1965). The existence of a two-whorled androecium in several *Tamarix* species was already discerned by Niedenzu (1895) and Arendt (1926), and was discussed for the whole genus by Warming & Potter (1932), but without adequate detail.

3. Configuration of the disk: Three main types of disk are distinguished:

   a. Hololophic: In a hololophic disk (Fig. 3) the primary five lobes (each lying between two adjacent stamens) are apically distinct, though always connate below and either free from or concrescent with the bases of the filaments. The lobes may be entire with obtuse, retuse or truncate apex, or slightly two-lobed, equi- or unequilateral.

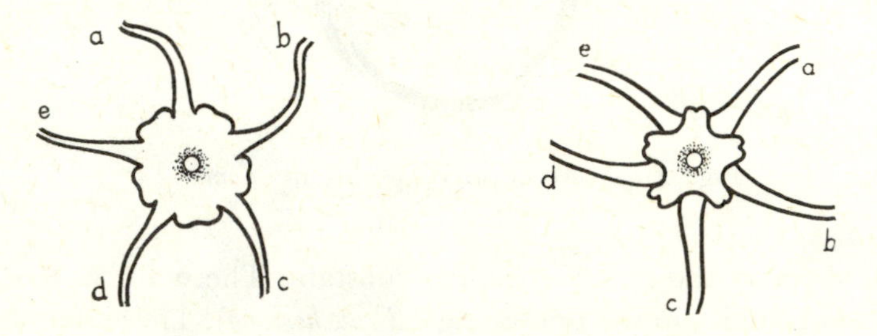

Fig. 3 Diagram of hololophic disks

b. Paralophic: In a paralophic disk (Fig. 4) the lobes are deeply bipartite and each half-lobe closely approaches the base of the adjoining filament and becomes concrescent with it.

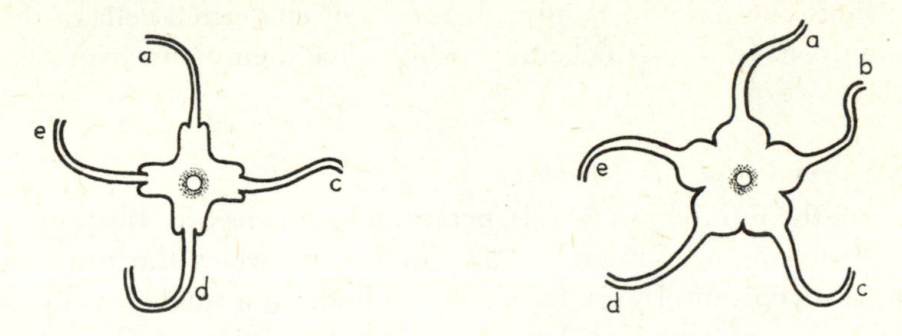

Fig. 4 Diagram of paralophic disks

c. Synlophic: In a synlophic disk (Fig. 5) the half-lobes of the bipartite discal lobes are very strongly confluent with the bases of the filaments, giving the impression that the filaments themselves have a broad base. (This broad base is, of course, formed by the connection of the filament with segments of the adjacent lobes from either side.) In other words, and better morphologically interpreted, when a paralophic disk has its lobes confluent with the base of the filaments the disk is synlophic.

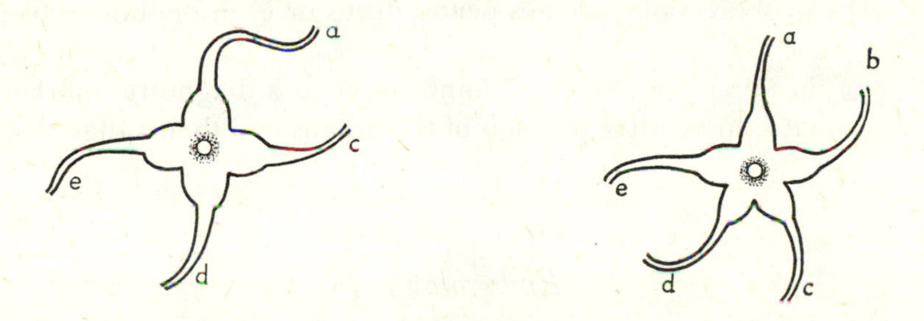

Fig. 5 Diagram of synlophic disks

Gynaecium

The ovary of *Tamarix* usually consists of three, sometimes of four and rarely of five, carpels. Accordingly, the number of stigmas is three, sometimes four or rarely five. Even in a single raceme the number of carpels varies (e.g., in *T. tetragyna*: 3–4; in *T. rosea*: 3–5); the shape of the ovary and stigmas also varies considerably within the species, and thus the diagnostic value of the gynaecium is small. For the vascular anatomy of the flower in general and of the carpels in particular, the reader is referred to Murty (1954), Puri (1939) and Saunders (1937–1939).

Seeds

The fruit of *Tamarix* is a many-seeded capsule. Neither the form nor the size of the seeds is diagnostically important. The seed bears an apical pappus. The apex is rostrate due to the twisted base of the unicellular hygroscopic hairs forming the pappus. Anatomical studies of the unicellular hairs were independently carried out by von Guttenberg (1926) and by Arendt (1926). Earlier studies, as well as those of this author, show the hairs to be protuberances of epidermal cells of the testa. The bases of the hairs have wavy thickenings which give them their hygroscopic properties.

Remark on the Number of Floral Parts

In some species the numbers of sepals, petals and stamens are constant and reliable. One of the problems in *Tamarix* is that in many species the number of certain floral organs is inconstant. In some species one finds in a single raceme anisomerous flowers with tetra-pentamerous calyx, a tetra-pentamerous corolla, a tetra-penta- or tetra-enneaandrous androecium and tri-tetracarpelled ovary, without any numerical correlation between the organs. The majority of species with anisomerous flowers are inconstant, especially with respect to the androecium. The rules prevailing in the androecium, and especially the distinction between antepetalous and antesepalous stamens, have already been described above. In a very general way one can point out that all species show a tendency towards pentamery, particularly in aestival racemes. Vernal racemes of tetramerous species have purely tetramerous flowers in their lower parts, tending towards pentamery in their upper parts. Thus, for instance, in a practically tetramerous species like *T. boveana* one occasionally finds in the apex of the vernal racemes pentandrous or even pentamerous-pentandrous flowers.

The numbers of floral parts alone thus cannot serve as a diagnostic marker without being correlated with the relative position of the organs and their range of numerical variability.

Embryology

The following species of *Tamarix* have been studied embryologically: *T. dioica*, Joshi & Kajale (1936); *T. aestivalis* (which the author was unable to identify), *T. africana*, *T. odessana* (which may be identical with *T. chinensis*), *T. pentandra* (which is surely *T. smyrnensis*) and *T. tetrandra* (which is doubtless *T. parviflora*), all by Mauritzon (1936); *T. ericoides*, Sharma (1939); *T. chinensis*, Puri (1939); *T. gallica* and Tamaricaceae in general, Paroli (1940) and Battaglia (1941, 1942); *Tamarix* in general, Johri & Kak (1954). From these studies it may be deduced that the development of the embryo sac in *Tamarix* is of the *Fritillaria* Type. Sharma (*op. cit.*) alleged that Mauritzon erred in attributing the species investigated to the *Adoxa* Type. Mauritzon used the term '*Lilium* Typus' for what is today known as the *Fritillaria* Type (according to Maheshwari, 1950, p. 122). Johri & Kak (*op. cit.*) generalized this phenomenon for the whole family Tamaricaceae. Their work indicates that a careful re-examination of the systematic position of the Tamaricaceae (or at least *Tamarix* and *Myricaria*) is necessary, since the *Fritillaria* Type of embryo sac is unknown in any other families of the Parietales, where Tamaricaceae are usually placed. Paroli (1940) observed two types of embryo-sac formation in *T. gallica*: (a) the *Euphorbia dulcis* Type, which is but a variation of the '*Lilium* Typus' or *Fritillaria* Type (Schnarf, 1929, p. 205), and (b) the *Adoxa* Type, where the egg often degenerates and the percentage of well-developed embryos, and consequently also that of viable seeds, is small. Sharma (1939) observed in *T. ericoides* double embryo sacs and the occurrence of polyembryony. This was also the first report of this phenomenon for the genus.

Palynology

Nair (1962) studied the pollen grains of the Indian species of *Tamarix*. Remarkable differences between species were observed. A key to the Indian species based on pollen dimensions and configurations was successfully elaborated. As an example, we give some data on the shape and size of the pollen grains from species studied by Erdtman (1952), Nair and the author:

T. aphylla: spheroidal, about 15 μ, reticulum regular
T. arabica: spheroidal, about 15 μ, reticulum regular
T. arborea: prolate, 27 x 18 μ, reticulum irregular
T. ericoides: spheroidal, 21 μ
T. dioica: spheroidal, 26 x 25–28 μ
T. gallica: subprolate, 19.5 x 15 μ
T. hampeana: prolate, 20 x 30 μ, no definite reticulum
T. indica: spheroidal, 14–17.5 μ
T. macrocarpa: prolate, 20 x 18 μ, reticulum regular
T. mannifera: prolate, 16 x 14 μ, reticulum regular
T. nilotica: prolate, 16 x 14 μ, reticulum irregular
T. palaestina: prolate, 18 x 16 μ, reticulum regular
T. parviflora: prolate, 25 x 18 μ, reticulum irregular
T. smyrnensis: prolate, 20 x 14 μ, reticulum regular
T. tetragyna: prolate, 22 x 16 μ, reticulum regular
T. tetrandra (which is probably *T. parviflora*): prolate, 24.5 x 16.5 μ.

Further investigation of this kind is needed in all the species. This will certainly yield more taxonomic evidence and support previous observations.

Chromosome Data

The following chromosome counts are recorded in the literature or were carried out by the author of this study:

T. aphylla,	2n = 24 (Bowden, 1945)
T. boveana,	2n = 24 (Reese, 1957)
T. dioica,	n = 12 (Malik, 1960)
T. ericoides,	n = 12 (Sharma, 1939)
T. gallica,	n = 12 (Bowden, 1945; Paroli, 1940)
T. hispida,	n = 12 (Bowden, 1945)
T. macrocarpa,	2n = 24
T. mannifera,	2n = 24
T. odessana,	n = 12 (Bowden, 1945)
T. palaestina,	2n = 24
T. parviflora,	n = 12; 2n = 24 (Bowden, 1945)
T. pentandra,	2n = 24 (Bowden, 1945)
T. smyrnensis,	2n = 24

The species mentioned above represent six series out of the nine included in this revision. It is evident that in this case chromosome counts do not offer any contribution to the solution of the taxonomic problem or offer any suggestion as to the evolutionary trend of the genus.

Paroli's observation (1940) that retarded chromosomes in the heterotypic anaphase sometimes give rise to small supernumerary nuclei in *T. gallica* is remarkable.

Palaeobotany

Of the papers dealing with fossils of *Tamarix*, the author has seen only the following: Krausel (1939) and Boureau (1951). The *Tamarix* fossil was named: *Tamaricoxylon africanum* (Krausel) Boureau, *Bull. Mus. Nat. Hist. Paris*, Ser. 2, 23: 468 (1951). Basionym: *Gynotrochoxylon africanum* Krausel, *Abh. Bayer. Akad. Wiss.*, 47: 97 (1939). According to Boureau the specimens from Somalia and Mauritania probably date from the Quaternary, and he attributes the specimens from Egypt that were found in the Lower Oligocene layer by Krausel to the same Quaternary layer, postulating that the specimens were fossils of roots which had penetrated into deeper layers. This is supported by the findings of Messeri (1938) on the xylem anatomy of some living *Tamarix* roots in the Fezzan.

Tamaricoxylon africanum cannot as yet be identified with one of the living species, but it undoubtedly belongs to the genus *Tamarix*.

Parasites

Angiosperm parasites of *Tamarix* and angiosperm competitors:
Cynomorium coccineum (Cynomoriaceae) has been reported on different *Tamarix* species in Yarkand (Henderson, 1873).
Plantations of *T. aphylla* withstand the invasion of the native grasses *Eragrostis cynosuroides* and *Imperata* (Deogun, 1939).

The following fungal parasites have been reported:
Coniothyrium caespitulosum on branches and branchlets of *T. gallica* in Italy (cf. Just, 1878, Vol. VI, p. 345).
Peniophora cremea f. *tamaricis* on *T. gallica* in France (Bourdot & Galzin, 1927).
Phellorinia gigantes on roots of *T. aphylla* in N. Africa (Maire, 1929).
Valesaria tamaricis on branches of *T. indica* in India (Mundkur & Ahmad, 1946).
Diplodia tamaricis on frozen branches of *T. chinensis* in Europe (Rabenhorst, 1873).
Cytospora tamaricis on *T. palaestina* in Israel (Rayss, 1943).
Sirodiplospora tamaricia on *T. aphylla* in Punjab (Sydow & Ahmad, 1939).
Phoma tamariscinum on *T. gallica*? in Austria (Thümen, 1877).

The following insect parasites have been reported:
Liothrips dampfyi (Thysanoptera) on *T. indica* in India (Ayyar, 1934).
Hemiberlesia megapore (Coccid.) on *T. aphylla* in N. Africa (Balachowsky, 1928).
Naiacoccus serpentinus var. *minor* (Coccid.) on *T. aphylla* and *T. canariensis* in Algeria (Balachowsky, 1930).
Chionaspis etrusca (Coccid.) on *T. gallica* in Corsica (Balachowsky, 1933).
Aphalara elegans (Psyllidae) on *T. aphylla* in N. Africa (Bergevin, 1932).
Pseudophlocus gestroi and *Artheneis chloratica* (Hemiptera) on *Tamarix* sp., in Giarabub (Bergevin, 1932).
Platygonatopus polychromus, a parasite of young nymphs of *Athysanus heydeni* on *Tamarix* sp. (Bernard, 1932).
Hyperopsis vinciguerrae, *Nephus tamaricis* and *Nephus tamaricis* var. *stamineus* (Coccinollids) all on *T. macrocarpa* in Giarabub (Capra, 1928–1930).
Salebria cingilella on *T. gallica* in France (Chrétien, 1926).
Ambliardiella tamaricum (Cecidomyi) on *T. africana* in N. Africa (Dieuzeide, 1931).
Agdistis tamaricis (Microlepidoptera) on *Tamarix* sp. in India (Fletcher, 1932).
Ascalani pachnodes larvae on stems of *T. indica*; *Odites spoliatrix* larvae on dry inflorescences of *T. indica*; *Scythis axenopa* larvae on *T. indica* in India (Fletcher, 1933).
Orthorrhinus cylindrostris larvae tunnel in the trunks of *Tamarix* spp. (French, 1933).
Ephestia glycyphloeas (Microlepidoptera) larvae feed on the exudation of *Tamarix* in Baluchistan (Meyrick, 1936).
Xerophilaphis tamaricophila, *Brevicorynella quadrimaculata* (Aphids) on *Tamarix* spp. in Central Asia (Nevsky, 1928).

Nanophyes muticus, *N. gyratus* and *N. aphyllae* (Coleoptera) on *T. aphylla* in N. Africa (Peyerimhoff, 1929).
Scraptia straminea on *Tamarix* sp., *Geranorrhinus certrator* on *T. trabutii*, *G. aphyllae* (Coleoptera) on *T. aphylla* (Peyerimhoff, 1931).
Julodis aequinoctialis subsp. *nemethi* on *T. africana* in N. Africa (Théry 1931).
Diplosis tamaricis on *T. aphylla* (Wachtl, 1886).
Euscelis stactogalus attacks all trees of the genus *Tamarix* in Hawaii (Zwaluwenburg, 1929).

Some of the insects mentioned above produce galls on *Tamarix*. The uses of galls in *T. africana*, *T. aphylla*, *T. indica*, *T. dioica* and *T. mannifera* for pharmaceutical purposes have been reviewed by Hartwich (1883). Galls and bud teratisms and their pathogens in *Tamarix* were studied by Slepyan (1962). According to him, *Tamarix* species are hosts to organisms of the families Itonididae, Psyllidae, Gelechiidae, Momphidae, Tortricidae, Curculionidae, Eriophyidae, which cause galls and bud teratisms. Itonididae are the most common ones. Vogl (1875) deals especially with *T. aphylla* galls.

Manna of the Jews, or Tamarisk Manna

Moldenke & Moldenke (1952) postulate three possible forms of biblical 'manna': (1) the resinous exudations of certain desert trees (particularly *Tamarix mannifera*); (2) algal manna (*Nostoc*); (3) lichen manna (particularly of the genus *Lecanora*). They were obviously not aware of the remarkable account of this subject by Bodenheimer (1929), who reviewed the historical literature referring to manna and added his own observations. According to him, manna is the sweet excretion of the following insects living on *Tamarix*: *Trabutina mannifera* (Ehrenb.) Bdhmr., *Najacoccus serpentinus* var. *minor* Green and perhaps also *Euscelis decoratus* Hpt. and *Opsius jucundus* Leth. The first two insects, as well as the sweet excretion, have been observed by the author also in the Dead Sea area. According to Bodenheimer, Bedouins living in the desert still collect the 'manna' early in the morning, before it is melted by the sun.

Uses

Tamarisk species can be used as efficient windbreaks; they also serve as hedges for dividing estates under desert conditions; e.g., *T. aphylla* (Brown, 1919; Woodbridge, 1931) or as dune fixers (Doney, 1945; F.H. 1928; Oliver, 1947; Sale, 1948), especially on sea shores (personal observations on the Mediterranean coast of Israel). Some species are tolerant of salt spray and are therefore useful as hedge plants near the sea-shore (Couppis, 1956); other species are highly effective in the fixation of river banks and ditches to prevent erosion (Leontiev, 1952; Russanov, 1950).

Smith (1941) mentions a variety of uses for *Tamarix* species. Heyer (1876) mentions its uses for heating.

Tamarix species are highly tanniniferous, the amount of tannin varying with the species. Kudriashev (1932) found 1.54% tannin substances in branches of *T. ramosissima*, and Sukhowkov (1929) found 12% tannin in young branches of two species of *Tamarix*. This should be taken into consideration when choosing species for grazing and browsing. Utilization experiments in grazing have been successful (Gary, 1960).

A number of species are known as ornamental or afforestation trees (Glading et al., 1945; Hawkins, 1958; Herriot, 1942; Lyon, 1924; Reich, 1891; Richardson, 1954; Vilbouchevitch, 1890 and 1892; Wilson, 1944).

Many species of *Tamarix* may become 'noxious' or troublesome due to their invasive habit and competition with neighbouring plants, and also because they impoverish water sources (Decker et al., 1962; Decker & Wetzel, 1957; Gary, 1963; Smueli, 1948; Waisel, 1960). *T. aphylla* and other species show a very high water output through transpiration (Waisel, 1960). Salt secretion, very common in many *Tamarix* species (Brunswick, 1920; Chapman, 1934; Decker, 1961; Marloth, 1887; Volkena, 1887), may lead to the desalinization of deeper soil layers, while increasing the salinity of the upper soil layers, the soil surface being covered each year by a layer of salty litter in the form of twigs (Litwak, 1957).

Although *Tamarix* is highly conditioned by edaphic factors, in many species the deciding factor in distribution is temperature. *T. aphylla*, *T. nilotica*, *T. dioica*, *T. stricta*, *T. indica*, *T. arabica*, *T. aucheriana*, *T. macrocarpa* and *T. senegalensis* demand high temperatures, while others inhabit temperate regions. The geographical distribution of the species no doubt indicates their thermal requirements. The species also differ markedly in their degree of salt resistance. While most tolerate salinity to a high degree, some are definitely not halophilous; e.g., *T. smyrnensis*, *T. chinensis* and *T. parviflora*. *T. hispida* and *T. parviflora* are extreme halophytes (Krupenikov, 1947; Zhemehuzhnikov, 1946). *T. aucheriana* has been observed by Zohary (1963, unpublished) and by the present author on the edges of the Iranian Kavirs, and *T. macrocarpa* has been observed in the salines of the Dead Sea, where the soil contains 8% chloride or more.

For further data on this subject see Gary (1963); Gimingham (1955); Kassas & Imam (1954); Marks (1950); Rivas (1945); Rivas & Amor (1945); and Walter & Walter (1953).

Once introduced, some species readily naturalize. They may also become invaders, as in the case of *T. pentandra* (*T. chinensis*) along mountain streams and reservoirs of the Wasatch Mountains (USA), as recorded by Christensen (1962).

Reproduction by seeds seems to be strictly limited (Waisel, 1960; Ware & Penfound, 1949) and is connected with special requirements (Horton et al., 1960; Wilgus & Hamilton, 1962).

Delimitation of the Genus Tamarix *and its Position in the Tamaricaceae Family*

Until Desvaux (1824) the genus *Tamarix* was considered a single taxon in its broad sense. Desvaux (1815, unpublished; 1824, published officially) was the first to divide *Tamarix* into *Tamarix* proper and his *Myricaria* (cf. *Historical Survey*). Link (1821) gave the first description of the Tamaricaceae. Desvaux included in Tamaricaceae the genera *Tamarix* and *Myricaria*, though Fabricius[6] had earlier pointed out the close relationship between *Reaumuria* and *Tamarix*. Ehrenberg (1827) gave two separate families: Reaumuriaceae and Tamaricaceae. In Reaumuriaceae he added *Hololachna*, a new genus formerly described under *Tamarix*. Ledebour (1831) founded *Eichwaldia* belonging to the Reaumuriaceae. Arnott (1834) established the genus *Trichaurus*, based on *Tamarix ericoides* Rottler.

Bentham & Hooker (1862) distinguished two tribes in the Tamariscineae: (a) the Tamarisceae and (b) the Reaumurieae.[7] Tamarisceae consisted of *Tamarix* (including *Trichaurus*, reduced as synonym) and *Myricaria*, while Reaumurieae consisted of *Hololachna* and *Reaumuria* (with *Eichwaldia* as synonym). This view is also accepted by Niedenzu (1895 and 1925) in both of his articles on the Tamaricaceae.

The delimitation of *Tamarix* and *Myricaria* has long been a matter of dispute. The present author's view is that these genera differ clearly only in their androecium. Ledebour et al. (1829) and Bentham & Hooker (*op. cit.*) rightly point out that in *Tamarix* the androecium consists of free stamens, while in *Myricaria* it is monadelphous (Fig. 6).

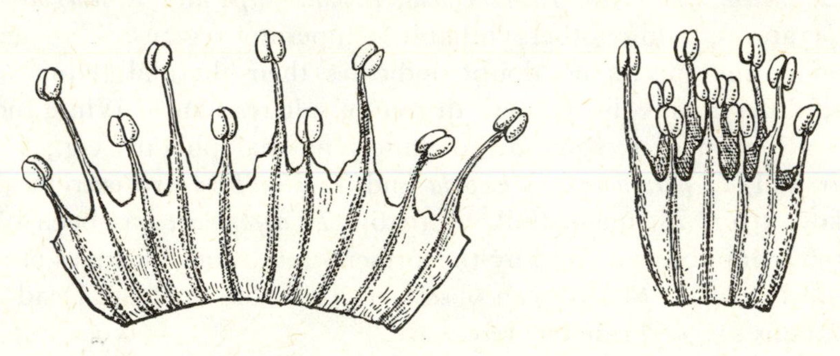

Fig. 6 Androecium of *Myricaria germanica* (on left — opened and flattened)

In addition, *Myricaria* has a membranous torus and a constant number of ten stamens, the antesepalous stamens being considerably longer than the antepetalous. One of the characteristics frequently considered essential (e.g., by Bunge, 1852; Desvaux,

6 The manuscript of Fabricius, a disciple of Linnaeus, was edited by Giseke and published in 1792 in a work entitled '*Caroli A. Linné, Praelectiones in Ordines Naturales Plantarum*,' Hamburg.

7 The tribe Fouquiereae does not concern us here.

1824; and Niedenzu, 1895, 1925) in distinguishing *Myricaria* from *Tamarix* is the stipitate pappus of the former as against the sessile pappus of the latter. This, however, was early proved by Bentham & Hooker (*op. cit.*) to be an unreliable characteristic in *Myricaria prostrata* Hook. f. & Thomas, who described the hairy tuft as 'coma sessilis Tamaricis', and by Dyer (1874) who, in his important remarks concerning *Myricaria*, also ascribed to it 'sessile plumes'. This has also been observed in other species.

As the distinction between the stipitate and the sessile pappus is not always obvious and sometimes requires careful examination, some species of *Tamarix* have often been transferred to *Myricaria* and vice versa (e.g., *T. ericoides*); this was certainly the reason for the establishment of the genus *Trichaurus*, to which a stipitate pappus was erroneously attributed. Bunge (1852) found that *T. ericoides*—the type species of *Trichaurus*—has a sessile pappus, and this has also been observed by the present author.

A comparison of some homologous floral organs in the four generally accepted genera of Tamaricaceae may elucidate the position of *Tamarix* within the family:

1. Inflorescence: In Reaumurieae the flowers are solitary, terminal or axillary; in Tamariceae they are arranged in more or less dense racemes.

2. Number of perianth segments: In *Reaumuria*, *Hololachna* and *Myricaria* the perianth is completely and constantly pentamerous. In *Tamarix* there is a marked trend towards tetramery.

3. The androecium: In *Tamarix* there is diplostemony and haplostemony; some species are partially diplostemonous and very few are triplostemonous or partially triplostemonous. In general, a whorled androecium can readily be traced. The antesepalous stamens are always present, while the antepetalous stamens are either present or abortive or replaced by staminodial nectariform lobes. The stamens are free.

In *Myricaria* diplostemony is the rule. The stamens are connate at the base and form a membranous tube.

In *Hololachna*, and especially in *Reaumuria*, the stamens seem to be pentadelphous, with the bundles presumably antepetalous. Antesepalous stamens are altogether missing; they have presumably been transformed into the ligules adnate to the petals at either side. Pleiomery of the antepetalous stamens has thus become the rule in Reaumurieae. This has happened partially and to a lesser extent in two *Tamarix* species. In Tamaricaceae the antesepalous stamens always persist, while the inner or antepetalous whorls in *Tamarix* (but not in *Myricaria*) become converted into staminodial nectary lobes, a general trend of abortion leading to haplostemony.

4. The ovary: *Reaumuria* has a pentacarpelled ovary, *Hololachna* and *Myricaria* a tricarpelled one. In *Tamarix* the ovary is generally tricarpelled, but in the vernal racemes of some species tetracarpelled or, very rarely (e.g., *T. rosea* and *T. dubia*), pentacarpelled ovaries may also occur.

5. Seeds: Seeds of all Tamaricaceae genera possess hygroscopic hairs. According to Brunner (1909) and the present author, the wall thickenings at the base of the unicellular hairs in Reaumurieae are sparser and less developed than those in

Tamariceae (see also Arendt, 1926, and von Guttenberg, 1926). Thus, in Reaumurieae the hygroscopic apparatus is less developed than in Tamariceae. In *Reaumuria* and *Hololachna* these hairs cover the whole surface of the seeds (Fig. 7a). *Tamarix* seeds have a sessile pappus at the apex, while most species of *Myricaria* possess a stipitate pappus (Figs. 7b–c, respectively).

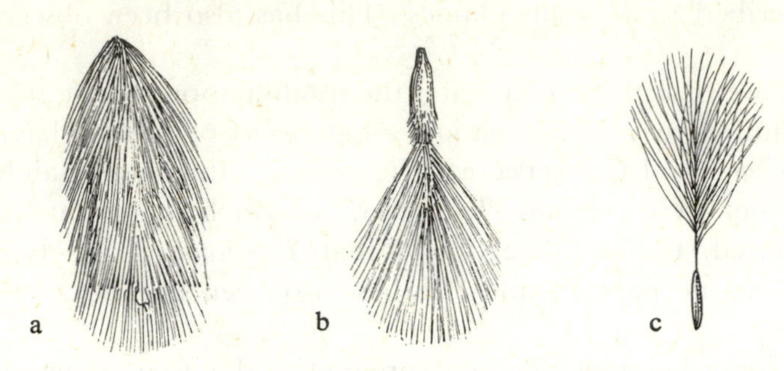

Fig. 7 Seeds of Tamaricaceae genera
a.–*Reaumuria;* b.–*Tamarix;* c.–*Myricaria*

The author's anatomical observations on the early stages of seed development of *T. macrocarpa* showed that the entire upper half of the seed is covered with such hairs and thus to some extent resembles that of *Reaumuria* (Fig. 8).

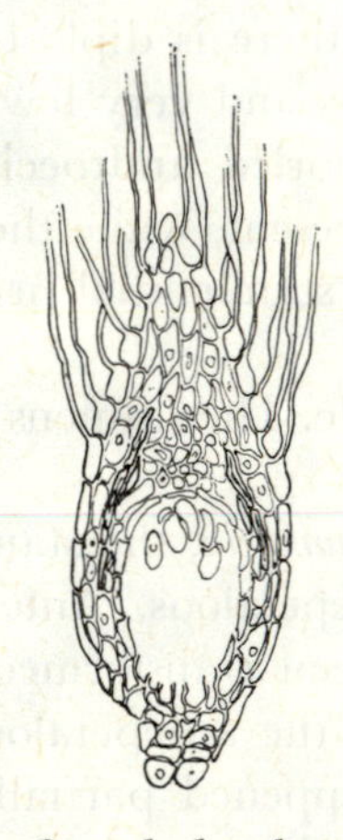

Fig. 8 Early stage of seed development in *T. macrocarpa*

The relation of *Tamarix* to the other genera of the family can thus be given as follows: The inflorescence of *Tamarix* is the most advanced in the family. *Tamarix* is the only genus in the family showing reductional trends towards tetramery. On the other hand, purely two-whorled androecia are still to be found in *Tamarix*, while in *Myricaria* the two whorls have been completely fused; in Reaumurieae the outer whorl has been transformed and fused with the petal whorl. Haplostemony is attained in *Tamarix* only. Seeds are most specialized for dispersion in *Myricaria*, somewhat less so in *Tamarix*, and even less so in Reaumurieae.

While *Tamarix* has reached a highly advanced evolutionary stage in some features, it has also preserved a number of the relatively primitive characteristics, and this supposedly gives *Tamarix* a more ancestral position within the family. Evolution, which presumably started from the 'archaic tamarisks' towards the more advanced and reduced congeneric species, also led through early ramification of the *Tamarix* stem to the other three highly advanced genera of the family (Fig. 9).

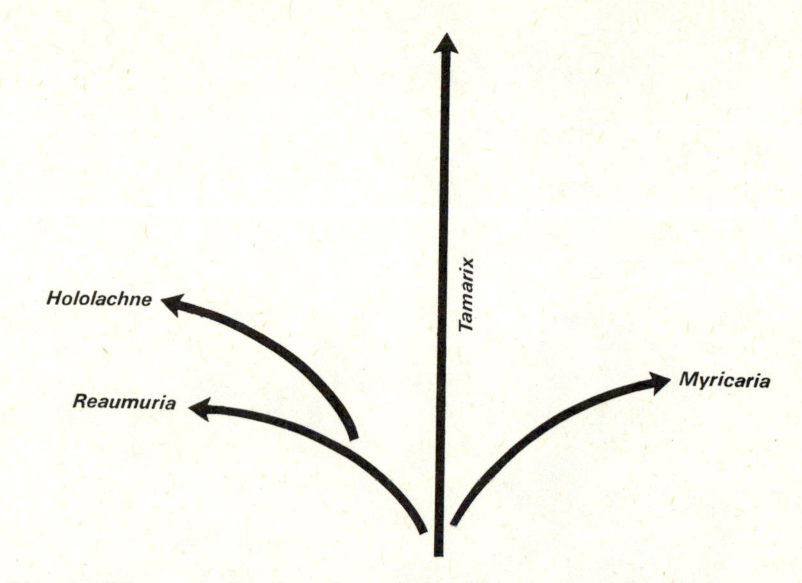

Fig. 9 Diagram showing the relationships among the genera of Tamaricaceae (explanation in text)

SYSTEMATIC PART

Description of the Genus

TAMARIX L.

Sp. Pl., 1:270 (1753); Gen. Pl., 5th ed., 131, No. 337 (1754).

Tamariscus P. Miller, Gard. Dict., abr. ed., 4 (1754).
Tamarix L., emend. Ledeb., Fl. Alt., 1:421 (1829).
Trichaurus Arn., in: Wight & Walker-Arnott, Prodr. Fl. Penins. Ind. Or., 1:40 (1834).
Myricaria Desv. Sectio *Parallelantherae* Ndz., in: Engler & Prantl, Nat. Pflanzenfam., III, 6:296 (1895).

Type species (lectotype): *T. gallica* L., designated by Britton, North American Trees, 702 (1908).

Shrubs, shrubby trees, or trees up to 10 m. Roots up to 30 m deep, more or less succulent. Branches usually many, sometimes forming very long stolons and then also water-storing, alternate, green, papillose, papillulose, hairy or glabrous when young, brownish, reddish-brown or blackish-brown, glabrous or glabrescent, with prominent lenticels when older, bark usually splitting. Leaves of various forms, usually small, scale-like, entire, alternate, not or more or less decurrent, glabrous or papillose or hairy, sessile with narrow base, amplexicaul or vaginate, with more or less deep salt-secreting glands. Inflorescence of simple or compound racemes, occurring as a rule on the branches of the previous year's growth and subsequently on the green branches of the current year or contemporaneously on both types of branches or on one type only. Flowers tetra- to pentamerous, monoecious or dioecious bracteate. Bracts usually single, occasionally two or more, scariously tipped or entirely herbaceous, glabrous, papillose or hairy, entire or denticulate, open or amplexicaul, shorter, of equal length or longer than pedicels and/or flowers. Calyx of 4 or 5 more or less unequal segments more or less connate at base, 2 outer and 2–3 inner, entire or denticulate in various patterns, of various shapes from suborbicular to triangular, glabrous, papillose or hairy. Corolla usually of 4 or 5 petals, white, pink or red in colour, elliptic, ovate or obovate, entire or emarginate, equi- or inequilateral, persistent, subpersistent or deciduous after anthesis. Androecium diplostemonous, the outer whorl of 4, 4–5 or 5 antesepalous stamens, the inner whorl of 4–10 antepetalous filaments usually somewhat shorter than the former or abortive, and then the androecium is haplostemonous with nectariferous antepetalous lobes. Stamens always free, emerging from a central, more or less fleshy, nectariferous, hypogynous torus having lobes especially when the androecium is haplostemonous; anthers apiculate or not. Pistil pyramidal or bottle-shaped, with 3–4 stigmas. Capsule many-seeded, conical-pyramidal, loculicidal. Seeds usually glabrous, sometimes hairy near the apical rostrate pappus of hairs; hairs unicellular, hygroscopic and emerging from an apical axis.

Synoptic Key to Sections and Series

1A Racemes 3–5 mm broad or in dioecious trees 5–7 mm broad; petals 1–2.25 mm long; stamens usually 5 (antesepalous); disk various Sect. **Tamarix** (p. 27)

2A No papillae present at all; leaves sessile with narrow bases Ser. **Gallicae** (p. 27)

2B Younger parts at least hairy, papillose or papillulose; leaves not vaginate Ser. **Leptostachyae** (p. 48)

2C No papillae present; leaves vaginate or appearing so Ser. **Vaginantes** (p. 78)

1B At least vernal racemes 5–12 mm broad, or 3–5 mm and then flowers tetrandrous; petals of various lengths; 4–5 of the 4–9 stamens antesepalous; disk with nectariferous lobes Sect. **Oligadenia** (p. 92)

3A Lowermost bracts of vernal racemes shorter than or equalling pedicels Ser. **Laxae** (p. 92)

3B Bract exceeding calyx, or shorter and then petals trullate-ovate; petals more than 2 mm long Ser. **Anisandrae** (p. 105)

3C Bracts longer than pedicels; petals 2–2.25 mm long Ser. **Arbusculae** (p. 134)

3D Bracts shorter than or equalling pedicels; racemes with fasciculate flowers at apex Ser. **Fasciculatae** (p. 141)

1C Racemes 6–10 (–15) mm broad; petals of various lengths; stamens 6–15 (mostly 10), of these 5 antesepalous and with slightly longer filaments; disk with no nectariferous lobes (except for *T. komarovii*) Sect. **Polyadenia** (p. 147)

4A In each raceme a few or several flowers have a pair of shorter antepetalous stamens between one or more pairs of longer antesepalous ones Ser. **Arabicae** (p. 147)

4B Only a single antepetalous shorter stamen, or none by abortion, between one pair of antesepalous longer ones Ser. **Pleiandrae** (p. 152)

Key to Species

1. Androecium triplostemonous or mostly or partially triplostemonous
   2. Antisepalous stamens with abruptly and distinctly broadening base, stamens mostly 12–13, rarely 15. 45. **T. pycnocarpa**
   2. Antisepalous stamens gradually and slightly broadening at base, stamens 10–11, rarely 12. 44. **T. aucheriana**
1. Androecium diplostemonous or partially diplostemonous or haplostemonous
   3. Androecium diplostemonous or partially diplostemonous
      4. Leaves vaginate
         5. Leaves vaginate all along; racemes 4–5 mm broad. 54. **T. stricta**
         5. Leaves vaginate at lower part; racemes 9–11 mm broad. 48. **T. ericoides**
      4. Leaves amplexicaul or sessile with narrow base
         6. Leaves sessile with narrow base; racemes of tetramerous flowers and/or also intermixed with pentamerous ones, rarely of pentamerous flowers only, but then some of the flowers subtended by 2–3 bracts; androecium always with less than 10 stamens
            7. At least bracts of the lower parts of racemes shorter than pedicels
               8. Racemes 6–7 mm broad; sepals less than 2 mm long. 25. **T. gracilis**
               8. Racemes 10–12 mm broad; sepals more than 2 mm long. 33. **T. hampeana**
            7. Bracts all longer than pedicels
               9. Bracts nearly equalling calyx, up to or exceeding flowers
                  10. All flowers of the raceme tetramerous; stamens (6–)8. 35. **T. octandra**
                  10. At least a part of the flowers of the racemes with 5 sepals or with 5 petals or both; stamens of most of the flowers 4–5 (6–9)
                     11. Petals trullate-ovate, shortly cuneate; rachis of racemes and younger parts usually papillose. 37. **T. tetragyna**
                     11. Petals obovate-unguiculate; rachis and younger parts not papillose. 31. **T. dalmatica**
               9. Bracts not exceeding calyx
                  12. Flowers usually tetramerous, each subtended by one bract only; insertion of filaments peridiscal. 38. **T. tetrandra**
                  12. Flowers usually pentamerous, at least some of them subtended by 2–3 bracts; insertion of filaments hypodiscal. 36. **T. rosea**
         6. Leaves amplexicaul, or leaves sessile with narrow base but then androecium of 10 stamens; racemes of pentamerous flowers only
            13. Leaves sessile with narrow base; younger parts of branchlets never papillose
               14. Leaves elliptic, over 10 mm long; petals obovate, 5.5–6 mm long. 50. **T. ladachensis**

14. Leaves scale-like, not over 5 mm long; petals broadly elliptic 3–3.5 mm long. 47. **T. dubia**

13. Leaves amplexicaul or nearly so; younger parts always papillose

15. Leaves strongly amplexicaul and densely imbricate. 46. **T. amplexicaulis** (see also *T. aucheriana*)

15. Leaves less amplexicaul to nearly sessile with narrow base, remote

16. All the flowers with diplostemonous androecium; sepals more or less acute

17. Disk with lobes; sepals more or less eroded-denticulate. 49. **T. komarovii**

17. Disk without lobes; sepals more or less entire. 52. **T. passerinoides**

16. Most of the flowers partially diplostemonous, rarely with some diplostemonous flowers intermixed on the same raceme; sepals more or less obtuse

18. Disk without lobes; stamens 6–9 (10). 51. **T. macrocarpa**

18. Disk synlophic; stamens usually 5, occasionally with an additional antipetalous stamen. 53. **T. salina**

3. Androecium haplostemonous

19. Flowers pentamerous

20. Leaves vaginate or pseudo-vaginate

21. Leaves vaginate

22. Dioecious trees

23. Bracts not vaginate; petals obovate; insertion of filaments hypodiscal. 22. **T. dioica**

23. Bracts vaginate; petals elliptic to ovate; insertion of filaments peridiscal. 23. **T. usneoides**

22. Monoecious trees

24. Bracts not vaginate, somewhat clasping; petals subpersistent or caducous; disk hololophic. 20. **T. aphylla**

24. Bracts vaginate; petals persistent; disk synlophic. 19. **T. angolensis**

21. Leaves, at least the younger ones, pseudo-vaginate

25. All leaves pseudo-vaginate and strongly adpressed for most of their length; sepals orbicular to broadly ovate. 21. **T. bengalensis**

25. Only younger leaves pseudo-vaginate, older leaves sessile with subauriculate to narrow base; sepals ovate to trullate-ovate. 13. **T. indica**

20. Leaves sessile with narrow base, sometimes auriculate but never vaginate or pseudo-vaginate

26. Racemes, at least the vernal ones, (5) 6–12 mm broad

27. At least the lower bracts of the vernal racemes shorter than, or about as long as, the pedicels
    28. Younger parts or at least rachis of raceme papillose; disk synlophic. 53. **T. salina**
    28. No papillose parts; disk hololophic or paralophic
        29. At least some flowers in each raceme subtended by 2 or 3 bracts; insertion of filaments hypodiscal. 36. **T. rosea**
        29. Each flower subtended by a single bract only
            30. Racemes 5–7 mm broad; sepals less than 2 mm long
                31. Petals persistent; insertion of filaments hypodiscal. 24. **T. chinensis**
                31. Petals caducous; insertion of filaments peridiscal. 25. **T. gracilis**
            30. Racemes 10–12 mm broad; sepals more than 2 mm long. 33. **T. hampeana**
27. Bracts longer than pedicels, usually about as long as or exceeding calyces
    32. Petals narrowly trullate-ovate or slightly obovate in outline, shortly clawed; younger parts and/or rachis of racemes papillose; disk hololophic to paralophic. 37. **T. tetragyna**
    32. Petals ovate to broadly trullate-ovate, not unguiculate; younger parts usually glabrous except for rachis of racemes; disk synlophic. 28. **T. africana**

26. Racemes usually less than 5 mm broad (occasionally 5 mm)
33. Petals persistent
    34. Sepals entire; at least lower bracts of vernal racemes more or less equalling pedicels. 24. **T. chinensis**
    34. Sepals more or less eroded-denticulate; bracts longer than pedicels
        35. Petals ovate to suborbicular, strongly keeled, especially in their lower part. 7. **T. smyrnensis**
        35. Petals obovate, not keeled. 6. **T. ramosissima**
33. Petals caducous
    34. Disk hololophic
        36. Leaves sessile with narrow base to subauriculate, divaricate, not imbricate
            37. Aestival racemes usually more than 7 cm long, up to 15 cm (see also *T. indica*)
                38. At least younger parts papillose; petals ovate; sepals finely denticulate. 18. **T. senegalensis**
                38. Entirely glabrous; petals obovate; sepals subentire to eroded. 3. **T. korolkowii**
            37. Aestival racemes usually less than 7 mm long
                39. At least rachis of racemes papillose; petals elliptic

40. Sepals entire; petals 1.75–2 mm long. 17. **T. nilotica**
40. Sepals densely and finely denticulate; petals 1.25–1.5 mm long. 14. **T. karakalensis**

39. Entirely glabrous; petals usually obovate
41. Sepals with papillose-denticulate margins, the outer 2 more obtuse than the inner; petals 1–1.5 mm long. 1. **T. arceuthoides**
41. Sepals entire, acute, the outer 2 more acute than the inner; petals 1–1.5 mm long. 5. **T. palaestina**

36. Leaves auriculate, adpressed, more or less imbricate, the younger amplexicaul
42. Sepals entire or subentire. 16. **T. mannifera**
42. Sepals densely and finely toothed
43. Younger leaves amplexicaul with touching margins, sometimes pseudo-vaginate; petals obovate. 13. **T. indica**
43. Younger leaves auriculate, not amplexicaul, with touching margins
44. Sepals obtuse; petals elliptic. 9. **T. aralensis**
44. Sepals acute; petals obovate. 8. **T. arabica**

34. Disk paralophic or synlophic
45. Petals ovate; disk paralophic
46. Sepals 1–1.5 mm long, entire, occasionally with a few irregular teeth; petals 1.75–2 mm long. 10. **T. arborea**
46. Sepals 0.75–1 mm long, finely, densely denticulate; petals 1.25–1.75 mm long. 4. **T. mascatensis**

45. Petals elliptic to obovate; disk synlophic
47. Leaves cordate-clasping; younger parts of plant hairy or papillose at least on rachis of racemes; petals 2 mm long. 12. **T. hispida**
47. Leaves not cordate-clasping; entirely glabrous; petals less than 2 mm long
48. Aestival inflorescences densely compound of racemes 7–15 cm long. 15. **T. leptostachya**
48. Aestival racemes not more than 7 cm long, usually shorter
49. Racemes usually with papillose rachis; sepals densely denticulate; petals obovate, 1.25–1.5 mm long. 11. **T. canariensis**
49. Rachis glabrous; sepals entire; petals more or less elliptic (rarely obovate), 1.5–1.75 mm long. 2. **T. gallica**

19. Flowers of each raceme tetramerous or tetra-pentamerous or pentamerous or all intermixed

50. Bracts equalling, up to, or much exceeding calyces
  51. Disk paralophic
    52. Petals obovate to elliptic-obovate, not unguiculate, 3–3.5 mm long, caducous; racemes 6–7 mm broad. 34. **T. meyeri**
    52. Petals narrowly obovate to narrowly elliptic-obovate, unguiculate or cuneate, usually 3.5–5 mm long, subpersistent; at least vernal racemes 8–10 mm broad
      53. Younger parts usually papillose; racemes with tetra- and penta- or tetra-pentamerous intermixed flowers. 37. **T. tetragyna**
      53. Entirely glabrous; racemes usually of tetramerous flowers (sometimes bearing pentamerous flowers at apex). 31. **T. dalmatica**
  51. Disk synlophic
    54. Bracts ending with a long, acute, diaphanous point; racemes 7–8 mm broad. 32. **T. elongata**
    54. Bracts ending with a short blunt point; racemes 8–10 mm broad
      55. Inner and outer sepals almost alike, obtuse, entire, entirely glabrous. 30. **T. brachystachys**
      55. The 2 outer sepals acute, entire, the inner 2 obtuse, more or less dentate; younger parts usually papillose. 29. **T. boveana**
50. Bracts not equalling calyces, always shorter
  56. Bracts shorter than or almost as long as pedicels
    57. The terminal 2–5 flowers of each raceme condensed as an umbel
      58. Bracts much shorter than pedicels; petals persistent; disk (holo-) paralophic. 43. **T. polystachya**
      58. Lower bracts about as long as pedicels, the others somewhat longer; petals caducous; disk (para-) synlophic. 42. **T. litwinowii**
    57. No condensed flowers at apices of racemes
      59. Disk synlophic to para-synlophic; lower bracts usually about as long as pedicels
        60. Racemes 6 mm broad, densely flowered; petals 2.5 mm long; younger parts often papillose. 27. **T. szowitsiana**
        60. Racemes 3–4 mm broad, loosely flowered; petals 2 mm long; no papillae present at all. 39. **T. androssowii**
      59. Disk holo-paralophic
        61. Racemes (8–) 10–12 mm broad; sepals more than 2 mm long. 33. **T. hampeana**
        61. Racemes 6–8 mm broad; sepals less than 2 mm long
          62. Racemes often with tetramerous, pentamerous and tetra-pentamerous intermixed flowers; petals elliptic to obovate. 25. **T. gracilis**
          62. Racemes only with tetramerous flowers; petals broadly elliptic to ovate. 26. **T. laxa**

56. Bracts longer than pedicels
   63. At least the first-appearing leaves amplexicaul with touching margins, decurrent, or leaves auriculate and somewhat clasping; petals obovate-elliptic to elliptic, sometimes slightly trullate-ovate
      64. Petals narrowly elliptic to obovate-elliptic; younger leaves amplexicaul with touching margins; bracts usually diaphanous. 40. **T. kotschyi**
      64. Petals elliptic to slightly trullate-ovate; leaves auriculate, bracts herbaceous. 10. **T. arborea**
   63. Leaves sessile with narrow base; petals parabolical (trullate-ovate)
      65. Bark black; bracts herbaceous in lower part, diaphanous in upper; petals (2.2–) 2.4–3.0 mm. 38. **T. tetrandra**
      65. Bark reddish-brown to purple; bracts entirely diaphanous; petals 1.8–2.0 mm. 41. **T. parviflora**

Enumeration of Sections, Series and Species

Section One. TAMARIX (See Appendix)

Decadenia Ehrenb., Linnaea, 2:253 (1827), pro subgenere, p.p.
Aestivales Bge., Tentamen, 6 (1852), p.p.
Sessiles Ndz., De Genere Tamarice, 4 (1895), pro subgenere, p.p.
Amplexicaules Ndz., *op. cit.*, 10, pro subgenere, p.p.
Primitivae Arendt, Beitr. Tamarix, 33 (1926), p. min. p.
Pentamerae Arendt, *op. cit.*, 36, p.p.
Eutamarix Gorschk., Not. Syst. Leningrad, 7:81 (1937), pro subgenere, p.p.

Type species: *T. gallica* L.

Leaves sessile with narrow base, subauriculate or vaginate. Racemes 3–5 mm broad except for a few dioecious species in which they are 5–7 mm broad. Bracts longer than pedicels. Flowers pentamerous or in a few species occasionally tetra-pentamerous or tetramerous in the vernal racemes only. Petals 1–2.25 mm long. Androecium haplostemonous, of (4–) 5 antesepalous stamens and of various discal structures, no antepetalous stamens present.

Series 1. GALLICAE Baum (ser. nov., see Appendix)

Paniculatae Bge., Mém. Acad. St. Pétersb., 7:294 (1851), pro sectione, p.p.
Xeropetalae Bge., Tentamen, 6 (1852), p.p.
Piptopetalae Bge., *loc. cit.*, p.p.
Epilophus Ndz., De Genere Tamarice, 8 (1895), pro subsectione, p.p.
Pentastemones Grex Sessiles Arendt, Beitr. Tamarix, 36 (1926), p.p.
Pentastemones Grex Semiamplexicaules Arendt, *op. cit.*, 39, p.p.
Ramosissimae Gorschk., in: Komarov, Fl. URSS, 15:311 (1949), p.p.

Type species: *T. gallica* L.

Entirely glabrous, no papillae present at all. Leaves usually sessile with narrow bases, except for *T. mascatensis*, where they are strongly clasping.

Included species: *T. arceuthoides* Bge., *T. gallica* L., *T. korolkowii* Regel & Schmalh. ex Regel, *T. mascatensis* Bge., *T. palaestina* Bertol., *T. ramosissima* Ledeb., *T. smyrnensis* Bge.

1. **T. arceuthoides** Bge., Mém. Acad. St. Pétersb., 7:295 (1851) [Plate I]

T. florida Bge. var. *kotschyi* Bge., Tentamen, 38 (1852).
T. askabadensis Freyn, Bull. Herb. Boiss., II, 3:1059 (1903).
T. karakalensis Freyn var. *verrucifera* Freyn, *op. cit.*, 1062, nom. illegit.
T. turkestanica Litw. f. *brachystachys* Litw. ex Gorschk., Not. Syst. Leningrad, 7:93 (1937).

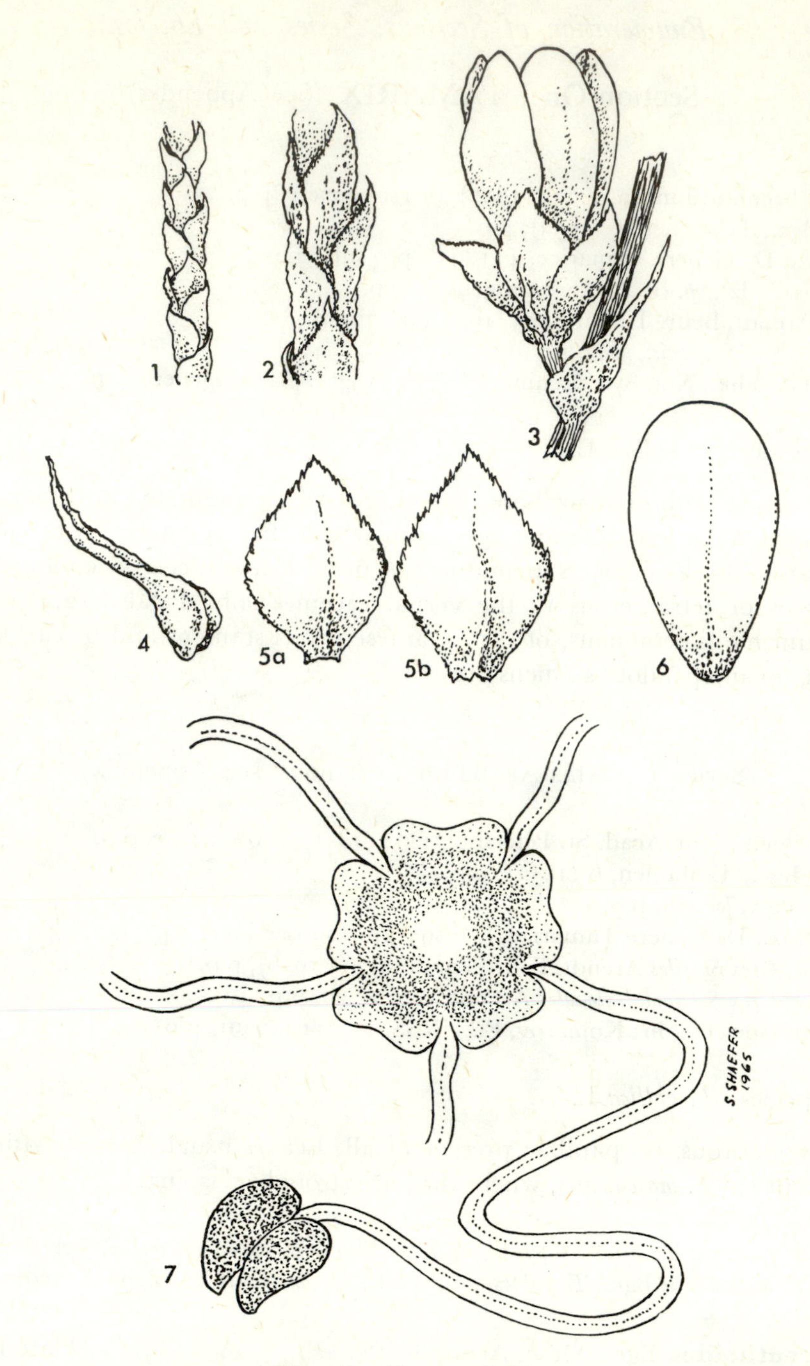

Plate I *T. arceuthoides*

1. Young twig (x 5); 2. id (x 10); 3. Flower (x 10); 4. Bract (x 20); 5a. Outer sepal (x 20); 5b. Inner sepal (x 20); 6. Petal (x 20); 7. Androecium (x 30).

Type: KAZAKH SSR: *Lehmann*, Häufig an den steinigen uferen des obern Sarafschan bis in den Karatau 6.9.1841 (holotype P, fragment of type K).

Tree, often shrubby, 2–3 (4) m tall, with reddish-brown bark, entirely glabrous. Leaves sessile with narrow base, usually with scarious papillose-scabridulous margins, 1–2.5 mm long. Inflorescences both vernal and aestival, composed of densely arranged racemes.[8] Racemes 1.5–5 cm long, 3–4 mm broad. Bracts as long as or scarcely longer than pedicels, entirely herbaceous, linear-triangular to narrowly ovate, acuminate with acute apex. Pedicel shorter than calyx. Calyx pentamerous. Sepals trullate-ovate, acute, 0.5–0.75 mm long, subentire (papillose-denticulate), the outer 2 more obtuse than the broader inner 3. Corolla pentamerous, caducous. Petals obovate, 1–1.5 mm long. Androecium haplostemonous, of 5 antesepalous stamens; insertion of filaments peridiscal; disk hololophic.

Flowering: April to August.

Habitat: Stony beds of temporary rivers, river banks.

Distribution: Kazakh SSR, Turkmen SSR, Uzbek SSR, Kirghiz SSR, Iraq, Iran, Afghanistan, Pakistan, Kashmir (see Map 1).

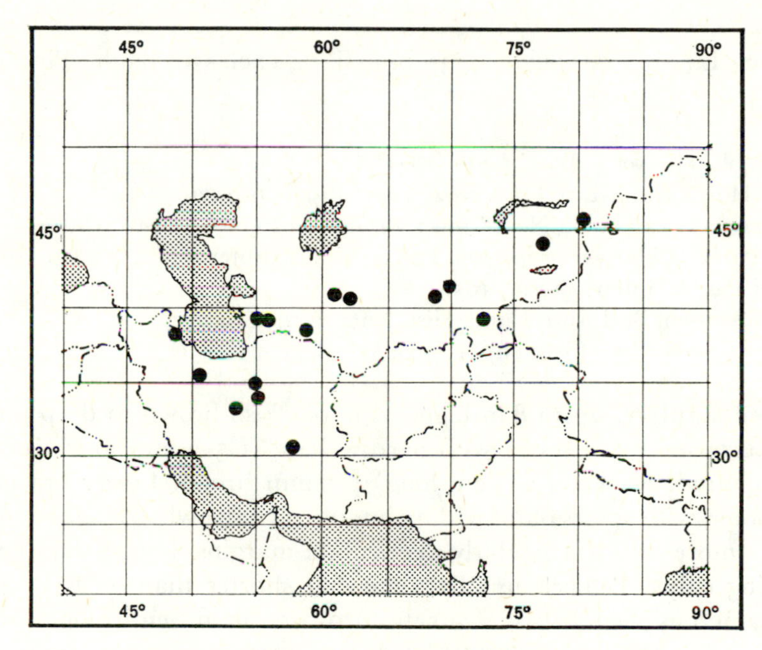

Map 1: *T. arceuthoides*

Selected specimens: TURKMEN SSR: *Sintenis* 468, Aschabad versus Besmen 4.6.1900 (B, S); *Litwinow* 981, Turcomania ad fl. Tedshen pr. Karry-Bent 4.7.1898 (B, E, LE); *Sintenis* 295, Aschabad pr. Kosen 19.5.1960 (holotype of *T. askabadensis* Freyn, G; isotypes G,

8 Specimens from higher altitudes and latitudes have mostly aestival inflorescences, those from lower altitutes and latitudes have mostly vernal ones.

P, W, WU); *Korshinsky* 2928, Roshan near outlet of Bartanya on humid soil in valley (holotype of *T. turkestanica* f. *brachystachys* Litw. ex Gorschk., LE). UZBEK SSR: *Bornmüller* 128, Prov. Samarkand ad ripas fluvii Serawschan prope pagum Gusar 1240 m, 18.7.1913 (B). KIRGHIZ SSR: *Igolkin* 527, Kirghizk Alatau Ig-Keletau ridge, Ulkun Kareka Canyon; stony bed of Taasl river 25.8.1931 (LE); *Iljin* 502, Talass Alatau, env. of Katmen-Tyube, river Tshitsh-Kemu conglomerate 8.8.1930 (LE). IRAQ: *Regel* 9, Diyala Liwa, Baquba on the banks of the Diyala river 4.4.1956 (B); *Haussknecht*, ad ripas fl. Tigris pr. Mossul .5.1867 (P). IRAN: *Bruns* 898, Teharan, Ufer des Djadje-Rud bei Gul-i-Djadje 22.5.1909 (B); *Aucher-Eloy* 4911, Ghilan (BM, P); *Bunge*, pr. Chabbis .4.1852 (FI, L, LE, P); *Kotschy* 157, inter Abushir et Schiras (holotype of *T. florida* var. *kotschyi* Bge., P; isotypes BM, G, K, OXF, S, W). AFGHANISTAN: *Giles*, South of Hindu Kush 7000′ (E); *Koelz* 12804, Iskarzir 8500 ft, river bed (E). PAKISTAN: *Siddiqui* & *Rahman* 26769, Chitral, Brimbret 30.7.1950 (UC). KASHMIR: *Stewart* 26462, Gilgit 16.7.1954 (UC).

Observations: (a) *T. karakalensis* var. *verrucifera* Freyn was based on a specimen with galls; thus, according to the rules, it is a nomen illegitimum. (b) The type of *T. turkestanica* Litw. f. *brachystachys* Litw. ex Gorschk. is mounted on one sheet together with the type of *T. turkestanica* Litw.

2. **T. gallica** L., Sp. Pl., 1:270 (1753), p. maj. p., excl. syn. Bauh., Pin., 485 (1623) [Plate II]

T. anglica Webb, Hooker J. Bot., 3:430 (1841).
T. algeriensis Hort., Hand-List Trees Kew, 1:35 (1894), pro syn.
T. pedemontana Savy ex Gand., Nov. Consp. Fl. Eur., 190 (1910), pro syn.
T. esperanza Pau & Villar var. *majoriflora* Pau & Villar, Broteria Bot., 23:101 (1927).
T. matritensis Pau & Villar, *op. cit.*, 105.
T. brachylepis Sennen, Butl. Inst. Catal. Hist. Nat., 32:90 (1932).
T. gallica L. var. *brachylepis* (Sennen) Sennen, *loc. cit.*

Tree, often shrubby, up to 8 m high, with blackish-brown to deep purple bark, entirely glabrous. Leaves sessile with narrow base, 1.5–2 mm long. Inflorescences loosely compound. Racemes 2–5 cm long, 4–5 mm broad. Lower bracts of vernal racemes oblong, with apices blunt with point, others narrowly triangular, acuminate, with usually more or less irregularly denticulate margins, longer than pedicels but not exceeding calyx. Pedicels usually somewhat shorter than or as long as calyx. Calyx pentamerous. Sepals trullate-ovate to ovate, acute, entire or subentire; the outer 2 somewhat smaller, slightly keeled, the inner somewhat longer and more obtuse, 0.75–1 mm long. Corolla pentamerous, caducous. Petals elliptic to slightly elliptic-ovate, 1.5–1.75 mm long. Androecium haplostemonous, of 5 antesepalous stamens; insertion of filaments peridiscal; disk synlophic; nectariferous tissue poor, i.e., torus almost membranous.[9]

9 In *T. canariensis*, which is closely related to *T. gallica*, there is more nectariferous tissue on both sides of the filaments and the disk tends to be paralophic.

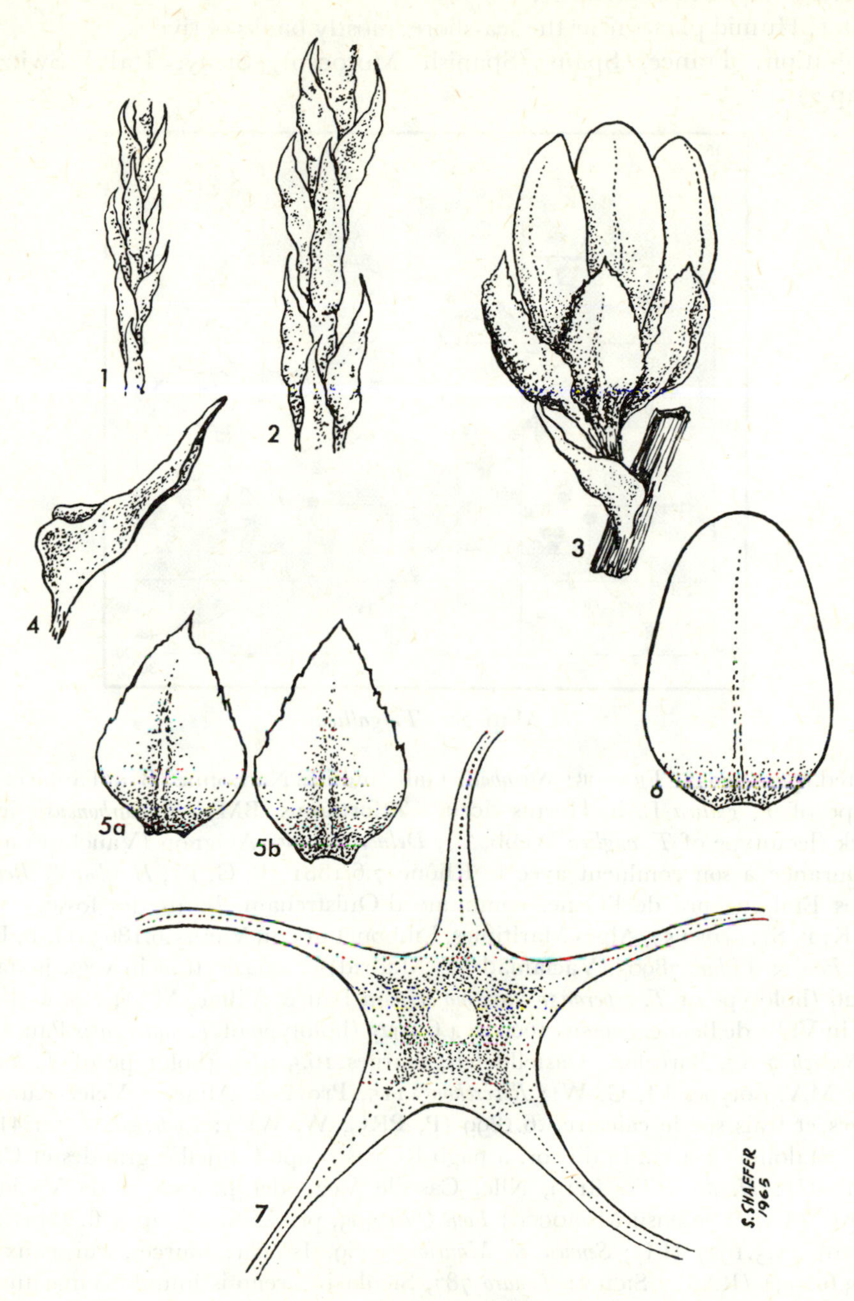

Plate II *T. gallica*
1. Young twig (x 5); 2. id (x 10); 3. Flower (x 10); 4. Bract (x 20);
5a. Outer sepal (x 20); 5b. Inner sepal (x 20); 6. Petal (x 20);
7. Androecium (x 30).

Flowering: April to September.

Habitat: Humid places near the sea-shore, mostly banks of rivers.

Distribution: France, Spain (Spanish Morocco), Sicily, Italy, Switzerland (see Map 2).

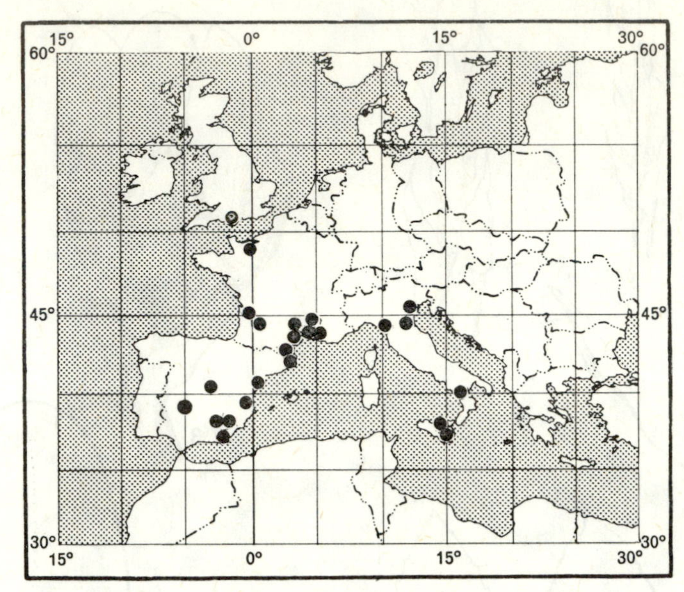

Map 2: *T. gallica*

Selected specimens: FRANCE: *Steinheil*, Gall. merid. Narbonne (B); 'Tamarix gallica' (lectotype of *T. gallica* L. in Hortus siccus Cliffortianus, BM); *T. narbonensis* in Herb. Lamarck (lectotype of *T. anglica* Webb, P); *Delacour* 3305, Avignon (Vaucluse) alluvions de la Durance à son confluent avec le Rhône 7.6.1881 (B, G, P); *Hardoin* & *Renou* 139, Calvados Embouchure de l'Orne, commune d'Ouistreham, bords des fosses 21.7.1853 (FI, G, K, P, S): *Cernat* 57, Alpes Maritimes, Embouchure du Var 25.6.1863 (G, P, US, W). SPAIN: *Pau* & *Villar*, 78905, Vaciamadrid (Madrid) Tamaricetum in vega juxta vicum 22.6.1926 (holotype of *T. esperanza* var. *majoriflora* Pau & Villar, MA); *Pau* & *Villar*, ad ripas fl. in Valle de Belmez, versus 1000 m 4.6.1925 (holotype of *T. matritensis* Pau & Villar, MA); *Sennen* 7017, Barcelone Castelldefels, marges 16.5.1929 (holotype of *T. brachylepis* Sennen, MA; isotypes FI, G, W); *Reverchon* 1115, Province Almeria Velez-Rubio lieux ombrages et frais sur le calcaire, .6.1899 (P, PRC, W, WU); *Rodrigues Lopes* 41, Costa Brava (Catalonia) 3-4 km in dir. or. a pago Roseas prope Canjellas grandes et C. petites 1.7.1961 (U); *H. del Villar* 8119, Nlle, Castille Vega del Jarama et de Vaciamadrid 29.6.1931 (MA). SPANISH MOROCCO: *Font Quer* 393, pr. Axdir ad ripas fl. Gius (Littore Chiphaeo) 15.5.1927 (FI); *Sennen* & *Mauricio* 9369, Ismoar, sources, ruisseaux (Benic Sicar) 3.6.1934 (RAB). SICILY: *Todaro* 787, Sicula in arenosis inundatis maritimis Terranova presso il fume Zapulla (FI, K, OXF, P, S, U). ITALY: *Dr Polch* 96, in salsuginosis maritimis insula Lido prope Venetiam dumetum efformans .8.1842 (syntype of *T. gallica* var. *virgata* Bge., W); *Krause* 201, Begroeide zandige kust van de Adriatische Zee bij Marina di Ravenne (Emilia) even ten N. van Garibaldi kanaal 11.9.1953 (U); *Minio*, Venezia 30.5.1942 (FI); *Moricand*, bords de la mer Livorne 1834 (FI, G, US). SWITZERLAND: *Lachenal*, in Helvetia (syntype of *T. gallica* var. *virgata* Bge., W).

Observations: a. *T. gallica* seems to be introduced in southern England. b. The author was able to see many voucher specimens of *T. anglica* Webb in K and FI.
c. Typification of *T. gallica*: Linnaeus' protologue and syntypes of *T. gallica* are:
1. Hort cliff. 111, which the author was able to see in London (BM);
2. Roy. Lugdb. 436, which is in L and which the author did not see. According to W. T. Stearn, there is a great probability that the Van Royen specimen is identical with the Hortus Cliffortianus one because of the closeness of time and material between the two;
3. Sauv. monsp. 45: the specimen does not exist;
4. Hort. Ups. 69: the author was able to see two specimens in Stockholm (S): one in the general herbarium with a mark 'Hort. Ups. 1.69.1', another which has some citations on the back of the sheet, in Linnaeus' special herbarium. According to Prof. Norlindth (personal communication) the citations were written by Dr Montin, who received specimens from Linnaeus' private gardener at Uppsala. Among other remarks, Dr Montin wrote on this sheet of *T. gallica* 'Specimen ex horto upsaliensi'. Both Dr Montin's specimen and the one in the general herbarium are identical with that in Hortus Cliffortianus;
5. Mat. med. 154 consists only of text and is not accompanied by a type specimen;
6. Bauh., Pin., 485, which the author was able to see in Burser's Hortus siccus at Uppsala (UPS). This specimen is an aestival inflorescene of *T. africana* Poir.;
7. Lob. ic. 211: from the drawing alone it is difficult to identify the species. The author was unable to find Lobel's topotype.

From the above-mentioned seven syntypes the author has chosen the one from *Hortus siccus Cliffortianus* as the lectotype of *T. gallica* L.; the others become automatically paralectotypes. It is now possible that the specimen in the Stockholm General Herbarium is an isoparalectotype. Another isoparalectotype might be the specimen No. 383.1 of the Linnaean herbarium at LINN; indeed, a number of Linnaean specimens of this herbarium were taken from Uppsala (W. T. Stearn, personal communication). The only paralectotype which should be excluded from the list is Bauhin's specimen. From Lobel's drawing alone no conclusion can be drawn. Another specimen of interest in the Stockholm General Herbarium is one identified by the son of Linnaeus, in his own handwriting, as *T. gallica*, and which the author could identify as an aestival inflorescence of *T. africana*.
d. See also observation (b) on *T. africana*.
e. The occurrence of *T. gallica* in Spanish Morocco is doubtful. The author regards most of the alleged *T. gallica* material which he saw from this area as *T. canariensis*. More and better material is needed to solve this problem.
f. See observation (a) on *T. canariensis*.

3. **T. korolkowii** Regel & Schmalh. ex Regel, Acta Horti Petrop., 5:582 (1877), ('korolkowi' orth. mut.) [Plate III]

T. florida Bge. var. *albiflora* Bge., Tentamen, 38 (1852).
T. montana Kom., Trav. Soc. Nat. St. Pétersb., 26:142 (1896).
T. turkestanica Litw., Trav. Mus. Bot. Acad. St. Pétersb., 7:72 (1910).

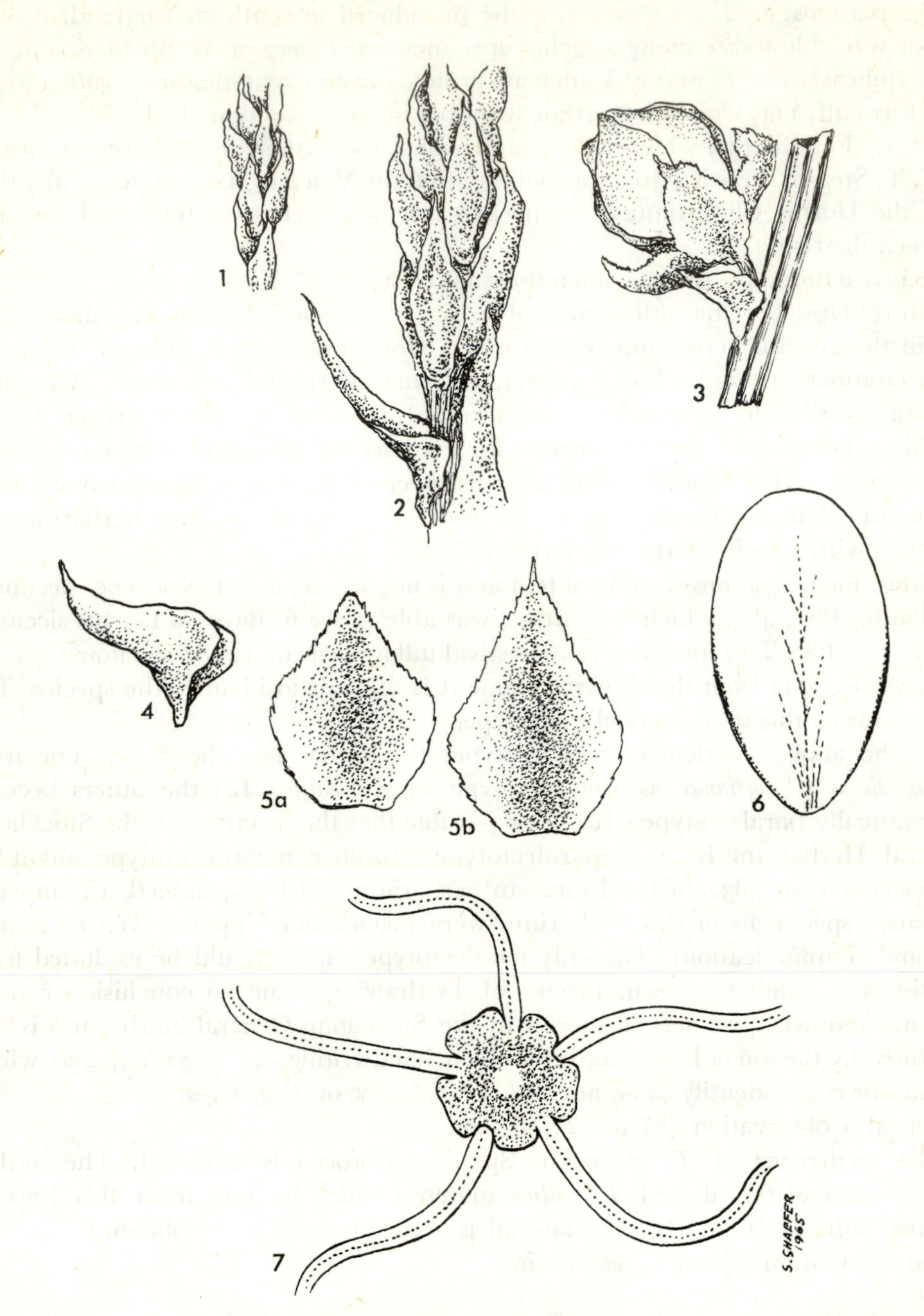

Plate III *T. korolkowii*
1. Young twig (x 5); 2. id (x 10); 3. Flower in bud (x 10); 4. Bract (x 20);
5a. Outer sepal (x 20); 5b. Inner sepal (x 20); 6. Petal (x 20);
7. Androecium (x 30).

Type: Uzbek SSR: *Korolkow & Krause*, Chiwa between Amu-Daria and Chiwa 25.5.1873 (holotype LE; isotype G, K).

Small tree or shrub with reddish-brown bark, entirely glabrous. Leaves sessile with narrow base, 1–3 mm long. Inflorescences simple when vernal, densely composed of racemes when aestival. Vernal racemes 2–7 cm, aestival ones 3–15 cm long, all 3–4 mm broad. Bracts longer than pedicels, except for the lowermost ones of the vernal inflorescences, which do not exceed the length of the pedicels and which are oblong and blunt, the others narrowly trullate, acuminate. Pedicels somewhat shorter than or as long as calyx. Calyx pentamerous. Sepals irregularly denticulate or erose-denticulate, 0.75–1.25 mm long, acute; the outer 2 ovate, slightly smaller than the broadly trullate inner 3. Corolla pentamerous, caducous. Petals 1.5–1.75 mm long, obovate-elliptic, somewhat inequilateral and emarginate at apex. Androecium haplostemonous, of 5 antesepalous stamens; insertion of filaments hypo-peridiscal (usually 1 hypo-, 4 peri-); disk hololophic.

Flowering: April to July, usually May–June.

Habitat: Beds of temporary rivers, near springs, moist places among conglomerates and on sands.

Distribution: Uzbek SSR, Kazakh SSR, Tadzhik SSR, Iran (see Map 3).

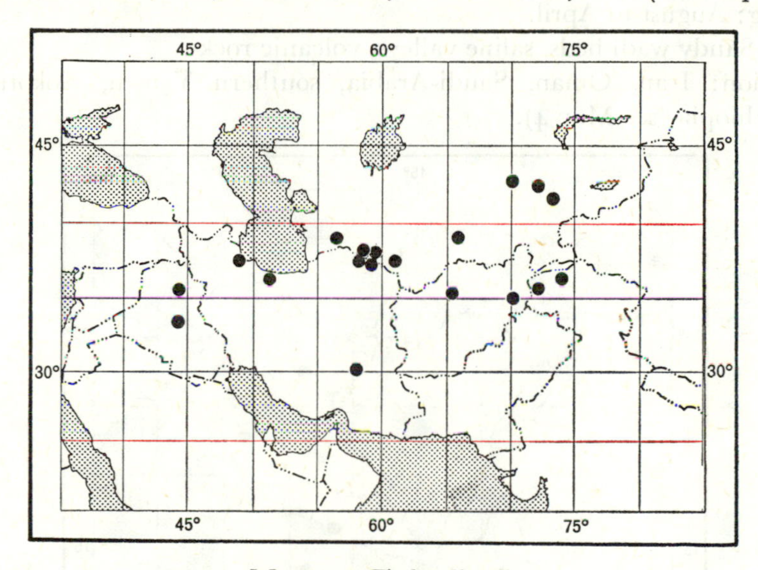

Map 3: *T. korolkowii*

Selected specimens: Uzbek SSR: *Litwinow* 3041, Uzbekistania ad ripam amnis prope pag. Andizhan 15.7.1913 (LE, US). Kazakh SSR: *Komarov*, Iskander Kul Alt. 7000 ped. (Zona Juniperus) 7.7.1892 (holotype of *T. montana* Kom., LE; isotype K); *Russanov* 161, Kazakstan Togai in middle of Ili riv. valley betw. Teresken and Tscharene 8.6.1929 (LE). Tadzhik SSR: *Korshinsky* 2928, Turkestania Roshan near outlet of Bartanga on humid soil in valley 1.8.1897 (holotype of *T. turkestanica* Litw., LE). Iran: *Strauss*, Pers. Occid. in monte Koh-Rud. .6.1908 (B); *Bunge*, inter Isfahan et Teheran 17.5.1859; *Buhse* 1210, prope Yesd (holotype of *T. florida* var. *albiflora* Bge., P).

Observation: On the type of *T. turkestanica* Litw. see observation (a) on *T. arceuthoides*.

4. **T. mascatensis** Bge., Tentamen, 60 (1852) [Plate IV]

Type: OMAN: *Aucher-Eloy* 4912, Regn. Mascate Secus Torrentem (holotype W; isotypes BM, FI, G, P).

Shrub or bushy tree with reddish-brown to deep brown bark, glabrous in all parts. Leaves amplexicaul with no coherent margins, 2–2.5 mm long. Vernal inflorescences simple and loose, usually aestival and densely compound. Racemes 1.5–3 cm long, 3–4 mm broad, densely flowered. Bracts longer than pedicels, linear, acuminate, with denticulate margins near their base. Pedicels shorter than calyx (flowers subsessile). Calyx pentamerous. Sepals 0.75–1 mm long, more or less finely denticulate,[10] the outer 2 ovate, acute slightly keeled, the inner trullate-ovate. Corolla pentamerous, early caducous. Petals 1.25–1.75 mm long, broadly ovate to ovate-elliptic. Androecium haplostemonous, of 5 antesepalous stamens; insertion of filaments peridiscal; disk para-synlophic.

Flowering: August to April.

Habitat: Sandy wadi beds, saline valleys, volcanic rocks.

Distribution: Iran, Oman, Saudi-Arabia, southern Yemen, Sokotra Island, Somalia, Ethiopia (see Map 4).

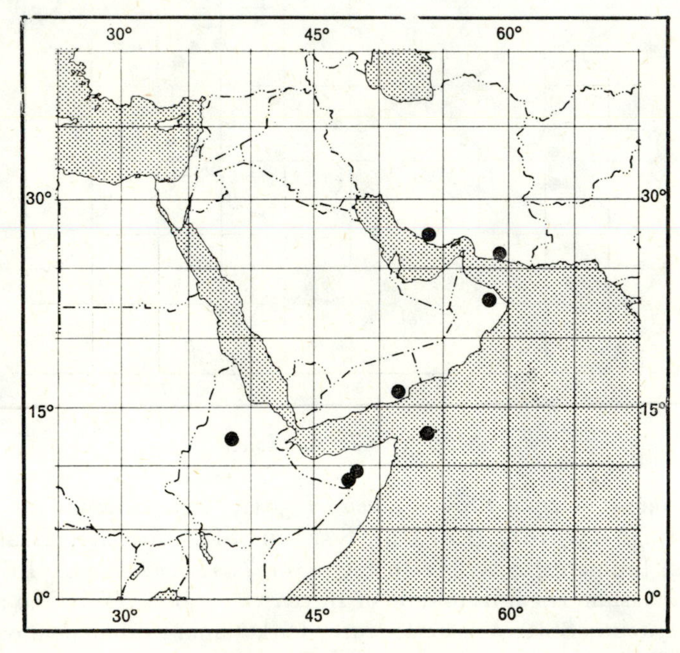

Map 4: *T. mascatensis*

10 Resembling to some extent those of *T. indica*.

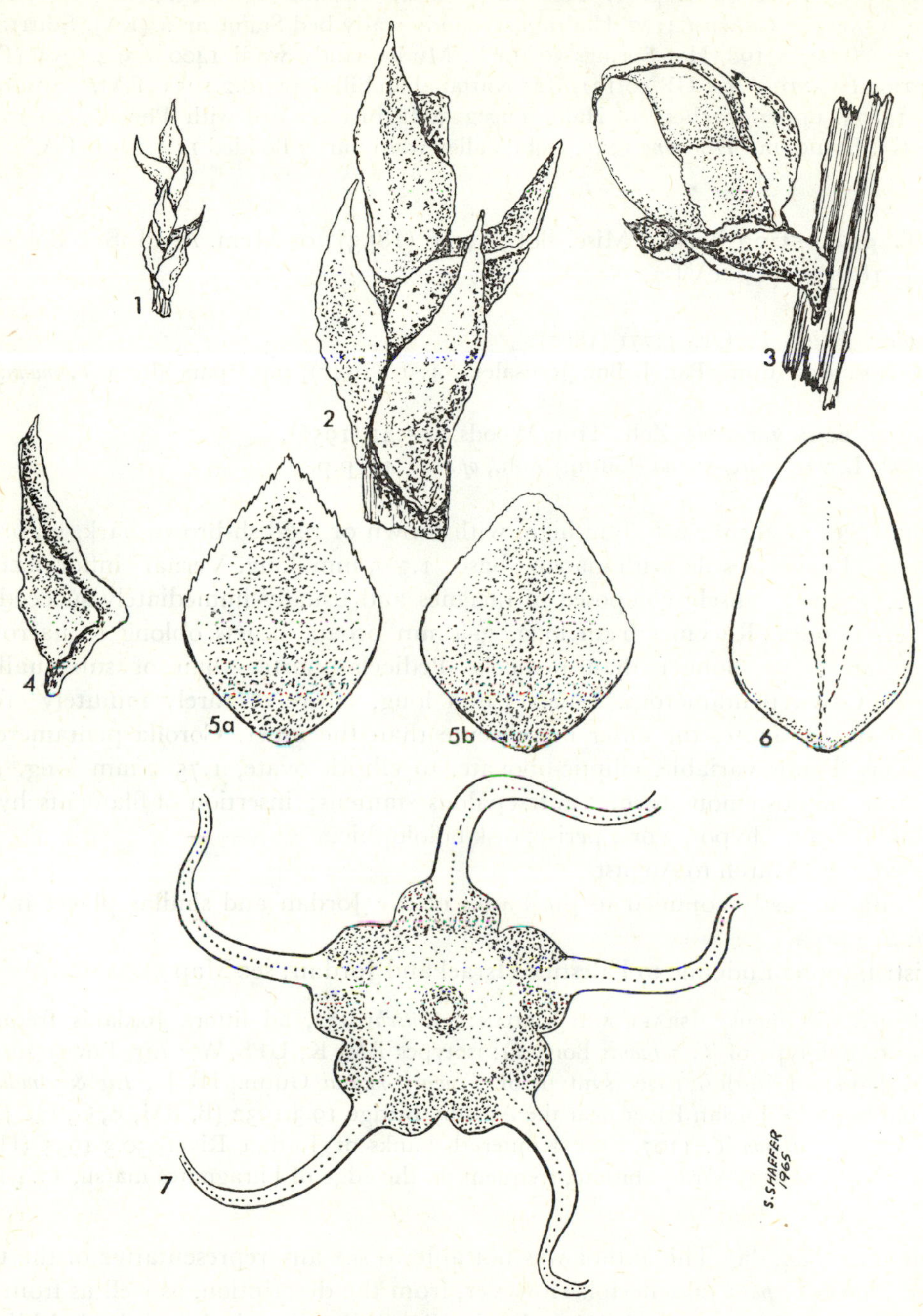

Plate IV *T. mascatensis*

1. Young twig (x 5); 2. id (x 10); 3. Flower in bud (x 10); 4. Bract (x 20); 5a. Outer sepal (x 20); 5b. Inner sepal (x 20); 6. Petal (x 20); 7. Androecium (x 30).

Selected specimens: IRAN: *Aucher-Eloy* 4510, desert. Sin. Persici (FI, G); *Bornmüller* 3353, Pers. austr. prov. Yesd ad Tschefta ditionis Agda 29.3.1892 (B, PRC). SAUDI-ARABIA: *Gilliland* 4127, Hadramaut sandy wady bed Sai'un area (EA). SOUTHERN YEMEN: *Grierson* 198, Hot Springs 30 km E. Mudia sandy wadi 1400 m 9.4.1953 (EA). SOKOTRA ISLAND: *Popov* GP/50/131, Ras Kattanahan hill slope 18.2.1953 (EA). SOMALIA: *Bally* 10875, upper reaches[11] of Halin Tug 2250 ft. in river bed with Phoenix 29.9.1956 (EA, K). ETHIOPIA: *Hemming* 1235, Dobi Valley wadi partly flooded 14.5.1957 (EA, K).

5. **T. palaestina** Bertol., Misc. Bot., 14:16 (1853); or Mem. Acad. Sci. Bologna, 4:424 (1853) [PlateV]

T. jordanis Boiss., Fl. Or., 1:771 (1867).
T. maris-mortui Gutm., Pal. J. Bot. Jerusalem, 4:50 (1947), p.p. (pars altera *T. mannifera* [Ehrenb.] Bge.).
T. jordanis Boiss. var. *typica* Zoh., Trop. Woods, 104:41 (1956).
T. gallica L. var. *maris-mortui* (Gutm.) Zoh., *op. cit.*, 44, p.p.

Small tree or shrub, 2–6(8) m high, with brown or reddish-brown bark, glabrous all over. Leaves sessile with narrow base, 1.5–5 mm long. Vernal inflorescences simple, aestival densely composed of racemes and usually immediately succeeding the vernal ones. Racemes 6 cm long, 3–4 mm broad. Bracts oblong to narrowly triangular, acute, longer than pedicels. Pedicels shorter than or subequalling calyces. Calyx pentamerous. Sepals 1 mm long, entire or rarely minutely erose, trullate-ovate, acute, the outer more acute than the inner. Corolla pentamerous, caducous. Petals variable, elliptic-obovate, to elliptic-ovate, 1.75–2 mm long. Androecium haplostemonous, of 5 antesepalous stamens; insertion of filaments hypo-peridiscal (1 or 2 hypo-, 3 or 4 peri-); disk hololophic.

Flowering: March to August.

Habitat: Mostly confined to the banks of the Jordan and similar places in the Dead Sea area.

Distribution: Endemic to Palestine (Israel and Jordan; see Map 5).

Selected specimens: ISRAEL AND JORDAN: *Kotschy* 432, ad littora Jordanis frequens 4.4.1855 (holotype of *T. jordanis* Boiss. G; isotypes BM, K, UPS, W); *Eig*, Lower Jordan Valley banks of Jordan 1926 (syntype of *T. maris-mortui* Gutm., HUJ); *Eig* & *Amdursky* 264, banks of the Jordan River near the Allenby Bridge 19.3.1932 (B, BM, E, G, HUJ, K, P, S, U, W); *Baum* T. 1107, near Kinnereth banks of Jordan River 30.3.1955 (HUJ, SMU); *Gillet* 15633, Azraq Shishan, frequent on the edge of Phragmites marsh, 17.4.1963 (K).

Observations: (a) The author was not able to see any representative of the type collection of *T. palaestina* Bertol. However, from the description, as well as from the locality where it was found—Habui ex valle Sidim in oris lacus Asphaltidis in

11 Transcribed 'weatches'.

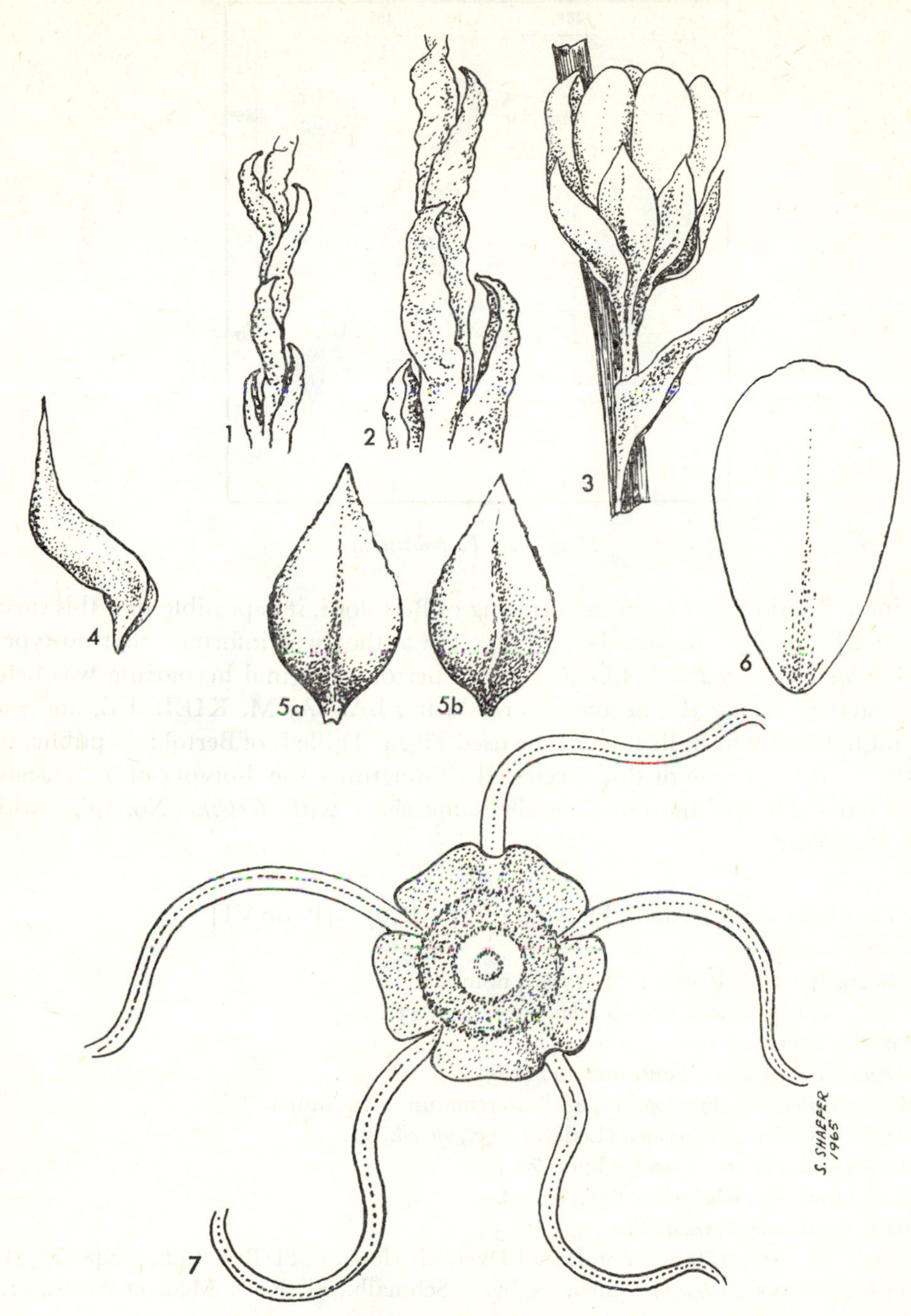

Plate V *T. palaestina*
1. Young twig (x 5); 2. id (x 10); 3. Flower (x 10); 4. Bract (x 20);
5a. Inner sepal (x 20); 5b. Outer sepal (x 20); 6. Petal (x 20);
7. Androecium (x 30).

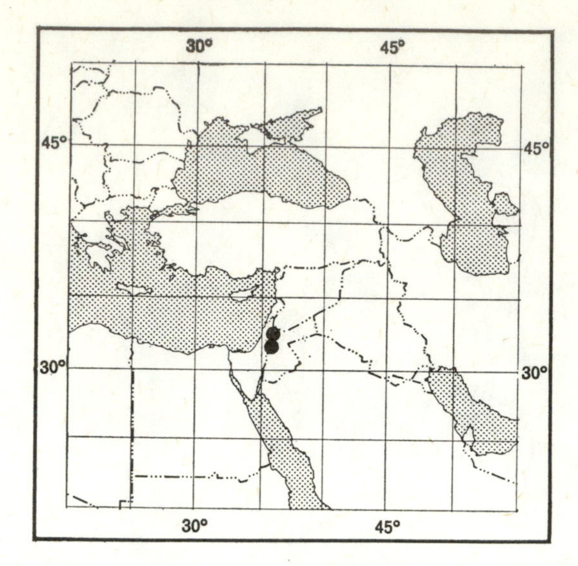

Map 5: *T. palaestina*

Palestina a Felisio—and from the drawing of Bertoloni, it is possible that this species is conspecific with *T. jordanis* Boiss. Since the author was informed that no type of *T. palaestina* exists in PAV, BOCO (where Bertol.'s original herbarium was before its destruction during the Second World War), BASSA, M, KIEL, LZ, and since he could not find it in B, BM, or FI, he used Fig. 4, Table I, of Bertoloni's publication (*loc. cit.*) as the neotype of this species. (b) Sometimes the isotypes of *T. jordanis*—*Kotschy* No. 432—are mounted on the same sheet with *Kotschy* No. 464, which is *T. smyrnensis*.

6. **T. ramosissima** Ledeb., Fl. Alt., 1:424 (1829) [Plate VI]

T. pendandra Pall., Fl. Ross., 2:72 (1788), nom. illegit.
T. gallica L. var. *micrantha* Ledeb., *ibid.*, 2:135 (1843), p.p.
T. eversmanni Presl ex Ledeb., *loc. cit.*, pro syn.
T. odessana Stev. ex Bge., Tentamen, 47 (1852).
T. eversmannii Presl ex Bge., *op. cit.*, 48 ('ewersmanni' orth. mut.).
T. pallasii Desv. var. *ramosissima* (Ledeb.) Bge., *op. cit.*, 51.
T. pallasii Desv. var. *brachytachys* Bge., *loc. cit.*
T. pallasii Desv. var. *minutiflora* Bge., *loc. cit.*
T. pallasii Desv. var. *tigrensis* Bge., *op. cit.*, 52.
T. gallica L. var. *pallasii* (auct. non Desv.) Dyer, in: Hook. f., Fl. Brit. Ind., 1:248 (1874).
T. pallasii Desv. var. *odessana* (Stev. ex Bge.) Schmalh., Fl. Ross. Med. et Austr., 1:169 (1895).
T. pentandra Pall. subsp. *tigrensis* (Bge.) Hand.-Mazz., Ann. Naturh. Mus. Wien, 27:57 (1913).
T. altaica Ndz., in: Engler & Prantl, Nat. Pflanzenfam., 21:287 (1925).
T. pallasii auct. plur. non Desv., p.p.

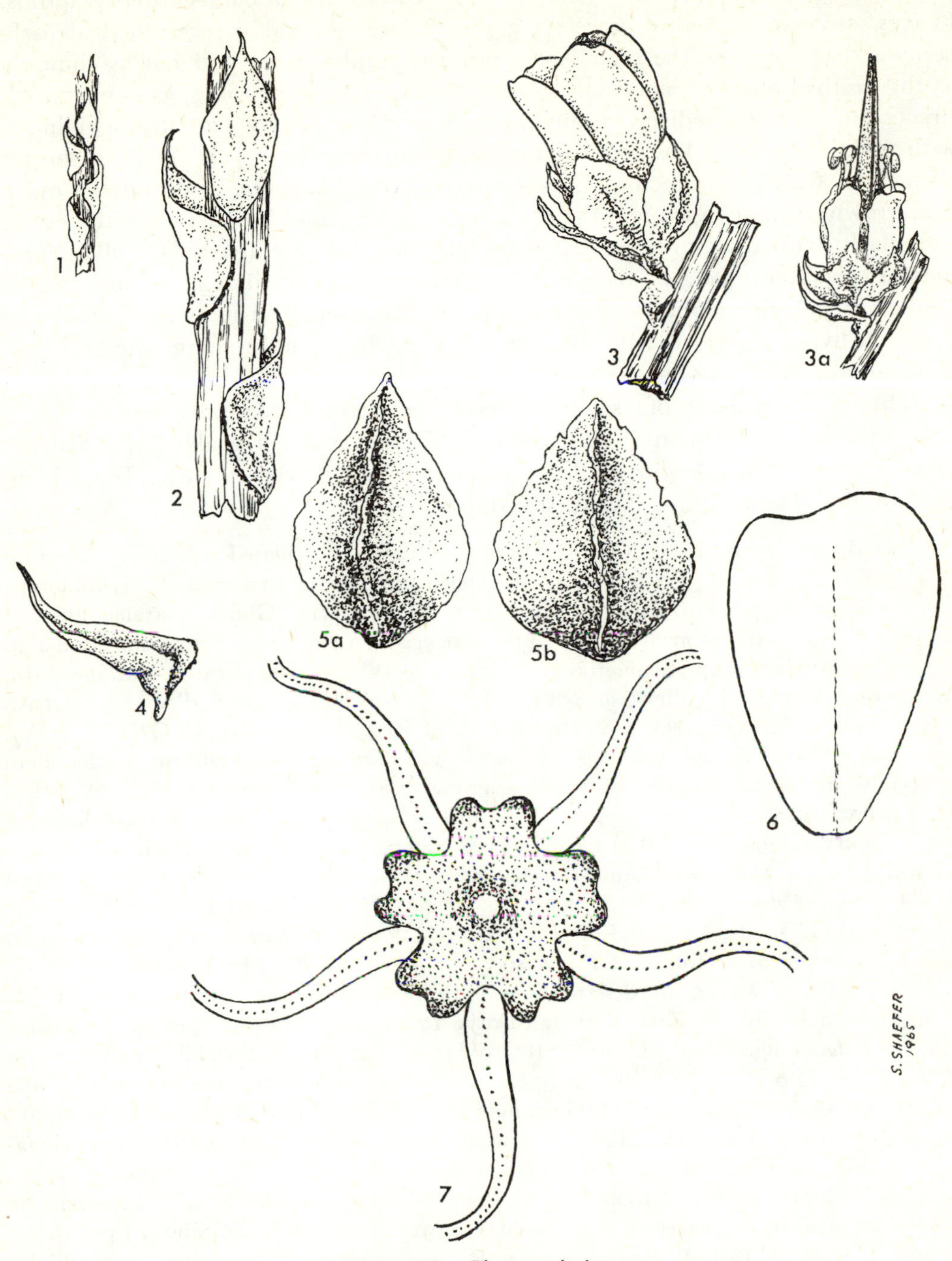

Plate VI *T. ramosissima*
1. Young twig (x 5); 2. id (x 10); 3. Flower (x 10); 3a. Young fruit (x 5); 4. Bract (x 20); 5a. Outer sepal (x 30); 5b. Inner sepal (x 30); 7. Androecium (x 30).

Shrub or shrubby tree, 1–5 (6) m high, with reddish-brown bark, entirely glabrous. Leaves sessile with narrow base, 1.5–3.5 mm long. Aestival inflorescences densely composed of racemes, the vernal ones usually simple, loose, and not as common as the aestival ones, often vernal-aestival. Racemes 1.5–7 cm long, 3–4 mm broad. Bracts longer than pedicels, triangular-trullate to narrowly trullate, acuminate, with margins more or less denticulate, mainly in their lower part. Pedicels shorter than calyx. Calyx pentamerous. Sepals narrowly trullate, acute, or the outer 2 ovate to narrowly trullate-ovate and the inner trullate-ovate and broader than the outer, irregularly denticulate to erose, 0.5–1 mm long, not connate at base. Corolla pentamerous, persistent. Petals 1–1.75 mm long, obovate to broadly elliptic-obovate, inequilateral. Androecium haplostemonous, of 5 antesepalous stamens; insertion of filaments hypodiscal; disk hololophic, its lobes usually strongly emarginate.

Flowering: May to October.

Habitat: Sandy shores of lakes and marshes, salty river banks, salty steppes.

Distribution: Ukrainian SSR, Russian SFSR, Mongolia, Kazakh SSR, Tadzhik SSR, Uzbek SSR, Kirghiz SSR, Azarbaijan SSR, Turkmen SSR, Iraq, Iran, Afghanistan, Tibet, China, Korea (see Map 6).

Selected specimens: IRAQ: *Chapman* 10730, Abughreib irrigated sub-desert alluvium 28.4.1948 (US); *Kotschy* [153] 453, in insulis Tigridis pr. Mossul sparsim 8.9.1841 (holotype of *T. pallasii* var. *tigrensis* Bge., P). IRAN: *Auchen-Eloy* 4511, Ghilan (paralectotype of *T. pallasii* Desv. var. *brachystachys* Bge., FI); *Rechinger* f. 1256, prov. Damghan-Seruna in desertis ad Sorcheh prope Semnan ca. 1600 m. (US, W). AFGHANISTAN: *Aitchison* 1028, Afghanistan Hari-Rud Valley 6.8.1858 (FI, W); *Pabot* A 809, 40 km S. d'Ankhoi 2.5.1958 (Herb. Pabot). TIBET: *Hook. f.* & *Thomson*, Tibet Reg. Trop. (CGE, G, OXF, P, S, W). UKRAINIAN SSR: *Szowits* (or *Lang* & *Szowits*) 146, Odessa circa salsum occidentalem in argillosis rarissima (holotype of *T. odessana* Stev. ex Bge., P; isotypes G, PRC, W, WU). AZARBAIJAN SSR: *Grossheim*, prov. Baku distr. Salijan dominium Kara-Tshala in steppa Shirvan 29.9.1925 (B, LE). TURKMEN SSR: *Androssow* 1870, Buchara in locis arenoso salsis pr. Farab ad fl. Amu-Darja 17.5.1905 (B, G, LE, PRC, S, WU). KAZAKH SSR-CHINA: *Politov*? (ex Herb. Petrop.), Songaria Chinensis ad lacum Saisang-Nor 1826? (BM, E, G, K, LE, OXF, P, PRC, S, W); *Karelin* & *Kiriloff* 718, in arenosis ad lacum Noor-Saissan (lectotype of *T. pallasii* var. *brachystachys* Bge., P; isolectotypes G, W). KAZAKH SSR: *Lehmann*, in deserto limoso ad fluvium Kuwan-Darja 14.7.1841, in collibus arenosis fluvios Kuwan et Jan-Darja 17.7.1841 (paralectotypes of *T. pallasii* var. *brachystachys* Bge., P). TADZHIK SSR: *Sidorenko et al.* 152, S. Tadzhikistan Tgrovayia Balka Tugai 12.6.1952 (LE). UZBEK SSR: *Popov* & *Vvedensky* 314, Deserta meridionalia Jaxartica in deserto, Mirza-Tschul (Goldanaja Step) in salsuginosis prope fontem Kamysty-Kuduk 17.5.1923 (B, E, G, K, LE, P, S). KIRGHIZ SSR: *Dobrov* 314b, Deserta septentrionalia Mujun-Kum in valle fl. Talas 21.7.1922 (B, E, G, HUJ, K, S). RUSSIAN SFSR: *Wizen*, Sarepta Wolgaschlucht 12.6.1909 (B, US); *Bornmüller* 1193, Turkestania Prov. Syr-Darja in desertis salsis inter Turkestan et Petrowsk 1.9.1913 (B); *P.S. Pallas*, Sibiria 14 d. (holotype of *T. pentandra* Pall., BM); *Eversmann*, in Russia (syntype of *T. eversmannii* Presl, PRC); *Klaus*, ad ostium Volgae (syntype of *T. eversmannii* Presl, P). MONGOLIA: *Przewalski*, Mongolia occidentalis Terra Ordos 1871 (E, G, K, P); *Potanin*, Gobi 1886 (G, K, LE, P, W). CHINA: *Chiao* 2853, Shantung prov. Hua Ying Sze 20.7.1930 (E); *Vorotnikov* 5774, Turkestania Sinensis, in deserto Takla-Makan Shor-Chagma-Bashi 24.6.1931 (S). KOREA: *Taquet* 132, in hortis Tjou Hjong Quelpaert .9.1907 (E, K).

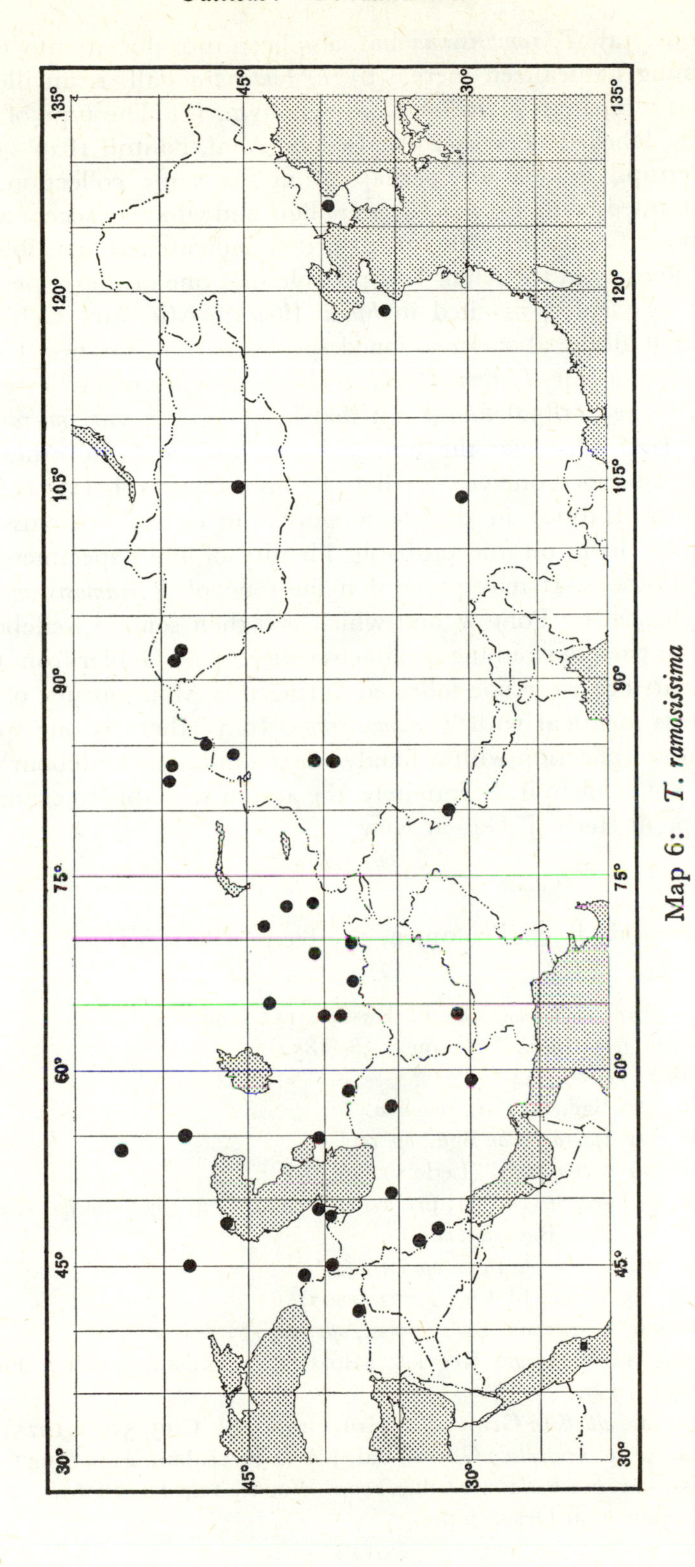

Map 6: *T. ramosissima*

Observations: (a) *T. ramosissima* has also been introduced into the New World and is becoming naturalized there. (b) *T. pentandra* Pall. is an illegitimate name because Pallas also cited *T. gallica* L. as synonym. (c) The type of *T. ramosissima*: In some sheets, labelled '*T. ramosissima* Ledeb., Altai, Politoff 1826' or: 'Saisang-Nor [ex Herb. Petrop.]' which are perhaps from the same collection, two different species are mounted, namely, *T. leptostachya* Bge. and what we accept as *T. ramosissima* Ledeb. From de Candolle (1880) p. 415, it is indicated that Gebler and Politow made a joint journey in the Altai. It is possible that one of these specimens was sent to Ledebour by Gebler, as cited in *Flora Altaica*: 'Mis. am. Gebler'. So far one may assume that all the above-mentioned specimens are isotypes. Ledebour (1843) in his *Flora Rossica* put *T. ramosissima* in the synonymy of his *T. gallica* var. *micrantha*. Later, Bunge (1852) found out that in *T. gallica* var. *micrantha* there were two species mixed. One was the true *T. ramosissima* and the other *T. leptostachya*, which he had described one year earlier. In his monograph (1852) Bunge remarks on both species 'Habitat in deserto songorico ad lacum Nor-Saissan (Politow!)'. This may throw light on the probable identity of both specimens of the above-mentioned collections. It may prove that the type of *T. ramosissima* is a specimen which was collected by Politow and which was then sent to Ledebour by Gebler. The author first tried to examine specimens collected by Gebler from CGE, L, OXF, P, and saw that they were not collected during this Altai journey of 1826 and that not even one is identical with *T. ramosissima*. In W there is one specimen of this assumed Politow collection with a handwriting similar to Ledebour's own, but the author could not confirm it. Fortunately, the author was able to examine a fragment of the holotype in Herb. J. Gay at Kew.

7. **T. smyrnensis** Bge., Tentamen, 53(1852) [Plate VII]

T. gallica L. var. *pycnostachys* Ledeb., Fl. Ross., 2:135 (1843).
T. florida Bge. var. *rosea* Bge., Tentamen, 38 (1852).
T. hohenackeri Bge., *op. cit.*, 44.
T. floribunda Stev. ex Bge., *loc. cit.*, pro syn.
T. pallasii Desv. var. *macrostemon* Bge., *op. cit.*, 50.
T. pallasii Desv. var. *pycnostachys* (Ledeb.) Bge., *loc. cit.*
T. lessingii Presl ex Bge., *op. cit.*, 51, pro syn. *T. pallasii* var. *pycnostachys* (Ledeb.) Bge.
T. pallasii Desv. var. *effusa* Bge., *loc. cit.*
T. pallasii Desv. var. *moldavica* Bge., *op. cit.*, 52.
T. bachtiarica Bge. ex. Boiss., Fl. Or., 1:772 (1867).
T. pallasii Desv. var. *smyrnensis* (Bge.) Boiss., *op. cit.*, 773.
T. hohenackeri Bge. var. *bungeana* Regel & Mlokoss., in: Kusn., Busch & Fomin, Fl. Cauc. Crit., 3(9):99 (1909).
T. pallasii Desv. var. *albiflora* Grint., Bull. Inf. Grad. Bot. Cluj, 5:95 (1925).
T. jordanis Boiss. subsp. *xeropetala* Gutm., Pal. J. Bot. Jerusalem, 4:49 (1947).
T. jordanis Boiss. var. *brachystachys* Zoh., Trop. Woods, 104:41 (1956).
T. pallasii auct. plur. non Desv., p.p.

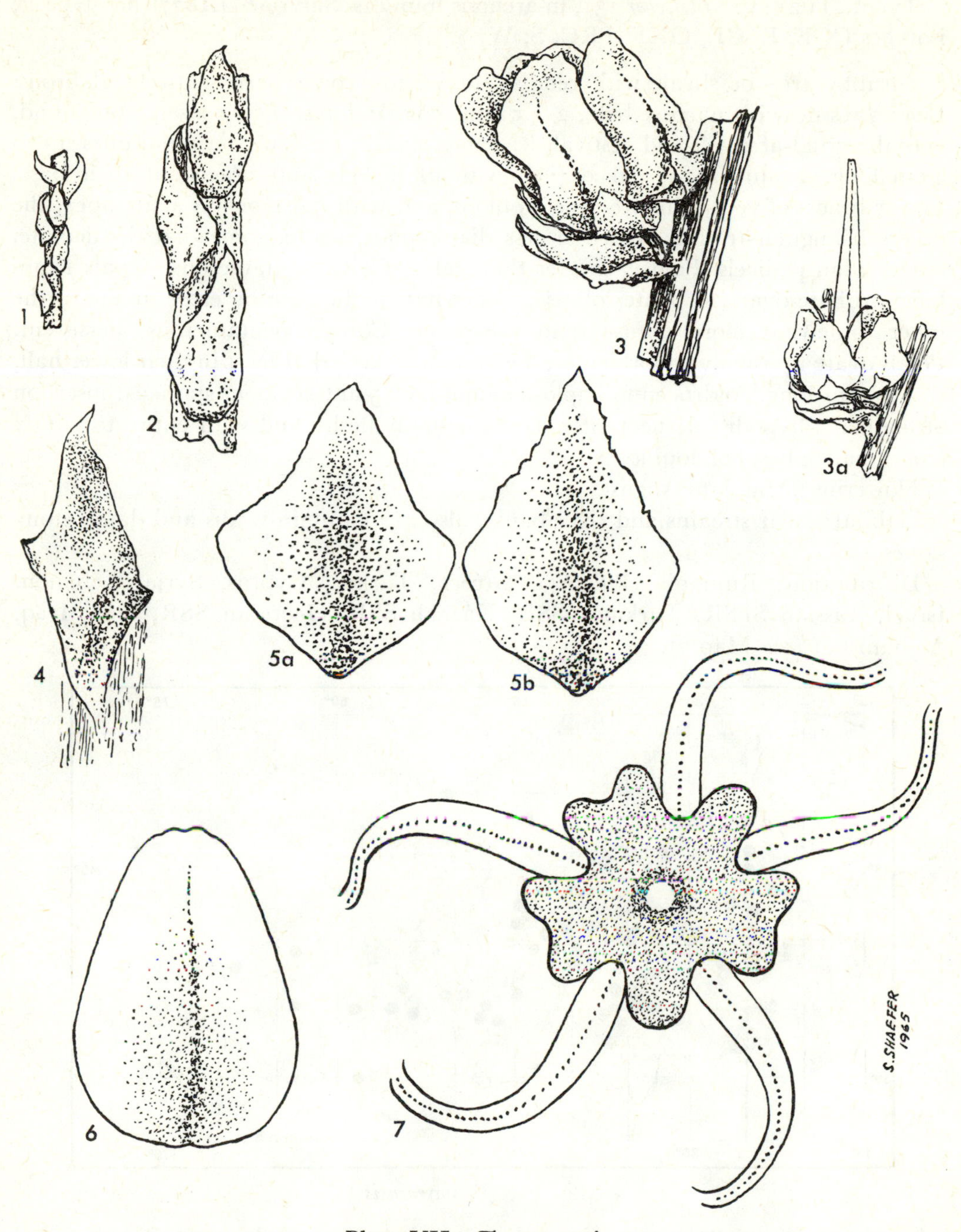

Plate VII *T. smyrnensis*
1. Young twig (x 5); 2. id (x 10); 3. Flower (x 10); 3a. Young fruit with persistent petals (x 5); 4. Bract (x 10); 5a. Inner sepal (x 40); 5b. Outer sepal (x 40); 6. Petal (x 20); 7. Androecium (x 35).

Type: TURKEY: *Fleischer* 131, in arenosis humidis Smyriae 4.1827 (holotype P; isotypes CGE, E, GL, OXF, PRC, S, W).

Shrubby tree or shrub with reddish-brown to brown bark, entirely glabrous. Leaves sessile with narrow base, 2–3.5 mm long. Inflorescences loosely compound, vernal, vernal-aestival, and aestival. Racemes 0.5–2.5 cm long, terminal ones up to 4 cm long, 4 mm broad, naked (i.e., without flowers and bracts) at their base. Lower bracts of vernal inflorescences oblong and with more or less acute apex, the others triangular-trullate, more or less diaphanous, acute, somewhat denticulate, longer than pedicels. Pedicel shorter than calyx. Calyx pentamerous. Sepals 1 mm long, trullate-ovate to ovate, obtuse, somewhat denticulate or erose at apex, the inner somewhat more obtuse than the outer. Corolla pentamerous, persistent. Petals ovate to usually suborbicular, inequilateral, keeled at least in their lower half, 2–2.75 mm long. Androecium haplostemonous, of 5 antesepalous stamens; insertion of filaments hypodiscal; nectariferous disk usually fleshy and with entire to faintly emarginate lobes, hololophic.

Flowering: March to August.

Habitat: Near streams and river banks, also in mountain wadis and damp stony slopes.

Distribution: Rumania, Bulgaria, Greece, Turkey, Cyprus, Syria, Lebanon, Israel, Russian SFSR, Turkmen SSR, Kazakh SSR, Georgian SSR, Iran, Iraq, Afghanistan (see Map 7).

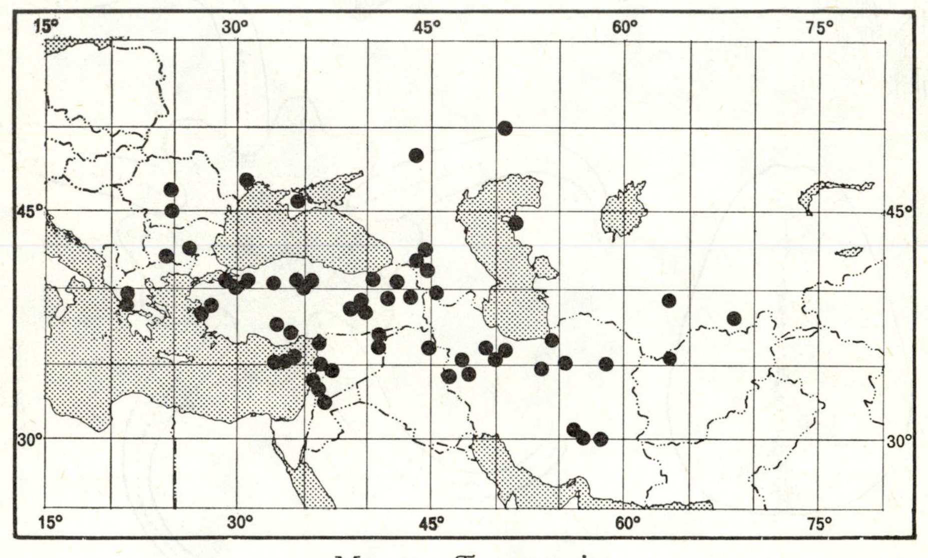

Map 7: *T. smyrnensis*

Selected specimens: RUMANIA: *Graescu*, in glarea ad margines fluviorum Teleagenu ad Scai'eni .6.1884 (PRC, W); *Guebhard* 51, in arenosis ad ripas fluvii Siret (holotype of *T. pallasii* var. *moldavica* Bge., W; isotype G). BULGARIA: *Stribrny*, in arenosis pr. Kricim 4.6.1897 (B, G, HBG, K, OXF, PRC, S, US, W). GREECE: *Bornmüller* 281, Peloponnesus Achaia ad Aegron (Tostitra) 31.5.1926 (B, G, S, W); *Bourgeau* 52, Rhodes près Salakos

31.5.1870 (BM, K, P, W). TURKEY: *Balansa* 132, marais d'eau saumâtre situés entre la Papeterie et la mer, près de Smyrne (BM, E, G, K, OXF, P); *Davis* & *Hedge* 29219, Prov. Tunceli, Tunceli Pulumur 10 miles from Tunceli 1000 m river banks 7.6.1957 (E, K). CYPRUS: *Kotschy* 572, ad alveas prope littora maris in viciniis Kuklia Amathus et aliis locis .3.1862 (G, K, L, P, S); *Sintenis* & *Rigo* 623 or 505?, entre Euriku et Galata 2.6.1880 (B, CGE, K, L, P, PRC, US, W, WU). SYRIA: *Pabot* 55, vallée du Yarmouk vers Hamme 27.1.1952 (Herb. Mouterde); *Mouterde*, bords du Tigre vers les ruines du Pont Romain 17.4.1955 (Herb. Mouterde). LEBANON: *Ehrenberg*, Syria Beirut .3.1820–1826 (syntype of *T. nilotica* Ehrenb., L, P); *Holleman-Haye*, Lattaquie 1936 (U). ISRAEL: *Post* 357, Palestine Huleh 6.4.1877 (E, FI); *Gutman*, Jordan Valley in wadis between Lisan hills of the Broken Land, 1–2 km west of Allenby Bridge 1941 (holotype of *T. jordanis* subsp. *xeropetala* Gutm., HUJ); *Waisel* 1238, Upper Jordan Valley Neoth Mordekhai 11.3.1953 (holotype of *T. jordanis* var. *brachystachys* Zoh., HUJ). RUSSIAN SFSR: *Lessing*, Rossia, hab. ad ripam dextram fluvii Ural (holotype of *T. lessingii* Presl, PRC; isotype P); *Meyer* 491 Caucasus Dagestan prov. near Temir-Chanchura between Petrowsk and Temir-Chan-Schura 18.5.1910 (UC); *Steven*, Kislar (syntype of *T. gallica* var. *pycnostachys* Ledeb., P); *Kolenati* 1157, ad fl. Alasan in prov. Saleki Trans.-Cauc. (syntype of *T. pallasii* var. *macrostemon* Bge., P). TURKMEN SSR: *Bornmüller* 1051, Turkestania Buchara prov. Kabadian Dsihili-Kul ad fluvii Wachsch ripas insularum 16.8.1913 (B). KAZAKH SSR: *Spiridonow* 719, Aktioubinskayia Gob. Tschelkarskyi Ou. (LE). GEORGIAN SSR: *Hohenacker*, ad rivulos pr. Helenendorf .6.1838 (lectotype of *T. hohenackeri* Bge., P; isolectotypes BM, FI, G, HBG, K, L, OXF, P, PRC, S, US, W); paralectotypes: *Szowits* 111, in sylvaticis ad fl. Chram 1.5.1829 (G, K, P); *Schuman*, Ufer der Kura Tiflis 11.7.1882 (PRC). IRAN: *Bunge*, Ssof inter Isfahan et Teheran 13.5.1859 (holotype of *T. bachtiarica* Bge.; G, isotypes FI, P); *Sintenis* 1334, Pers. bor. prov. Asterabad Bender Ges in maritimis 25.11.1900 (B, E, G, WU); *Bornmüller* 1220, Kurdistania Riwandous ad fines Pers. in infer. monte Helgrud 25.6.1893 (B, G); *Buhse* 1349/2 (b!) (holotype of *T. florida* var. *rosea* Bge., P); *Szovits*, Iberia and/or Azerbeidshan (holotype of *T. floribunda* Stev., P); *Szowits* 364, in glareosis ad torrentem Arvin prope Badalan dicto Khoi prov. Azerbeidzan 7.6.1928 (syntype of *T. pallasii* var. *macrostemon* Bge., P); *Szowits*, Karababa, Juni 1829 (holotype of *T. pallasii* var. *effusa* Bge., P). IRAQ: *Chapman* 11201, Haji Omran igneous metamorphic rocks near streams 3.6.1948 (YS). AFGHANISTAN: *Griffith* 960 & 961, 'Affghanistan' (L); *Pabot* 782, Chikargan 5 km N. W. 1.5.1958 (Herb. Pabot).

Observations: (a) *T. florida* var. *rosea* Bge.—Typification: The sheet bearing Buhse's label with No. '1349/2 (b!)' bears also the Greek letter *delta*, which means var. [*delta*] *rosea*. This agrees also with the general description. (b) All syntypes of *T. pallasii* var. *macrostemon* Bge. are conspecific with *T. smyrnensis*, and all of them are kept in Paris. (c) *T. smyrnensis* was often confused with *T. ramosissima*, from which it mainly differs in its ovate to orbicular and keeled petals; *T. ramosissima* (petals not obovate and not keeled) seems to be much more salt-tolerant than *T. smyrnensis*, which is common along mountain river banks.

Series 2. LEPTOSTACHYAE (Bge.) Baum (comb. nov.)

Parviflorae Ehrenb., Linnaea, 2:257 (1827), p. maj. p.
Leptostachyae Bge., Mém. Acad. St. Pétersb., 7:293 (1851), pro sectione.
Piptopetalae Bge., Tentamen, 6 (1852), p. maj. p.
Epidiscus Ndz., De Genere Tamarice, 7 (1895), pro subsectione, p.p.
Mesodiscus Ndz., *op. cit.*, 8, pro subsectione, p.p.
Pentastemones Grex *Sessiles* Arendt, Beitr. Tamarix, 36 (1926), p.p.
Pentastemones Grex *Semiamplexicaules* Arendt, *op. cit.*, 39, p.p.
Hispidae Gorschk., in: Komarov, Fl. URSS, 15:308 (1949).
Micranthae Gorschk., *op. cit.*, 310.
Ramosissimae Gorschk., *op. cit.*, 311, p.p.

Type species: *T. leptostachya* Bge.

At least younger parts hairy, papillose or papillulose. Leaves sessile with narrow bases to auriculate or more or less clasping.

Included species: *T. arabica* Bge., *T. aralensis* Bge., *T. arborea* (Sieb. ex Ehrenb.) Bge., *T. canariensis* Willd., *T. hispida* Willd., *T. indica* Willd., *T. karakalensis* Freyn, *T. leptostachya* Bge., *T. mannifera* (Ehrenb.) Bge., *T. nilotica* (Ehrenb.) Bge., *T. senegalensis* DC.

8. **T. arabica** Bge., Tentamen, 55[12] (1852) [Plate VIII]

T. scebelensis Chiov., Fl. Somala, 93 (1929).
T. gallica L. var. *longispica* Zoh., Trop. Woods, 104:46 (1956).
T. gallica L. var. *eilathensis* Zoh., *loc. cit.*

Type: YEMEN: *Botta*, Arabiae felicis in Tehama, Souera bords du ruisseau 10.1837 (holotype P; isotypes G, K, P).

Bushy tree with brown or reddish-brown bark, younger parts entirely papillose. Leaves sessile with narrow base, cordate-auriculate, slightly amplexicaul when young, 1.75–2.5 mm long. Aestival inflorescences compound, vernal ones rare. Racemes 2–5 cm long, 3–4 mm broad, densely flowered. Bracts longer than pedicels, linear-triangular, acute, with somewhat dentate margins near base. Pedicel slightly shorter than calyx. Calyx pentamerous. Sepals 1 mm long, trullate-ovate, erose-denticulate, with many fine, thin, acute, minute teeth, more or less acuminate but not very acute at apex, the 2 outer more acute and slightly keeled. Corolla pentamerous, caducous. Petals obovate, 1.5 mm long, equi- to inequilateral. Androecium haplostemonous of 5 antesepalous stamens; insertion of filaments hypo-peridiscal (1 to 2 hypodiscal and the other 3 to 4 peridiscal); disk hololophic.

Flowering: July to October, sometimes also March, April.

Habitat: Dry river beds, sands.

12 *T. arabica* Pall., Nova Acta Acad. Sci. Imper. Petrop., 10:376 (1797) is a nomen nudum and therefore does not invalidate Bunge's name of this species.

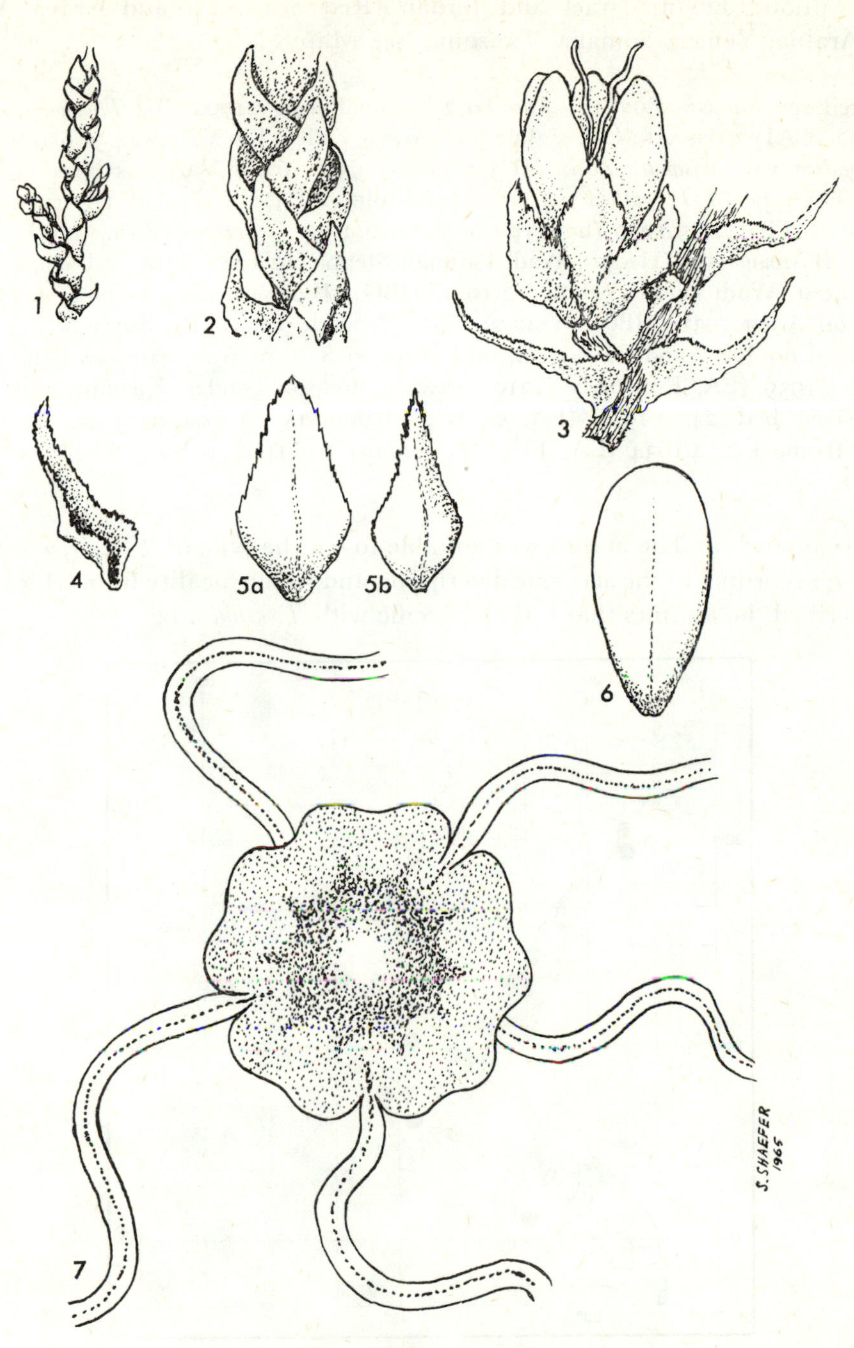

Plate VIII *T. arabica*
1. Young twig (x 5); 2. id (x 10); 3. Flower (x 10); 4. Bract (x 20); 5a. Inner sepal (x 20); 5b. Outer sepal (x 20); 6. Petal (x 20); 7. Androecium (x 40).

Distribution: Egypt, Israel and Jordan (Red Sea coast and Arava Valley) Saudi Arabia, Yemen, Somalia, Tanzania (see Map 8).

Selected specimens: EGYPT: *Kneucker* 275, Ismailia 22.3.1902 (B); *Täckholm*, Helwan 3.10.1952 (CAI). ISRAEL: *D. Zohary* 895, Arava Valley, Ein Yahav 24.3.1950 (holotype of *T. gallica* var. *anisandra* Zoh., HUJ); *Tadmor* 934, Arava Valley, salines near Eilat .4.1949 (holotype of *T. gallica* var. *eilathensis* Zoh., HUJ); *Tadmor* 961, Arava Valley environs of Eilat 26.9.1950 (holotype of *T. gallica* var. *longispica* Zoh., HUJ). SAUDI ARABIA: *Wissmann* 565, Hedjas, Wadi Fatimah Steppe .12.1927 (HBG); *Wissmann* 1121, Hadramaout, Wadi Hadjer (Hafir) .4.1941 (HBG); *Wissmann* 2414, Gebirge des Hinterlandes von Aden 1931 (HBG). YEMEN: *Bové*, Arabiae felicis (P). SOMALIA: *Hemming* 1570, Bihen on Las Anod-Gave road, bushy tree to 8 ft in wadi cut through gypsum plain 10.9.1959 (EA, K); *Bally* 11210, between Bosaso (Bender Kassim) and Elayu, in dry river bed 24.10.1956 (EA, G, K). TANZANIA: *Greenway* 3943, W. Usambaras, Mkomazi 22.4.1934 (EA, FHO, K); *Fraser* 61894 & 61895, Singila 17.7.1936 (FHO, K).

Observations: (a) The author was not able to see the type of *T. scebelensis* Chiov. However, according to the accurate description and to the locality from which it was first described, he assumes that it is conspecific with *T. arabica* Bge.

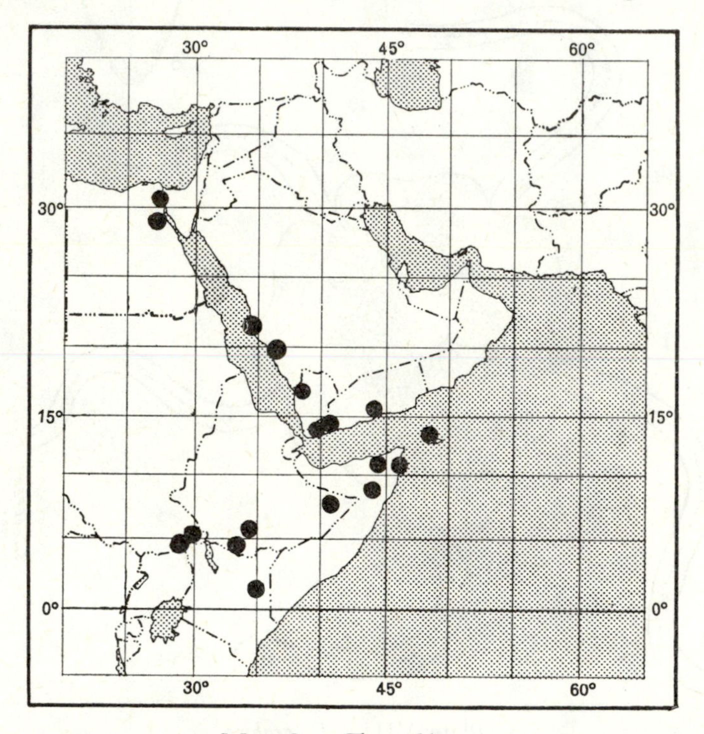

Map 8: *T. arabica*

9. **T. aralensis** Bge., Tentamen, 59 (1852) [Plate IX]

T. florida Bge. var. *rigida* Bge., *op. cit.*, 38.
T. florida Bge. var. *aucheri* Bge., *op. cit.*, 38.
T. mannifera (Ehrenb.) Bge. var. *persica* Bge., *op. cit.*, 64.
T. bungei Boiss., Fl. Or., 1:774 (1867).
T. ninae Gorschk., Soviet. Bot., 4:117 (1936).

Type: KAZAKH SSR: *Lehmann*, Nordküste der Aralsee 30.6.1841 (holotype P; isotype P; fragmentum typi in Herb. J. Gay, K).

Tree, often shrubby, with brown to reddish-brown bark, younger parts more or less papillose all over or at least on rachis. Leaves sessile with narrow base, subauriculate, sometimes faintly amplexicaul, 1–2 mm long. Vernal inflorescences simple, aestival ones densely compound. Racemes densely flowered, 2–6 cm long, 3.5–5 mm broad, rachis minutely papillose. Bracts narrowly trullate, subulate, longer than pedicels, margins near base only more or less denticulate. Pedicel shorter than calyx. Calyx pentamerous. Sepals strongly, deeply and irregularly denticulate, trullate-ovate, 0.75–1 mm long, all more or less obtuse, but the 2 outer more acute than the inner. Corolla pentamerous, caducous. Petals elliptical, sometimes obovate or somewhat ovate, 1.75–2 mm long. Androecium haplostemonous of 5 antesepalous stamens; insertion of filaments peridiscal; disk hololophic.

Flowering: March to August.

Habitat: Salty steppes and sands.

Distribution: Kazakh SSR, Turkmen SSR, Tadzhik SSR, Uzbek SSR, Iran (see Map 9).

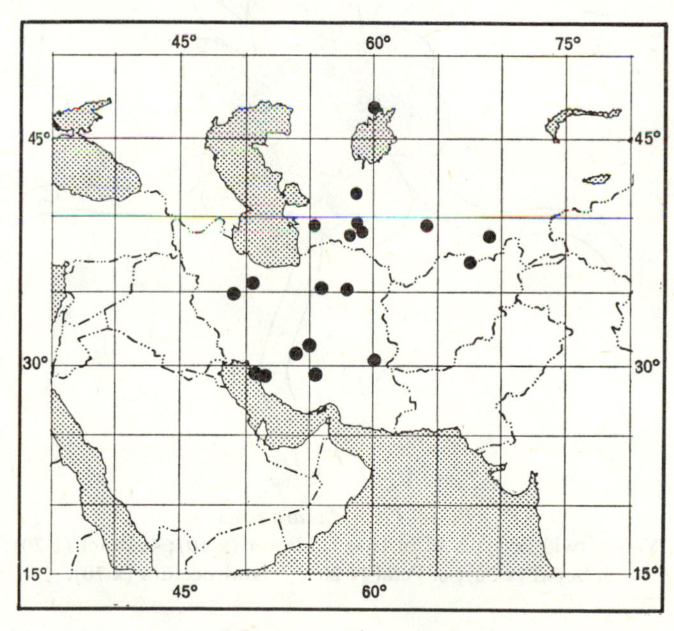

Map 9: *T. aralensis*

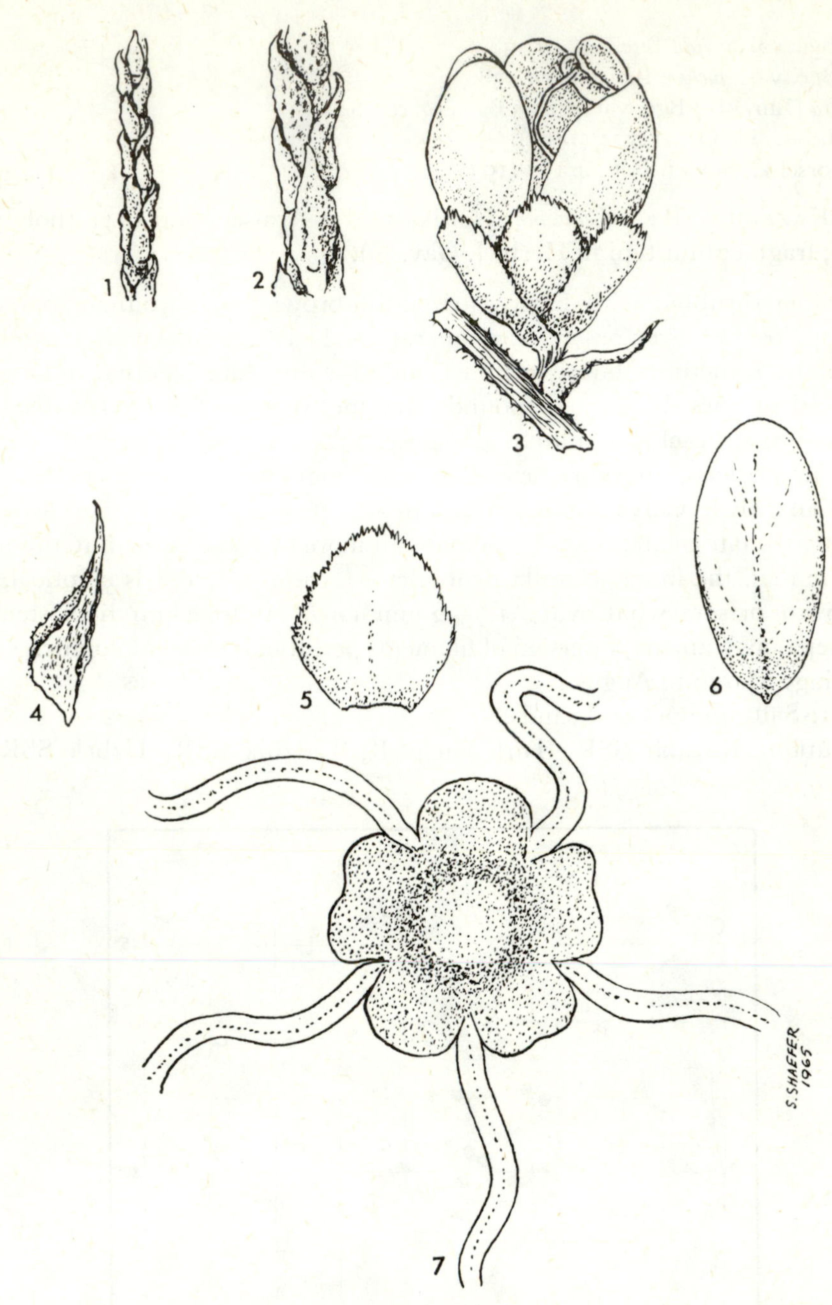

Plate IX *T. aralensis*
1. Young twig (x 5); 2. id (x 10); 3. Flower (x 10); 4. Bract (x 20); 5. Sepal (x 20); 6. Petal (x 20); 7. Androecium (x 70).

Selected specimens: TURKMEN SSR: *Sintenis* 1591, Kasandschik, in steppis 28.4.1901 (BM, E, G, K, L, P, WU); *Blinovsky* 144, SE of Kopet Dagh, Khodja Bulan Canion 22.7.1948 (LE); *Sintenis* 467, Aschabad versus Besmen 4.6.1900 (G) pro parte (L, PRC); *Androssow* 1419, Turkestania dominatio Buchara, in arenosis pr. Farab (ad fl. Amu-Darja) 20.5.1902 (B, G, LE, PRC); *Dubiansky* & *Basilevskaja,* sands of Kara Kum, Hatab border between sands and cultivated land, valley with loam (isotype of *T. ninae* Gorschk., LE). TADZHIK SSR: *Regel,* Turkestania, in deserto pr. Kurgan-Tyube ad fl. Wachisch ripam sinistram 1300' 22/7–3/8.1883 (LE). UZBEK SSR: *Bornmüller* 1095, Buchara prov. Kabadian, ad ripas prope Kakaity 18.8.1913 (B); *Russanov* 427, delta of Amu-Darja Kuvaschi Djorma 28.7.1928 (LE). IRAN: *Bornmüller* 3346, prov. Yesd, ad Kermandschahan 4.4.1892 (B, OXF); *Bunge,* prope Kefter-Chan inter Kerman et Yesd 21.4.1859 (holotype of *T. bungei* Boiss., G; isotypes FI, G, K, P); *Kotschy* No. O, Persia austr. (holotype of *T. mannifera* var. *persica* Bge., P; isotypes B, CGE, FI, G, K, L, OXF, P, POM, S, UPS, W); *Aucher-Eloy* 4510, desert. Sinus Persici (holotype of *T. florida* var. *aucheri* Bge., P; isotype FI); *Buhse,* prope Dshendak 10.4.1849 & *Buhse* 1210A(3), Husseinon bei Rishm .4.1849 (syntypes of *T. florida* var. *rigida* Bge., P).

10. **T. arborea** (Sieb. ex Ehrenb.) Bge., Tentamen, 67 (1852) [Plate X]

T. gallica L. var. *arborea* Sieb. ex Ehrenb., Linnaea, 2:269 (1827).
T. arborea (Sieb. ex Ehrenb.) Bge. var. *subvelutina* Boiss., Fl. Or., 1:776 (1867).
T. arborea (Sieb. ex Ehrenb.) Bge. var. *fluviatilis* Sickenb., Mem. Inst. Egypt., 4:189 (1901).
T. sokotrana Vierh., Oesterr. Bot. Zeitschr., 54:62 (1904).
T. pseudo-pallasii Gutm., Pal. J. Bot. Jerusalem, 4:51 (1947), p.p. (pars altera = *T. nilotica* [Ehrenb.] Bge.).
T. sodomensis Zoh., Trop. Woods, 104:35 (1956).
T. negevensis Zoh., *op. cit.*, 38.
T. jordanis Boiss. var. *negevensis* Zoh., *op. cit.*, 42.
T. jordanis Boiss. var. *sodomensis* Zoh., *loc. cit.*
T. gallica L. var. *pachybotrys* Zoh., *op. cit.*, 47.

Lectotype: EGYPT: *Sieber,* ad Cairo Aegypti (PRC, isolectotypes G, K, OXF, P, PRC, W).

Tree, often bushy, with brown to reddish-brown bark, younger parts more or less papillose. Leaves auriculate, slightly clasping, 2–3 mm long. Aestival inflorescences composed of racemes; vernal inflorescences simple. Racemes 2–6 cm long, 5 mm broad. Bracts longer than pedicels, linear-triangular, long acuminate, with more or less denticulate margins near base, acumen entire. Pedicel shorter than calyx. Calyx [13] pentamerous. Sepals 1–1.25 mm long, the outer 2 trullate-ovate, acute, the inner broadly trullate-ovate, more obtuse, entire or occasionally with very sparse irregular teeth. Corolla [13] pentamerous, caducous. Petals elliptic to usually somewhat trullate-ovate, 1.75–2 mm long. Androecium haplostemonous, of 5 antesepalous stamens (occasionally also with 1–2 abortive antepetalous stamens); insertion of filaments peridiscal; disk fleshy, nectariferous, para-synlophic.

13 Vernal racemes occasionally have tetramerous or tetra-pentamerous flowers intermixed with the pentamerous ones.

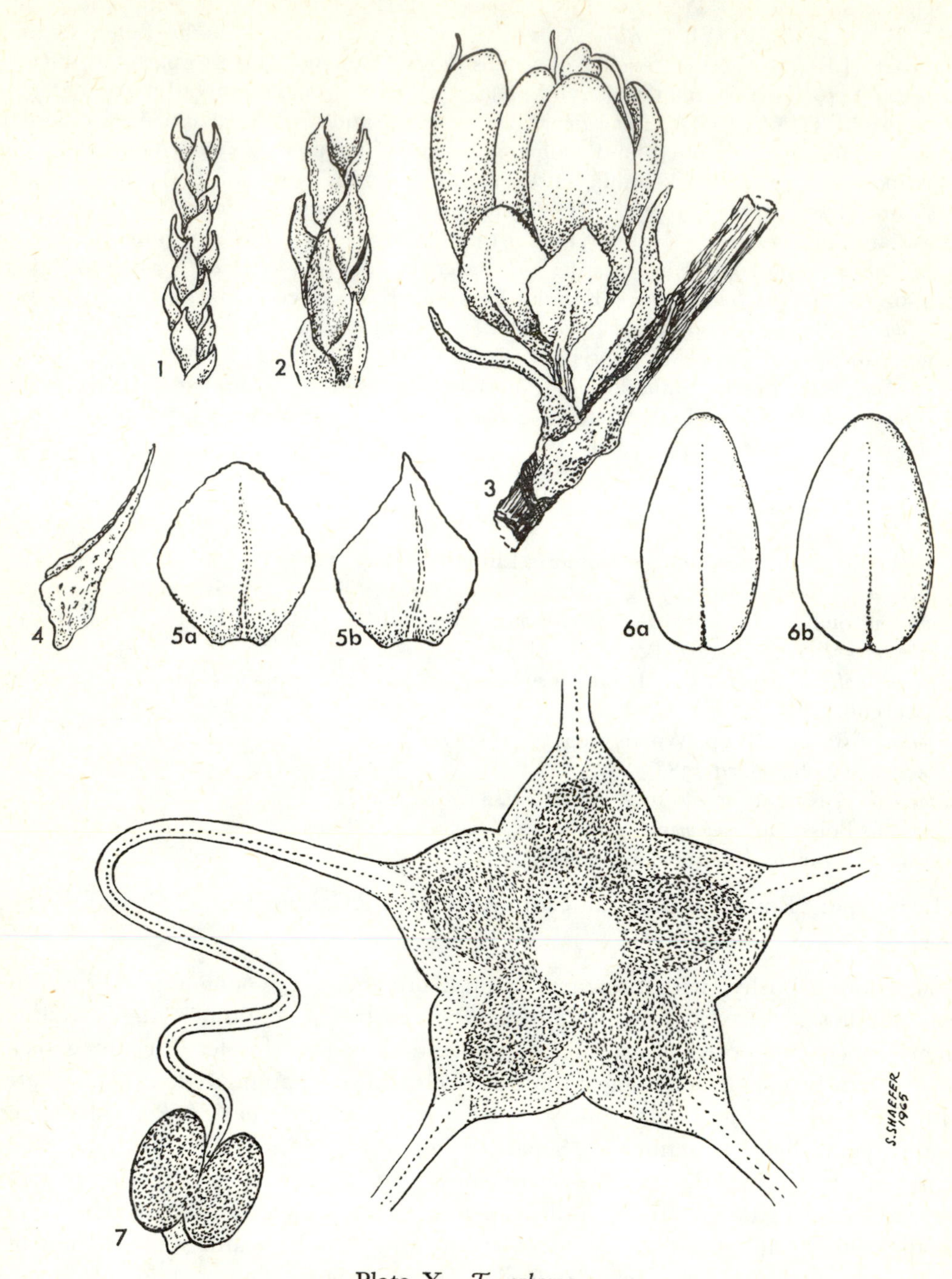

Plate X *T. arborea*

1. Young twig (x 5); 2. id (x 10); 3. Flower (x 20); 4. Bract (x 20); 5a. Inner sepal (x 20); 5b. Outer sepal (x 20); 6a–b. Petals (x 20); 7. Androecium (x 40).

Var. **arborea.** Bark reddish-brown; younger parts almost glabrous to papillose. Racemes not longer than 6 cm. Bracts not longer than calyx.

Var. **subvelutina** Boiss. Bark blackish-brown, very papillose all over. Racemes longer than type, up to 10 cm long. Bracts usually exceeding calyces. Intergrades with var. *arborea.*

Flowering: September to April, rarely also May, August.

Habitat: Salty banks of rivers, oases, coast (Red Sea), dunes and wadis.

Distribution: Tunisia, Libya, Egypt, Israel, Jordan, Sudan, Sokotra Island (see Map 10).

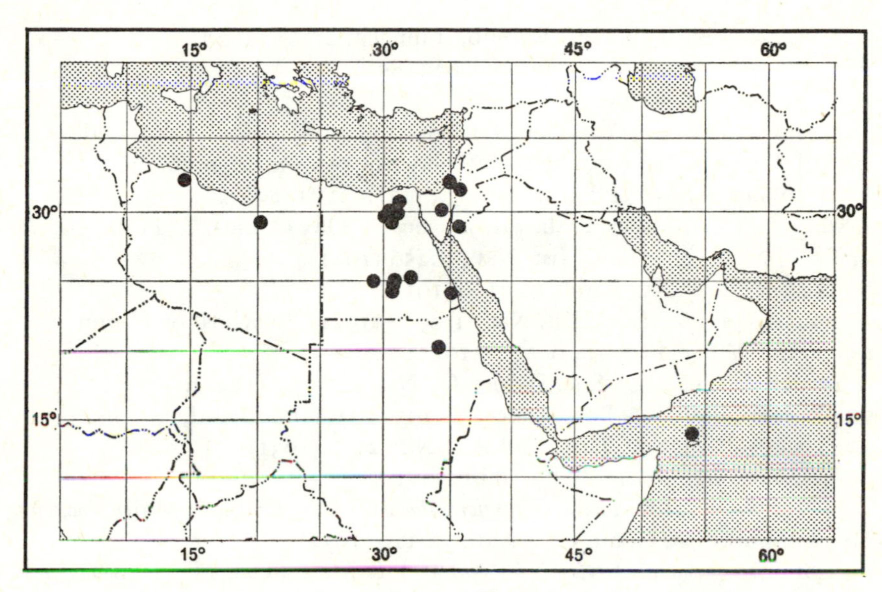

Map 10: *T. arborea*

Selected specimens: TUNISIA: *Bisseling* 955126116, Tunis, Kheredinne .4.1951 (L); *Ross*, Tunis .5.1884 (HBG). LIBYA: *Bornmüller* 729, Tripolis in desert pr. Gargaresh 21.4. 1933 (B, BM, K, US); *Ascherson* 678, Lybische Wüste, beim Brunnen Kerani 28.12.1873 (P). EGYPT: *Letourneux* 233b, in fossis secus vias prope Ismailia .4.1880 (B, FI, P, PRC, W); *Kneucker* 66, Sinai Wadi Hebran 29.3.1904 (B); *Ehrenberg*, Sinaigebirge (fragmentum typi of *T. gallica* var. *mannifera* f. *effusa* Ehrenb.) Herb. J. Gay (K); *Samaritani* 3609, pr. Alexandrian ad lacum Ramle 31.10.1858 (holotype of *T. arborea* var. *subvelutine* Boiss., G); *Täckholm*, Kharga oasis near the town 15.1.1928 (CAI); *Hassib*, Dakhla oasis 10.2.1931 (CAI); *Ehrenberg*, Aegyptus Fayum 1821 (P), *T. gallica arborea* ad Cahiram (S), paralectotypes of *T. gallica* var. *arborea* Sieb. ex Ehrenb. ISRAEL: *Eig*, Tel Aviv 1923; *Eig*, Gaza Coast sands 1924; *Zohary* id. 1927; *Gutmann*, id. 1941 (syntypes of *T. pseudo-pallasii* Gutm., HUJ); *Tadmor* 602, Arava Valley, Wadi Um Tarfa near Sodom 19.3.1950 (holotype of *T. sodomensis* Zoh., HUJ); *Tadmor* 1321, S. shore of Dead Sea wadi Um Tarfa near Sodom 18.3.1950 (holotype of *T. jordanis* var. *sodomensis* Zoh., HUJ); *D. Zohary* 601, Arava Valley, Ein Hotzev, near spring 24.3.1950 (holotype of *T. negevensis* Zoh., HUJ); *Waisel* 1269, Negev Ein Murra 8.4.1955 (holotype of *T. jordanis* var. *negevensis* Zoh., HUJ); *Tadmor* 809,

Arava Valley, Ein Hotzev 5.12.1952 (holotype of *T. gallica* var. *pachybotrys* Zoh., HUJ). JORDAN: 68*c*, on calcareous Ram sandstone below the edge of the escarpment west of Ail (E). SUDAN: *Kassas* 846, Kha Arbaat 17.12.1954 (CAI). SOKOTRA ISLAND: *Paulay*, im Dünensande des Strandgebietes langs der ausseren Grenze der Avicennien-Bestande von Gubbet-Shoab 8.12.1899 (holotype of *T. sokotrana* Vierh., WU).

Observations: (a) *T. arborea* var. *subvelutina* is confined to Egypt. (b) Many isotypes of *T. gallica arborea Sieb.* ex Ehrenb. are in G, PRC and W.

11. **T. canariensis** Willd., Abh. Akad. Berlin Physik, 1812-1813: 79 (1816) [Plate XI]

T. gallica L. var. *canariensis* (Willd.) Ehrenb., Linnaea, 2:268 (1827).
T. gallica L. var. *agrigentina* Bge., Tentamen, 62 (1852).
T. gallica L. var. *sardoa* Bge., *loc. cit.*
T. brachystylis J. Gay ex Coss., Ann. Sci. Nat. Bot., IV, 4:283 (1855), nom. nud.
T. brachystylis J. Gay ex Coss. var. *sanguinea* J. Gay ex Coss., *loc cit.*, nom. nud.
T. brachystylis J. Gay ex Batt. & Trab., Fl. Alg., 1(2):321 (1889).
T. brachystylis J. Gay ex Batt. & Trab. var. *sanguinea* J. Gay ex Batt. & Trab., *loc. cit.*
T. lagunae Cab., Mem. Soc. Esp. Hist. Nat., 8:256 (1911).
T. geyrii Diels, Bot. Jahrb., 54 Beibl. 120:100 (1917).
T. weylerii Pau, Mem. Soc. Esp. Hist. Nat., 1:293 (21921), 'weyleri' orth. mut.
T. esperanza Pau & Villar, Broteria Bot., 23:101 (1927).
T. riojana Sennen & Elias, Bol. Soc. Iberica Ci. Nat., 27:133 (1928).
T. murbeckii Sennen, Bull. Soc. Bot. France, 78:193 (1931), nom. nud.
T. leucocharis Maire, Bull. Soc. Hist. Nat. Afr. N., 22:30 (1931).
T. brachystylis J. Gay var. *geyrii* (Diels) Maire, *loc. cit.*
T. balansae J. Gay ex Batt. & Trab. var. *micrantha* Maire & Trab. ex Maire, *loc. cit.*
T. sireti Sennen, Butl. Inst. Catal. Hist. Nat., 32:110 (1932).
T. gallica L. var. *lagunae* (Cab.) Maire, in: Jahand. & Maire, Cat. Pl. Maroc, 2:488 (1932).
T. muluyana Sennen & Maur., Cat. Fl. Rif Or., 42 (1933), nom. nud.
T. valdesquamigera Sennen & Maur. *loc. cit.*, nom. nud.
T. gallica L. subsp. *leucocharis* (Maire) Maire, Bull. Soc. Hist. Nat. Afr. N. 26:184 (1935).
T. gallica L. subsp. *epidiscina* Maire var. *submutica* Maire & Trab. ex Maire, *loc. cit.*

Type: CANARY ISLANDS: *Broussonet*? Tenerife? (holotype Herb. Willd. No. 6062 B; isotypes: *Broussonet* "Tamar. an a gallica differt?" (Canaries 1801), BM, HUJ).

Bushy tree with reddish-brown bark, younger parts papillose to sparsely papillulose, rachis of raceme always papillose. Leaves sessile with narrow base, 1.5–2.5 mm long. Vernal inflorescences simple, aestival densely compound. Racemes 1.5–5 cm long, 4–5 mm broad, densely flowered. Bracts linear-triangular, long acuminate to subulate, entire, adaxial side and margins often papillulose, almost equalling to somewhat exceeding calyx. Pedicel more or less equalling calyx. Calyx pentamerous. Sepals incised-denticulate with many narrow and small dense teeth, 1.5–1.75 mm long, acute, the outer 2 trullate-ovate, the inner trullate. Corolla pentamerous, caducous. Petals obovate to narrowly obovate, 1.25–1.5 mm long. Androecium haplostemonous of 5 antesepalous stamens; insertion of filaments peridiscal; disk synlophic.

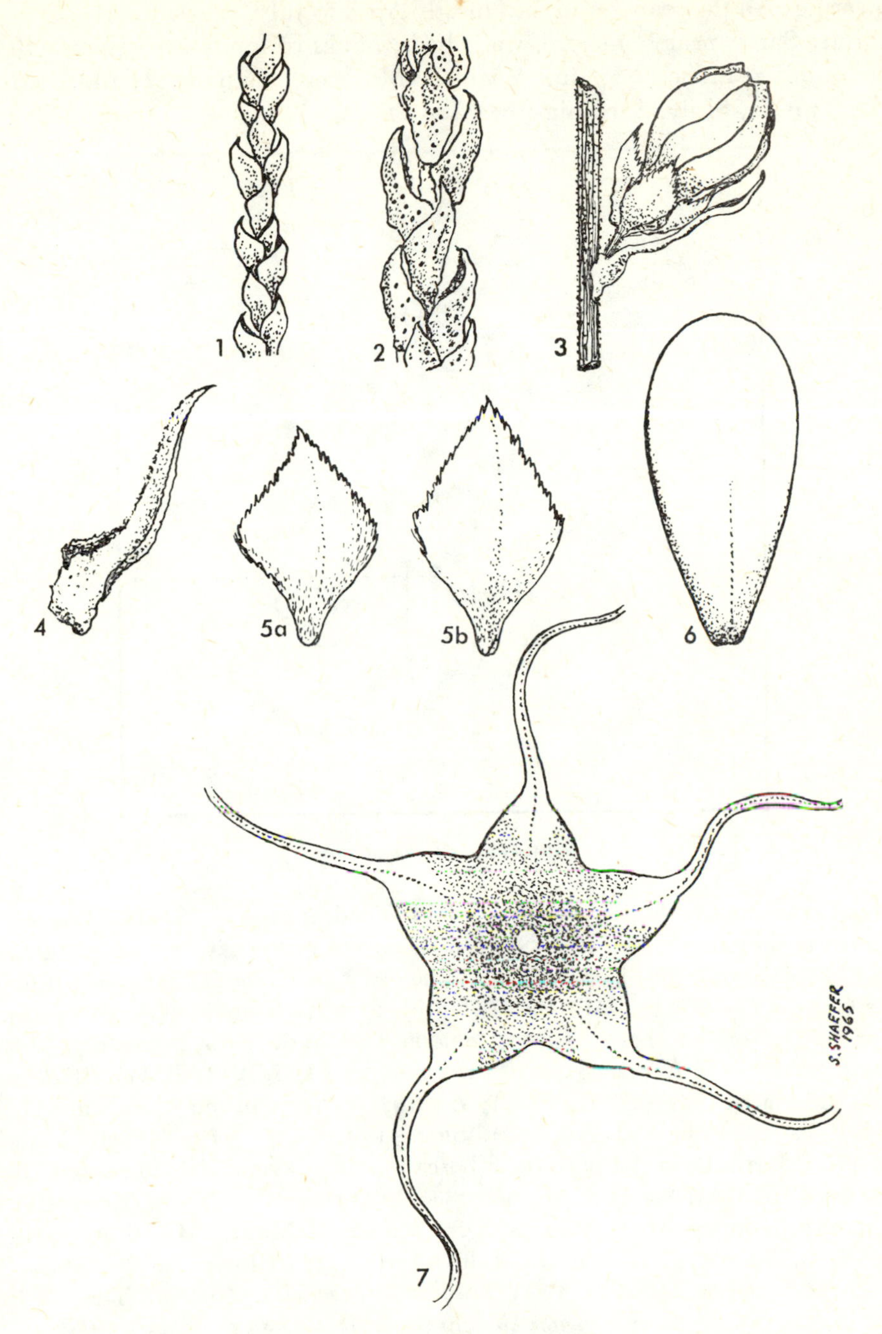

Plate XI *T. canariensis*
1. Young twig (x 5); 2. id (x 10); 3. Flower (x 10); 4. Bract (x 20); 5a. Outer sepal (x 20); 5b. Inner sepal (x 20); 6. Petal (x 25); 7. Androecium (x 30).

Flowering: All the year round, but mostly April to July.

Habitat: Sands near sea-shore, wadi beds, banks of temporary rivers.

Distribution: France, Spain, Spanish Morocco, Canary Islands, Morocco, Algeria, Tunisia, Sicily, Sardinia (see Map 11).

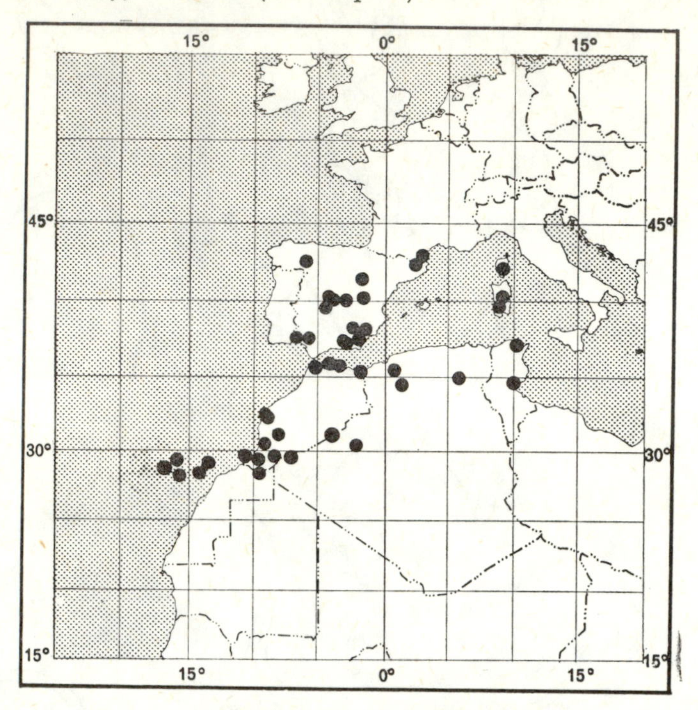

Map 11: *T. canariensis*

Selected specimens: FRANCE: (7961), Port Vendres (RAB). SPAIN: *Bourgeau* 2295, bords de la rivière à Aranjuez 26.6.1854 (B, G, K, OXF, P, PRC); *Elias* 4283, Logrono, Alcanale (Rioja) bords de l'Ebre 29.8.1921 (holotype of *T. riojana* Sennen & Elias, MA; isotypes G, W); *Pau* (78910), Aranjuez (Madrid) 21.5.1887 (holotype of *T. esperanza* Pau & Villar, MA); *Sennen* & *Jeronimo* 7250, Amleria, Cuevas de Vera, lierges de l'Almanzora 14.7.1929 (holotype of *T. sireti* Sennen, MA; isotypes FI, G, P, RAB, W); *Willkomm* 389, Granada Alpujares 1827 (G, K, PRC); *Boissier*, ad littus prope Malacam (G, P, W, syntypes of *T. gallica* var. *paniculata* Bge.). SPANISH MOROCCO: *Pau* (78856), Mauritania, Tetuan, Rio Martin Cava del paco de la borca margen daviche del rio .5.1921 (holotype of *T. weyleri* Pau, MA); *Sennen* & *Mauricio* 7846, lit et marges du Nekor, Route d'Alhucer-mas 7.7.1931 (holotype of *T. valdesquamigera* Sennen & Maur., MA; isotypes BM, G); *Mauricio* 8784, Ulad-Settut bords de la Muluya 12.10.1933 (holotype of *T. muluyana* Sennen & Maur., MA; isotypes BM, G, P, W); *Sennen* & *Mauricio* 7566, Beni Tuzin à Rastataf ruisseau 12.7.1930 (isotype of *T. murbeckii* Sennen, BM). CANARY ISLANDS: *Bornmüller* 613, Gran Canaria Caldera de Bandarna in aridis 21.5.1900 (G, P, PRC, W, WU); *Bourgeau* 285, Teneriffa ad littora maris 1845 (BM, CGE, G, K, OXF, P, US, W). MOROCCO: *Faure*, massif des Beni-Snassen, environs de Berkane lieux humides près la Moulouya 22.6.1932 (E, S, U); *Maire*, Marrakech bords de l'Oued Tensift alluvoins un peu salés 16.7.1924 (RAB). ALGERIA: *Chevallier* 190b, Sahara, Biskra in alluv. amnis 9.10.1903

(B, HBG, P); *Balansa* 990, bords de l'Oued-Biskra à Biskra 5.11.1853 (FI, G, K); *Desfontaines*, Barbarie (P); *Maire* 249, Mouydir Tigelgemin 450 m 29.2.1928 (lectotype of *T. leucocharis* Maire, P; isolectotypes FI, K, RAB), paralectotypes: *Maire* 252, Mouydir gorges d'Arak 600–650 m (P, RAB, UC); *Wilczek*, ad ripas fluminis Acif-n-Ait-Amer prope castellam Tauri 1.4.1934 (holotype of *T. gallica* subsp. *epidiscina* var. *submutica* Maire & Trab., P); *Balansa*, Bords de l'Oued Biskra à Biskra 6.4.1853 (holotype of *T. brachystylis* var. *sanguinea* J. Gay, K); *Maire* 1609, Temassinin, lieux humides salés 360 m (holotype of *T. balansae* var. *micrantha* Maire & Trab., P). Tunisia: *Pitard* 341, Gabès in arenosis salsuginosis .2.1908 (BM, G, K, L, P); *Letourneux*, Oued Tomerza 8.5.1887 (FI, P). Sicily: *Splitgerber*, Sicilia Maccalubbi prope Agrigentum .5.1833 (holotype of *T. gallica* var. *agrigentina* Bge., W; isotype S). Sardinia: *Müller* 229, in humidis prope Cagliari .7.1828–1829? (holotype of *T. gallica* var. *sardoa* Bge., W; isotypes E, GL, K, PRC).

Observations: (a) It is not always easy to distinguish between the aestival forms of *T. canariensis*, *T. gallica* and *T. africana*. *T. africana* and *T. canariensis* are always papillulose or papillose at least on the rachis of their racemes, while *T. gallica* is always glabrous. The petals of *T. africana* are trullate-ovate to ovate, while those of *T. canariensis* are obovate and those of *T. gallica* are elliptic. There are also slight differences in the length of petals. (b) The author did not see the type of *T. geyrii* Diels, which was lost during World War II in Berlin. The identity of this species is certain from its distribution and adequate description.

12. **T. hispida** Willd., Abh. Akad. Berlin Physik, 1812–1813: 77 (1816) [Plate XII]

T. pentandra var. — Pall., Fl. Ross., 2:72 (1788).
T. tomentosa Smith, in: Rees, Cyclop., 35 (1): No. 2 (1817).
T. canescens Desv., Ann. Sci. Nat. Bot., I, 4:348 (1824).
T. karelini Bge., Mém. Acad. St. Pétersb., 7:294 (1851).
T. hispida Willd. var. *genuina* Bge., Tentamen, 69 (1852).
T. hispida Willd. var. *pleiandra* Bge., *op. cit.*, 70.
T. karelini Bge. var. *densior* Trautv., Bull. Soc. Imper. Nat. Moscou, 39:310 (1866).
T. serotina Bge. ex Boiss., Fl. Or., 1:773 (1867).
T. karelini Bge. var. *malomae* Trautv., Acta Horti Petrop., 1:273 (1872).
T. karelini Bge. var. *hirta* Litw., Sched. Herb. Fl. Ross., 5:79 (1905).
T. lipskyi Gand., Bull. Soc. Bot. France, 65:27 (1918).
T. hispida Willd. f. *internodiis abbreviatis* Korov., Sched. Herb. Fl. Asiae Med., 13:44 (1927), nom. illegit.
T. hispida Willd. f. *internodiis elongatis* Korov., *loc. cit.*, nom. illegit.

Type: Russian SFSR: *Pallas*, habitat ad mare caspium locis salsis (holotype Herb. Willd. No. 6060, B; isotypes BM, PRC, and '*T. pubescens* Pall'., BM).

Shrub or small tree, 1–5 m tall, with reddish-brown bark, usually very hairy on all parts to almost glabrous, with at least papillulose rachis of racemes (calyx sometimes glabrous and other floral parts usually glabrous). Leaves sessile, cordate-auriculate at base, 1.5–2.25 mm long. Aestival inflorescences densely composed of racemes, the terminal raceme usually much longer. Racemes 1.5–7 cm, also up to

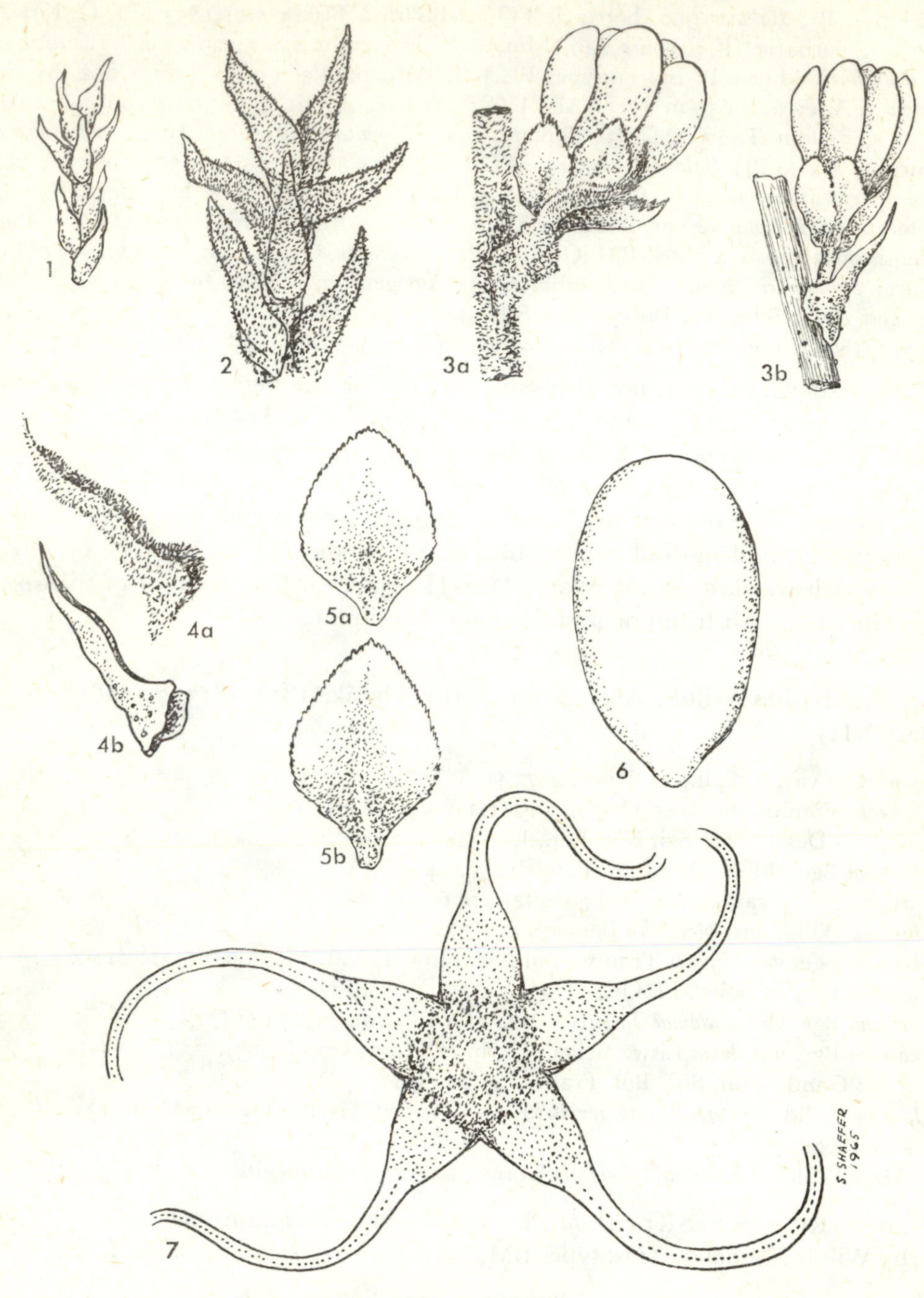

Plate XII *T. hispida*

1. Young twig (x 5); 2. id (x 10); 3a. Flower of var. *hispida* (x 10); 3b. Flower of var. *karelini* (x 10); 4a. Bract of var. *hispida* (x 20); 4b. Bract of var. *karelini* (x 20); 5a. Inner sepal (x 20); 5b. Outer sepal (x 20); 6. Petal (x 20); 7. Androecium (x 30).

15 cm long, 3–5 mm broad. Bracts longer than pedicels, narrowly triangular, acuminate, entire. Pedicel shorter than calyx. Calyx pentamerous. Sepals 0.75–1 mm long, densely and irregularly denticulate with fine teeth especially towards apices, the outer 2 more acute and slightly keeled. Corolla pentamerous, caducous. Petals 1.5–2 mm long, usually obovate, occasionally elliptic. Androecium haplostemonous, of 5 antesepalous stamens; insertion of filaments peridiscal; disk synlophic, filaments mostly with enlarged and nectariferous bases, or disk almost membranous.

Var. **hispida.** Very hairy on all parts to (rarely) inconspicuously papillose. Sepals 1 mm long, densely and irregularly denticulate with fine teeth especially towards apices. Petals 2 mm long, usually obovate. Filaments with enlarged and nectariferous bases.

Var. **karelini** (Bge.) Baum. As a rule, inconspicuously papillose, sometimes also hairy. Sepals subentire, 0.75 mm long. Petals 1.5 mm long, ovate-elliptic. Filaments with rudiments of nectariferous tissue at their bases; disk more membranous. Intergrades with var. *hispida* to some extent.

Flowering: June to December.

Habitat: Sandy hills, salty flats, salty banks of rivers and salty marshes.

Distribution: Kazakh SSR, Mongolia, Russian SFSR, Chinese Turkestan, Uzbek SSR, Turkmen SSR, Afghanistan, Iran (see Map 12).

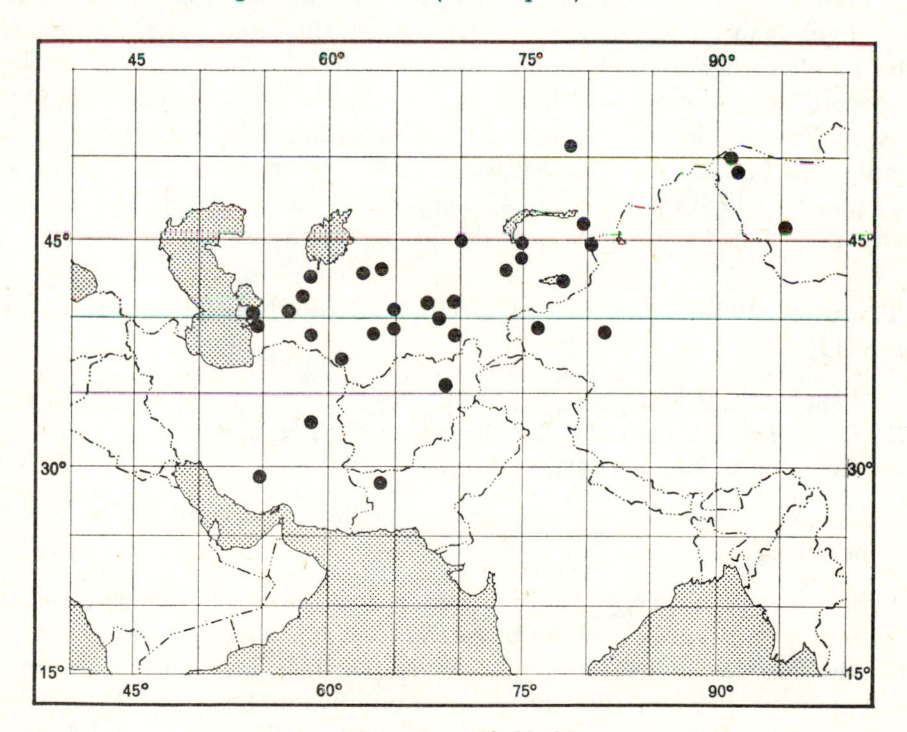

Map 12: *T. hispida*

Selected specimens: CHINA AND KAZAKH SSR: *Schrenk*, Songoria ad fl. Ili and/or fl. Tschu 6.8.1842 (B, E, G, K, L, OXF, P, S, US, W); *Vorotnikov* 5739 & 5761, Turkestania Sinensis, Takla Makan Mazar-Tagh 11.6.1931 (S). KAZAKH SSR: *Vvedensky* 311b, deserta meridionalia Jaxartica ad ripas salsa Tuz-kane 7.10.1923 (B, E, G, HUJ, K, LE, P, S, W); *Tamkon* 17, Kzyl-Orvodsk, Khofman-Kala along the river of Aoula 7.10.1940 (LE). MONGOLIA: *Przewalski*, Mongolia occidentalis Terra Khalkha 1873 (E, P); *Pobedimova* 626, Gobi desert Chargin-Godi 9.1.1930 (LE, UC). RUSSIAN SFSR: *Politov*, Sibiria Ulatau, flussufer des Tschu (LE, PRC, WU). UZBEK SSR: *Razorosky* 100, Klovasky reg. (Dzungarian Steppes) N. of Ursantiaevsk, sandy hills 15.9.1935 (LE); *Meyer* 661, Samarkand pr. on saline places in the desert near Skostakos 24.9.1910 (UC). TURKMEN SSR: *Bornmüller* 1023, Turkestania Buchara Kurban-Tube in salsis planitiei fluvii Warosin ad Kurban-Tube 14.8.1913 (B); *Karelin*, Turcomania ins. Ogurtschinsk (holotype of *T. karelini* Bge., P; isotypes P, G, LE); *Androssow* 1420, Turkestania Bucharae and fl. Amu-Darja pr. Farab. in salsis 24.8.1901 (fl.) 15.9.1901 (fr.) (holotype of *T. karelini* var. *hirta* Litw., LE; isotypes B, G, K, PRC, S, WU); *Lipsky* 866, Boukhara 21.7.1896 (holotype of *T. lipskyi* Gand., LY). AFGHANISTAN: *Pabot* 1145, 20 km N.W. Haybak 30.7.1958 (Herb. Pab.); *Neubauer* 140, Dschelangir zw. Baghlan und Aliabad bei Baghlan, am steilen und felsigen Flussufer 5.10.1960 (W). IRAN: *Bornmüller* 3348, Pers. austr. prov. Farsistan Niris ad lacum salsum 6.10.1892 (B, P); *Aellen* & *Sharif*, Kachan-Kaviz 11.11.1948 (W) (Esf. Herb.); *Bunge*, in septentrione urbis Birdschand, ad fontes salinas in humidis inter Herat et Tebes .11.1858 (holotype of *T. serotina* Bge., G).

Observations: (a) A fragment of the specimen cited by Bunge as '*T. pycnocarpa* Karel., Enum. Turcom. n. 336 (ex parte)', which is the holotype of *T. hispida* var. *pleiandra* Bge., is in Paris. This specimen also bears the name '*T. turcomanica* m.' in Bunge's handwriting. (b) The author was not able to see the type of *T. canescens* Desv. According to the original description, and following the opinions of previous students of *Tamarix*, its identity with *T. hispida* Willd. is beyond question. (c) *T. serotina* Bge. ex Boiss.: Bunge himself was aware of the close relations of *T. serotina* with *T. karelini* (which is but a form of *T. hispida*), as can be seen from his notes and observations attached to the type specimen in Herbier Boissier at Geneva.

13. **T. indica** Willd., Ges. Naturf. Freunde Berlin Neue Schr., 4:214 (1803) [Plate XIII]

T. epacroides Smith in: Rees, Cyclop., 35: No. 4 (1817).
T. gallica L. var. *indica* (Willd.) Ehrenb., Linnaea, 2:268 (1827).
T. troupii Hole, Indian Forester, 45:248 (1919).

Type: INDIA: *Klein*, *T. articulata* Ind. 1799 (holotype Herb. Willd. No. 6063, B; isotype PRC).

Bushy shrub or small tree, 3–8 m high, with reddish-brown to greyish-brown bark, glabrous, except for papillose rachis of racemes, or entirely papillose. Leaves sessile with narrow base to amplexicaul (without coherent margins), with all intermediate stages on the same specimen (i.e., the youngest usually amplexicaul, others auriculate to sessile with narrow base), 1–1.5 mm long. Vernal inflorescences simple, aestival

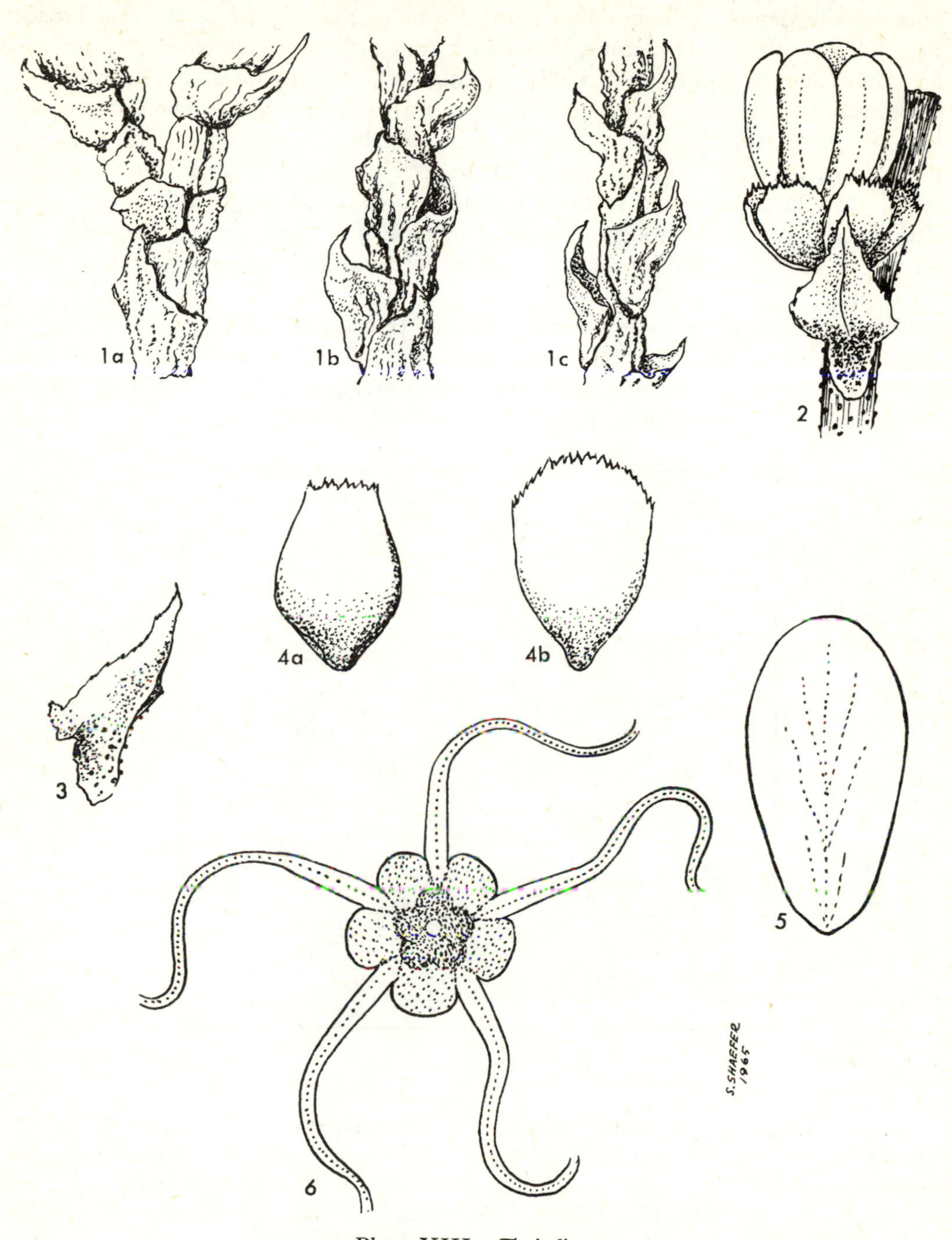

Plate XIII *T. indica*

1a. Young twig with amplexicaul leaves (x 10); 1b. id, with half-clasping leaves (x 10); 1c. id, with simple leaves (x 10); 2. Flower and papillose rachis of raceme (x 15); 3. Bract (x 20); 4a. Outer sepal (x 20); 4b. Inner sepal (x 20); 5. Petal (x 20); 6. Androecium (x 30).

ones usually densely compound racemes. Racemes 4–15 cm long, 3–5 mm broad. Bracts longer than or sometimes equal in length to pedicels, narrowly triangular, acuminate, acute, slightly keeled, with slightly and minutely papillulose denticulate margins. Pedicel somewhat shorter than calyx. Calyx pentamerous. Sepals 0.75–1 mm long, ovate or trullate-ovate, somewhat truncate, the outer 2 more acute than the inner ones, all deeply incised-denticulate especially towards apex, with fine, thin teeth. Corolla pentamerous, caducous. Petals 1.5–2 mm long, elliptic to obovate-elliptic or obovate. Androecium haplostemonous of 5 antesepalous stamens; insertion of filaments peridiscal (inconspicuously somewhat hypodiscal); disk tiny, hololophic.

Flowering: August to December; occasionally January, February.

Habitat: Forming almost pure scrub on saline soil; it seems, however, that two ecotypes may be distinguished according to salt tolerance.

Distribution: Ceylon, India, Pakistan, Afghanistan (see Map 13).

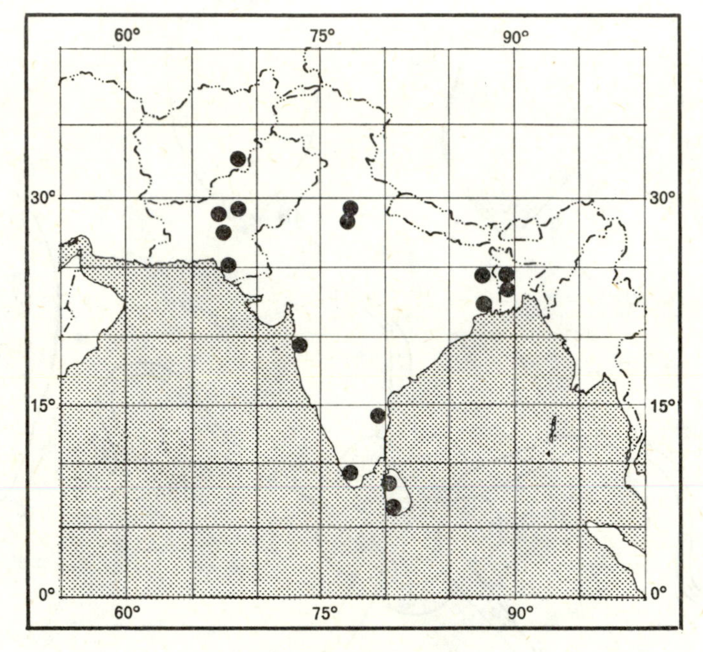

Map 13: *T. indica*

Selected specimens: CEYLON: *Thwaites* 1254, Ceylon 1854 (BM, CGE, G, K, P, W); *Walker*, Ceylon (E, K). INDIA: *Koenig*, *Tamarix* habitat copiose prope flumen Colloram maximum non ad junctionem 'flumina Iny' nominis (holotype of *T. epacroides* Sm., LINN. No. 383.3; isotypes? BM, FI, OXF, S); *Wight* 950, peninsula Ind. or. 1834 (BM, CGE, E, G, GL, K, P, US); *Leschenault*, Bengal 1821 (GDC, P); *Gaudichaud* 7, Calcutta, bords du Gange .4.1837 (FI, G, P); *Heinig*, Bengal Sunderbuns .12.1892 (U, W); *Elliot*, Madras 1852; *Walt* 3606, N.W. Himalayas (E). PAKISTAN: *Lace* 3605, Baluchistan N.W. Sibi 6.12.1887 (E); *Burtt* 15619, Karachi Tatta 60 miles E. of Karachi 30.11.1958 (E); *Troup* 16344, Sibi. AFGHANISTAN: *Griffith* 958, Afghanistan (FI, K).

Observation: The author was able to examine the holotype of *T. troupii* Hole from a small fragment which was kindly preserved by Mrs Schieman-Czeika in Vienna (W).

14. **T. karakalensis** Freyn, Bull. Herb. Boiss., II, 3:1060 (1903) [Plate XIV]

T. karakalensis Freyn var. *scoparia* Freyn, *op. cit.*, 1062.
T. karakelensis Freyn var. *myriantha* Freyn, *loc. cit.*

Type: TURKMEN SSR: *Sintenis* 1966, Kisil-Arwat, Karakala in valle fluvii Sumbar 24.6.1901 (isotypes E, W, WU).

Shrubby tree, 2–3 m high, with reddish-brown bark; younger parts glabrous to minutely papillose, except for rachis of raceme which is usually papillose. Leaves sessile with narrow base, 1.25–2.25 mm long. Inflorescences usually densely composed of aestival racemes. Racemes 0.5–4 cm long, 3–4 mm broad. Bracts equalling or longer than pedicels, with more or less finely denticulate margins, triangular, acute. Pedicel shorter than calyx. Calyx pentamerous. Sepals 0.5–0.75 mm long, trullate-ovate, acute, much (finely) incised-denticulate, especially at apex, the outer 2 slightly keeled. Corolla pentamerous, caducous. Petals 1.25–1.5 mm long, obovate-elliptic or rarely elliptic to obovate. Androecium haplostemonous, of 5 antesepalous stamens, insertion of filaments hypodiscal;[14] disk hololophic.
Flowering: June to August.
Habitat: Salty places on banks of rivers and in wadis.
Distribution: Turkmen SSR, Iran (see Map 14).

Selected specimens: TURKMEN SSR: *Sintenis* 467, Aschabad *versus* Besmen 4.6.1900 (holotype of *T. karakalensis* var. *myriantha* Freyn, G; isotypes B, E, K, P, S, W); *Sintenis* 646b Aschabad, Suluklu (Saratowka) ad fines Persise, inter Mergen Ulja et Kalkulap 26.7.1900 (holotype of *T. karakalensis* var. *scoparia* Freyn, G; isotypes BM, E, K, P, S, W); *Blinovsky* 66, Geok-Tepinsk Central Kopet-Dag on Geok-Tepe-Germab rd. 21.7.1950 (LE); *Andrusczenko* 3046, ad ripam lacus salsi Malla-Kara prope stationem viae ferreae Dshebel 22.7.1912 (BM, G, K, LE, S, US, W). IRAN: *Bornmüller* 3360a, Pers. interior, prov. Yesd, in vallibus montium ad Taft 6.4.1892 (B); *Bruns* 896, Abarrabad 12.5.1904 (HBG).

Observations: (a) The Sintenis labels and their numbers are not always consistent in the case of *Tamarix*. The Type collection of *T. karakalensis* is thus heterogeneous. Even in G different specimens are mounted on herbarium sheets bearing identical labels and collection numbers. (b) *T. karakalensis* Freyn is similar to *T. remosissima* in its general habit. It differs from the latter, among other characteristics, in its papillosity and its caducous petals (in *T. ramosissima* the younger parts are glabrous and the petals are persistent).

14 Sometimes the insertion is not easily distinguished here, and has to be examined carefully because of the small disk which is less fleshy and more membranous.

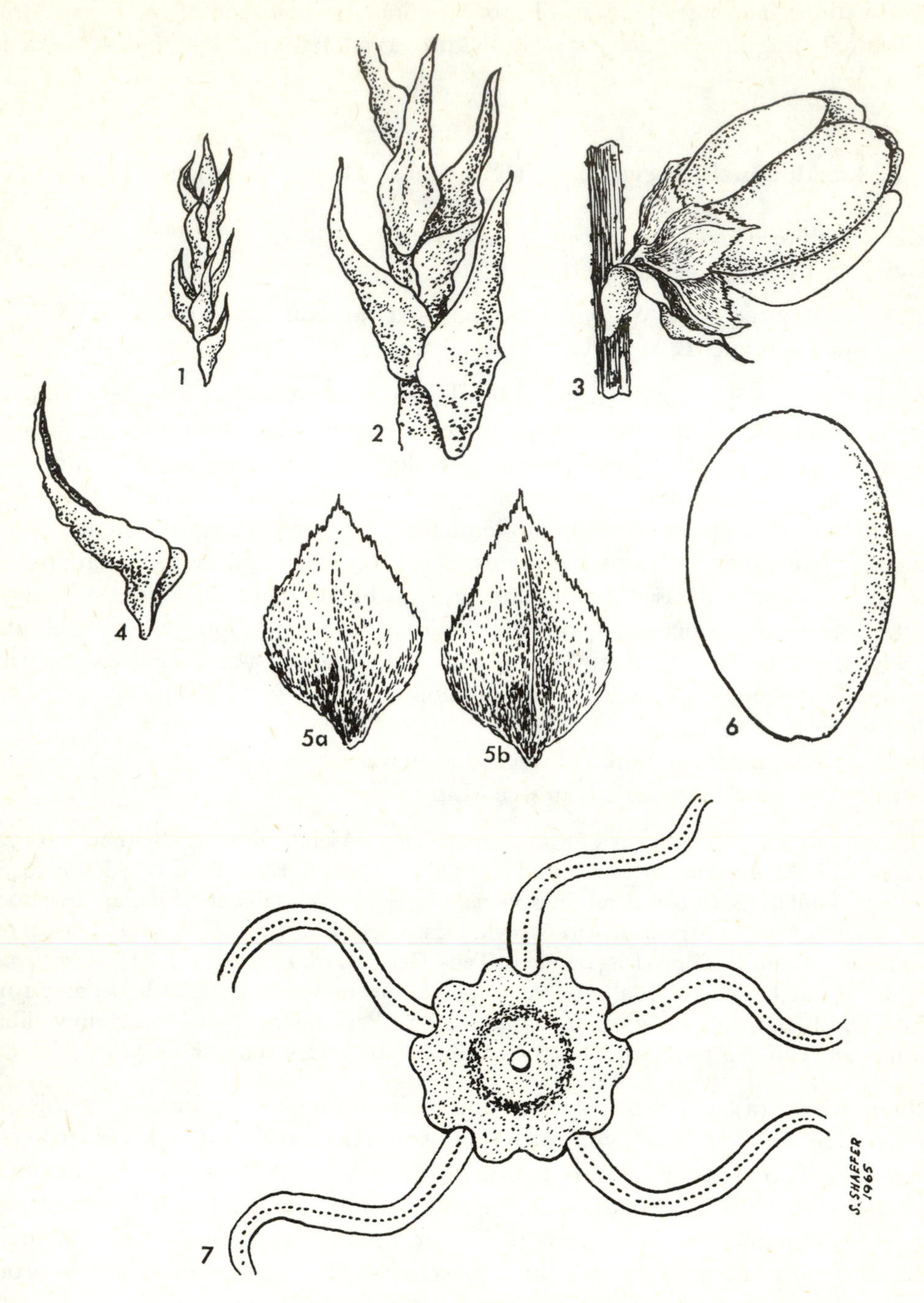

Plate XIV *T. karakalensis*
1. Young twig (x 5); 2. id (x 10); 3. Flower (x 10); 4. Bract (x 20); 5a. Inner sepal (x 20); 5b. Outer sepal (x 20); 6. Petal (x 20); 7. Androecium (x 30).

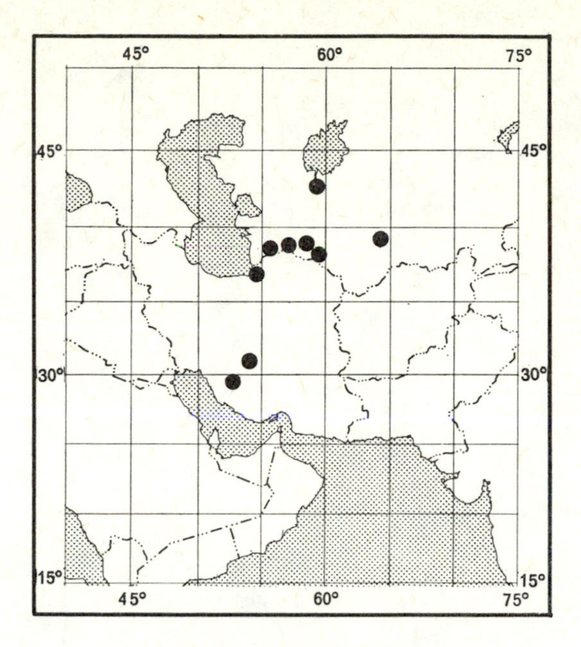

Map 14: *T. karakalensis*

15. **T. leptostachya** Bge., Mém. Acad. St. Pétersb., 7:293 (1851) [Plate XV]

T. gallica L. var. *micrantha* Ledeb., Fl. Ross., 2:135 (1843), p.p.
T. kasakhorum Gorschk., Not. Syst. Leningrad, 7:91 (1937).

Lectotype: Russian SFSR: *Syssow*, in der Kirghisensteppe 'Barssuki' am Aralsee 8.1840 (P, isolectotype LE).

Shrub or very small tree, 1–4(6) m, with brown bark; younger parts glabrous, rarely sparsely muricate-papillulose. Leaves sessile with narrow base, 2–3 mm long. Aestival inflorescences densely composed of racemes. Racemes 7–15 cm long, 3 mm broad. Bracts entirely herbaceous, narrowly trullate-triangular, longer than pedicels, acuminate, those subtending racemes shorter, cordate and blunt with a broad obtuse point. Pedicel as long as calyx. Calyx pentamerous. Sepals 0.5–0.75 mm long, acute, entire or subentire, trullate-ovate, more or less obtuse. Corolla pentamerous, caducous. Petals obovate, 1.5 mm long. Androecium haplostemonous, of 5 antesepalous stamens; insertion of filaments peridiscal; disk synlophic.

Flowering: May to July.

Distribution: Russian SFSR, Kazakh SSR, Turkmen SSR, Tadzhik SSR, Kashmir, Mongolia (see Map 15).

Var. **leptostachya.** Bracts narrower than to about as broad as rachis of racemes. Sepals usually entire. Disk tiny, nearly membranous.

Habitat: Mainly along streams.

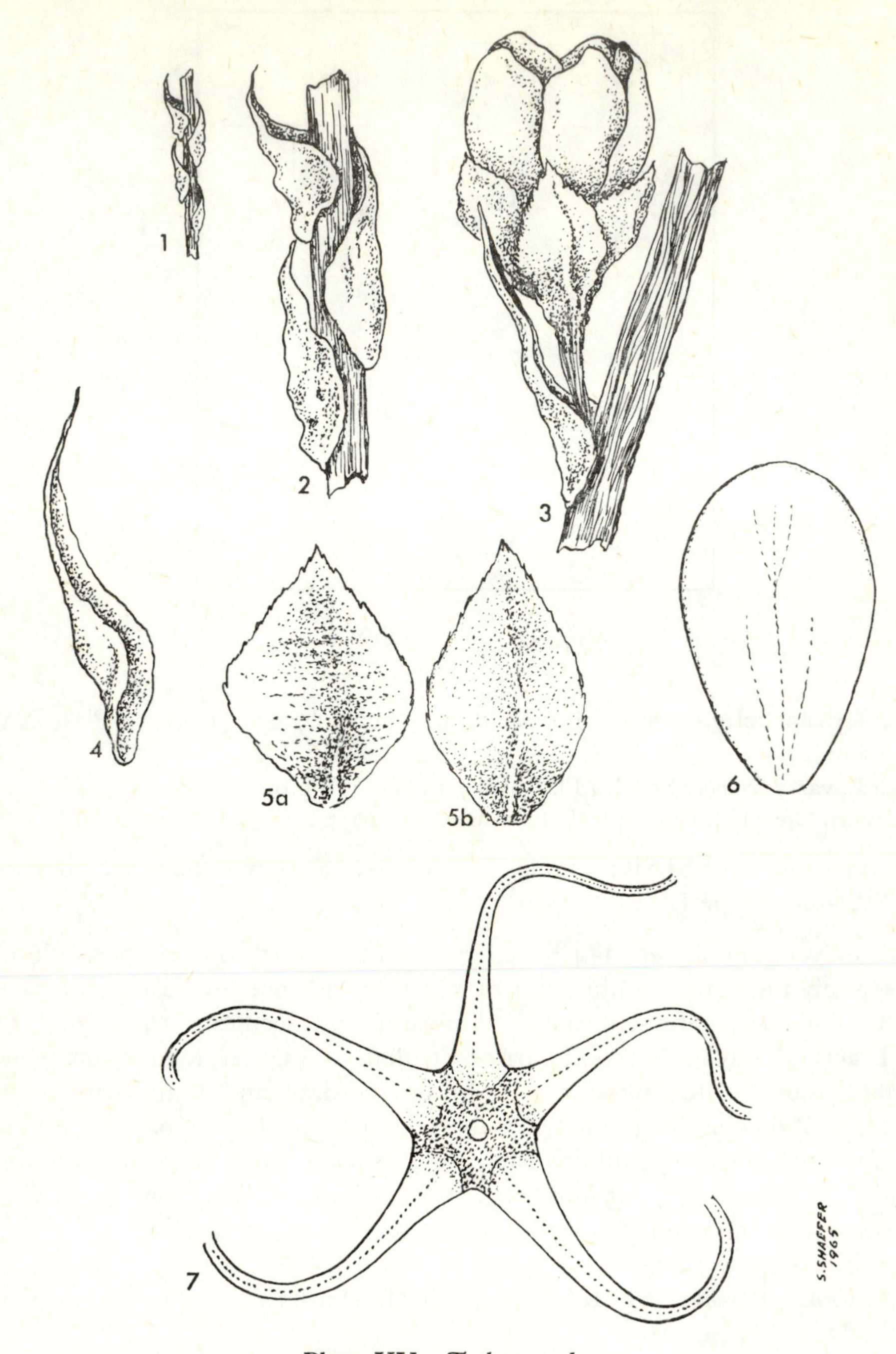

Plate XV *T. leptostachya*
1. Young twig (x 5); 2. id (x 10); 3. Flower (x 10); 4. Bract (x 20); 5a. Inner sepal (x 20); 5b. Outer sepal (x 20); 6. Petal (x 20); 7. Androecium (x 30).

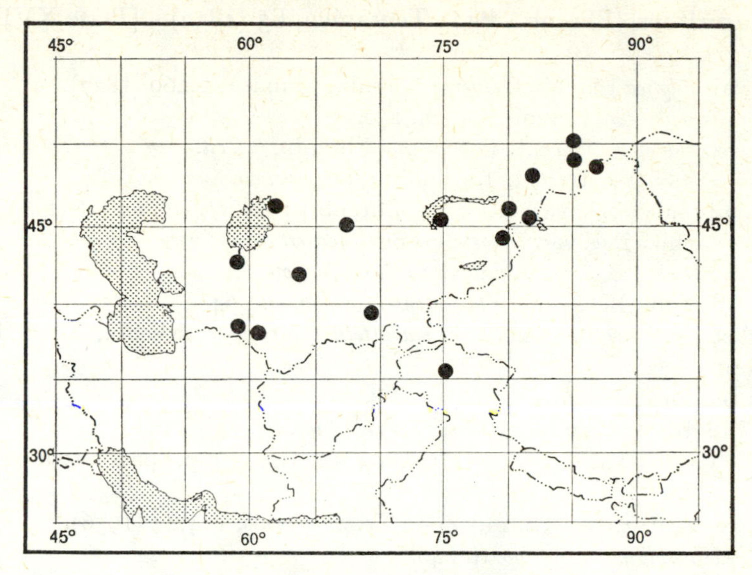

Map 15: *T. leptostachya*

Selected specimens: KAZAKH SSR: *Politoff*, Altai desert. dzongar. ad lacum Nor-Saissan 1826 (syntypes of *T. gallica* var. *micrantha* Ledeb., E, P); *Lintschevsky* & *Lintschevsky* 529, Bakhash-Alakul plain Lepsa river valley 31.7.1934 (LE); *Kuznezov* 407, delta Amu-Darja 13.7.1928 (LE). RUSSIAN SFSR and KAZAKH SSR: *Politov*, Irtysch (LE); *Lehmann*, Lehmgrunde am Kuwan-Darja 14.7.1841 (paralectotype of *T. leptostachya* Bge., P); *Lehmann*, Sandstrecken zwischen den Kuwan und Jan-Darja 19.7.1841 (P). TADZHIK SSR: *Fedtschenko* & *Tschernova* 153, Dolina riv. Varzob. 12.7.1933 (LE). KASHMIR or WESTERN TIBET: *Winterbottom* 832, Shangosi, Valley of Indus 12.10.1847 (K); *Paffen* 294, Karakorum Hunza und Nayar Gebiet, Sarat: Ufer- und Grundwassergehölze der *Artemisia* (Wüsten) Steppe 2450 m (M, W). MONGOLIA: *Hsia* 3167, Inner Mongolia near Wula-shan 26.8.1931 (K).

Var. **kasakhorum** (Gorschk.) Baum. Bracts broader and shorter than in var. *leptostachya*. Sepals sometimes faintly denticulate. Disk somewhat fleshier.

Habitat: Seems to be confined to sand dunes.

Selected specimens: KAZAKH SSR: *Pavlov* 126, Asia Media Kazakstania centralis Reg. Kzyl-Ordinsk in steppa salsuginosa areonsa pr. pag. Sadr-Sarai 14.6.1929 (holotype LE); *Pavlov* 53, Kzyl-Ordinsk sand dunes near Aman Tash near lake of Ashtshi-Kuduk (LE); *Janorusmy*, Songaria Tabulgaly-Koyum 1840–1843 (B). TURKMEN SSR: *Androssow* 1746, Zacasp. obl., Askhabad (LE).

Observation: The species has been introduced into the New World to a very limited extent; it is occasionally cultivated as an ornamental plant.

16. **T. mannifera** (Ehrenb.) Bge., Tentamen, 63 (1852) [Plate XVI]

T. gallica L. var. *nilotica* Ehrenb. f. *cinerea* Ehrenb., Linnaea, 2:269 (1827).
T. gallica L. var. *mannifera* Ehrenb., *op. cit.*, 270.
T. gallica L. var. *mannifera* Ehrenb. f. *divaricata* Ehrenb., *loc. cit.*
T. gallica L. var. *mannifera* Ehrenb. f. *effusa* Ehrenb., *loc. cit.*
T. mannifera (Ehrenb.) Bge. var. *divaricata* (Ehrenb.) Bge., *op. cit.*, 64 (1852).
T. mannifera (Ehrenb.) Bge. var. *leptostachys* Bge., *loc. cit.*
T. mannifera (Ehrenb.) Bge. var. *leucanthera* Bge., *loc. cit.*
T. mannifera (Ehrenb.) Bge. var. *purpurascens* Bge., *op. cit.*, 64.
T. arborea (Sieb. ex Ehrenb.) Bge. var. *mannifera* (Ehrenb.) Sickenb., Mem. Inst. Egypt, 4:189 (1901).
T. aeruginosa Sickenb., *loc. cit.*
T. nilotica (Ehrenb.) Bge. var. *desertorum* Sickenb., *loc. cit.*
T. maris-mortui Gutm., Pal. J. Bot. Jerusalem, 4:50 (1947), p.p. (pars altera = *T. palaestina* Bertol.).
T. gallica L. var. *maris-mortui* (Gutm.) Zoh., Trop. Woods, 104:44 (1956).
T. gallica L. var. *abiadensis* Zoh., *op. cit.*, 46.
T. gallica L. var. *brevispica* Zoh., *loc. cit.*
T. gallica L. var. *divergens* Zoh., *op. cit.*, 47, nom. illegit.[15]
T. gallica L. var. *microcarpa* Zoh., *loc. cit.*
T. gallica L. var. *ascalonica* Zoh., *op. cit.*, 48.
T. gallica L. var. *erythrocarpa* Zoh., *loc. cit.*

Small tree or shrub with reddish-brown bark, younger parts usually papillose, sometimes only sparsely (not entirely) papillulose. Leaves sessile with narrow base, subauriculate, 2–3 mm long, more or less adpressed-imbricate. Vernal inflorescences simple and dense, aestival inflorescences very densely compound. Racemes 1–5 cm long, except for terminal raceme in each inflorescence which may reach 8 cm, 4 mm broad. Bracts more or less triangular, slightly longer than or rarely equalling pedicels in vernal racemes. Pedicel shorter than to more or less equalling calyx. Calyx pentamerous. Sepals 0.75–1 mm long, ovate to trullate-ovate, practically entire or minutely, sparsely and slightly denticulate. Corolla pentamerous, caducous. Petals 1.75–2 mm long, elliptic to obovate, equi- or inequilateral, entire or emarginate. Androecium haplostemonous, of 5 antesepalous stamens; insertion of filaments peridiscal; disk hololophic.

Flowering: August to May, sometimes also July.

Habitat: Salty flats and deserts, wadis, oases, coast (Red Sea), fields.

Distribution: Egypt, Israel, Jordan (see Map 16).

Selected specimens: EGYPT: *Schimper* 329, Sinai, in Arabia petraea loco 'Nafch' 16.5.1835 (syntypes of *T. mannifera* var. *purpurascens* Bge., B, CGE, E, FI, G, GL, K, L, OXF, P, S, UPS, US, W, WU); *Täckholm*, Kharga Oasis near the town 15.1.1928 (CAI); *Täckholm* & *Kassas*, Dakhla Oasis 13.2.1952 (CAI); *Ehrenberg*, in vallibus Sinaiticis 1823 (isotypes of *T. gallica* var. *mannifera effusa* Ehrenb., E, L, P, S, 'Mons Sinai' PRC); *Ehrenberg*, Sinai 1823

15 See Zohary (*loc. cit.*) and compare with Art. 67 — a name based on a monstrosity (galls!).

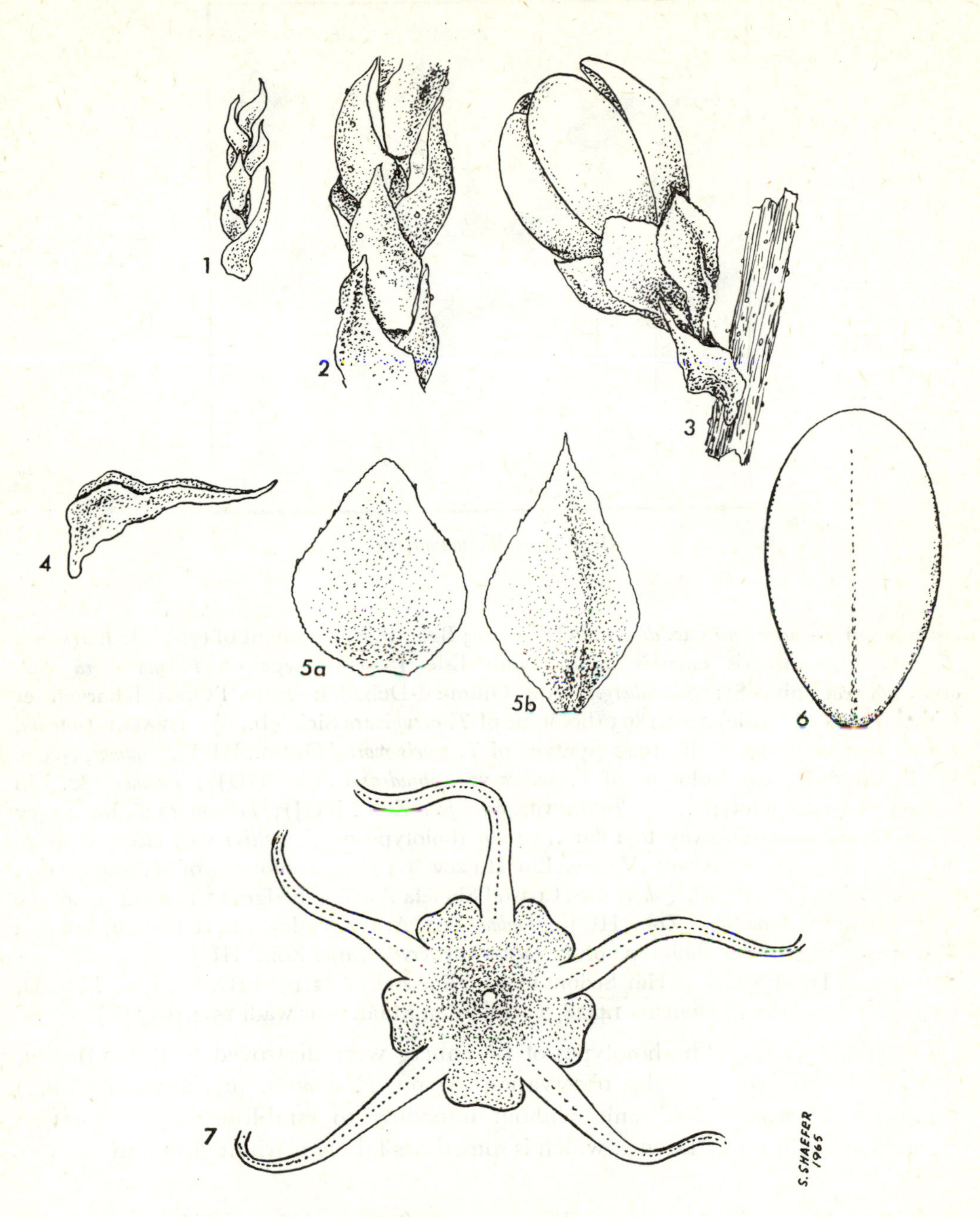

Plate XVI *T. mannifera*
1. Young twig (x 5); 2. id (x 10); 3. Flower (x 10); 4. Bract (x 20); 5a. Inner sepal (x 20); 5b. Outer sepal (x 20); 6. Petal (x 20); 7. Androecium (x 30).

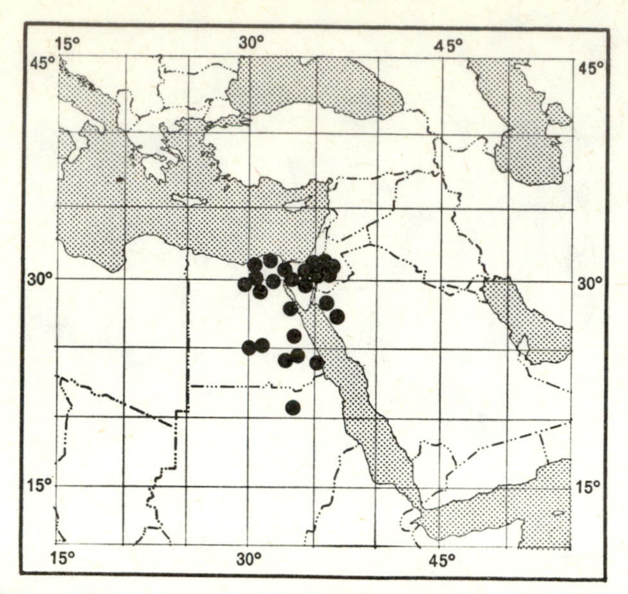

Map 16: *T. mannifera*

(isotypes of *T. mannifera* var. *divaricata* [Ehrenb.] Bge., P, S; fragment of type K); *Ehrenberg*, *T. mannifera* antheris carneis, Sinai Wadi Esle 1823 (isotype of *T. mannifera* var. *divaricata* Ehrenb., S); *Sickenberger*, Ain Onim-ed-Debadeb entre l'Oasis Khargeh et Okhel (Oasis Abbassie) 20.2.1893 (holotype of *T. aeruginosa* Sickenb., S). ISRAEL: *Gutman*, Lower Jordan Valley Kalia 1940 (syntype of *T. maris-mortui* Gutm., HUJ); *Tadmor*, Negev Wadi Abiad .8.1949 (holotype of *T. gallica* var. *abiadensis* Zoh., HUJ); *Tadmor* 880, Ein Hotzev .3.1950 (holotype of *T. gallica* var. *brevispica* Zoh., HUJ); *Tadmor* 1346, far Negev Wadi Hyani near crossway to Eilat .12.1949 (holotype of *T. gallica* var. *divergens* Zoh., HUJ); *Tadmor* 1330, Arava Valley, Ein Hotzev 3.3.1950 (holotype of *T. gallica* var. *microcarpa* Zoh., HUJ); *M. Zohary* 918, Coastal Shefela Ascalon hedges 17.10.1954 (holotype of *T. gallica* var. *ascalonica* Zoh., HUJ); *Tadmor* 662, Arava Valley, Ein (Ghadian) Yotvata near spring 29.12.1952 (holotype of *T. gallica* var. *erythrocarpa* Zoh., HUJ); *Zohary* & *Rayss* 752, env. of Dead Sea near Har Sedom moist saline soil 18.12.1938 (G, HUJ, K, L, S, U, UC, US). JORDAN: *Dinsmore* 14428, Transjordan Maan near wadi 15.4.1937 (E).

Observations: (a) The holotypes of Ehrenberg were destroyed in Berlin during World War II. (b) See also observation (a) on *T. nilotica*. (c) Decaisne (1835) speaks of *T. mannifera* Ehrenb. without intending to establish the combination *T. mannifera* (Ehrenb.) Decne., which is sometimes falsely attributed to him.

17. **T. nilotica** (Ehrenb.) Bge., Tentamen, 54 (1852) [Plate XVII]

T. gallica L. var. *nilotica* Ehrenb., Linnaea, 2:269 (1827).
T. gallica L. var. *nilotica* Ehrenb. f. *glaucescens* Ehrenb., *loc. cit.*, p.p. (excl. pl. *canariensi*).
T. gallica L. var. *heterophylla* Ehrenb., *op. cit.*, 270.

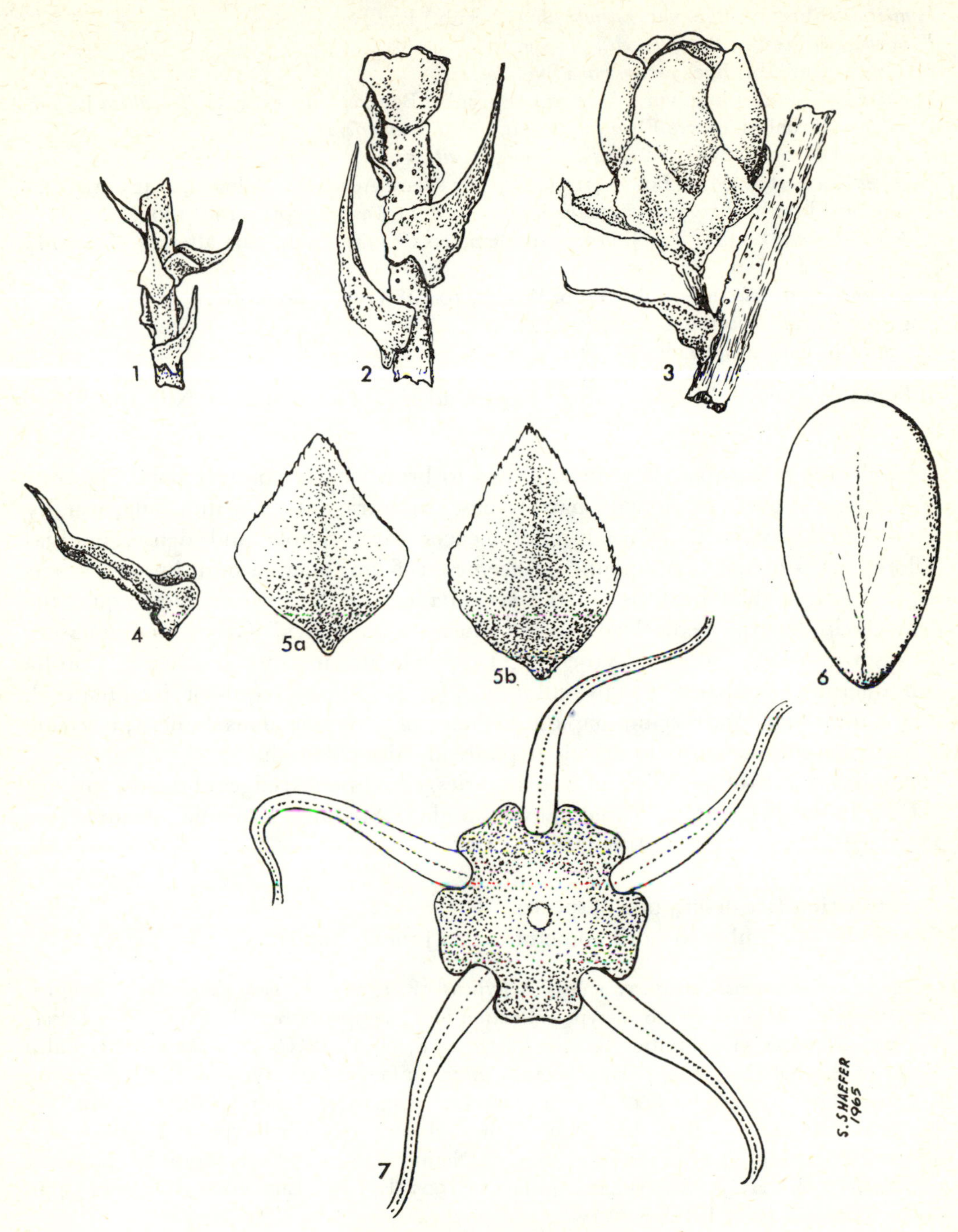

Plate XVII *T. nilotica*
1. Young twig (x 5); 2. id (x 10); 3. Flower (x 10); 4. Bract (x 20); 5a. Inner sepal (x 20); 5b. Outer sepal (x 20); 6. Petal (x 20); 7. Androecium (x 30).

T. nilotica (Ehrenb.) Bge. var. *heterophylla* (Ehrenb.) Bge., *loc. cit.*
T. ehrenbergii Presl ex Bge., *loc. cit.*, pro syn.
T. nilotica (Ehrenb.) Bge. var. *genuina* Bge., *loc. cit.*
T. nilotica (Ehrenb.) Bge. var. *glaucescens* (Ehrenb.) Bge., *op. cit.*, 55, excl. *T. gallica* L. var. *nilotica* Ehrenb. f. *cinerea* Ehrenb., *loc. cit.* (= *T. mannifera*).
T. nilotica (Ehrenb.) Bge. var. *abyssinica* Bge., *op. cit.*, 55.
T. aegyptiaca Bertol., Misc. Bot., 14:15 (1853); or Mem. Acad. Sci. Bologna, 4:423 (1853).
T. nilotica (Ehrenb.) Bge. var. *fluviatilis* Sickenb., Mem. Inst. Egypt, 4:189 (1901).
T. pseudo-pallasii Gutm., Pal. J. Bot. Jerusalem, 4:51 (1947), p.p. (pars altera = *T. arborea* [Sieb. ex Ehrenb.] Bge.).
T. gallica L. var. *micrantha* Zoh., Trop. Woods, 104:48 (1956), nom. illegit. [16]
T. gallica L. var. *subpatens* Zoh., *op. cit.*, 49.
T. gallica L. var. *tenuior* Zoh., *loc. cit.*

Type: EGYPT: *Ehrenberg*, prov. Fayum ad rivula et in insula Nili 1820–1826 (syntypes K, L).

Small tree or shrub with reddish-brown to brown bark, younger parts glabrous to papillose. Leaves sessile with narrow base, subauriculate, 2–3 mm long, usually remote and divaricate. Vernal inflorescences rare, simple and dense, aestival inflorescences densely compound. Racemes 1–6 cm long, 4 mm broad. Bracts slightly longer than pedicels; in the vernal racemes the lowest bracts equal the pedicels in length. Pedicel somewhat shorter than calyx. Calyx pentamerous. Sepals 1 mm long, entire or subentire, trullate-ovate, more or less acute. Corolla pentamerous, caducous. Petals elliptic-ovate to elliptic, equi- or inequilateral, 1.75–2 mm long. Androecium haplostemonous, of 5 antesepalous stamens; insertion of filaments conspicuously to slightly hypodiscal; disk hololophic.

Habitat: Banks of the Nile and its tributaries, canal banks, edges of ponds, springs.

Distribution: Lebanon, Israel, Egypt, Sudan, Somalia, Ethiopia, Kenya (see Map 17).

Var. **nilotica** (see description of species).
Flowering: November to April, sometimes also June to August.

Selected specimens: EGYPT: *Savi*, Aegyptus 1843, id. '*T. senegalensis* DC.' lungo il Nilo versio il villagio di Sciubra (type collection of *T. aegyptiaca* Bertol., FI, G, W); *Letourneux* 30, ad fossas in paludibus Ramle 5.4.1877 (E, G, P, PRC, W); *Davis* 6005, Ealfu banks of the Nile 6.12.1943 (E). ISRAEL: *Eig*, Haifa 1921 (syntype of *T. pseudo-pallasii* Gutm., HUJ); *Gutman* T. 860, Kiryat Motzkin 30.3.1941 (holotype of *T. gallica* var. *micrantha* Zoh., HUJ); *Boyko* 867, Acco Plain Galia 2.7.1949 (holotype of T. *gallica* var. *subpatens* Zoh., HUJ); *M. Zohary* 841, Sharon Netanya 26.11.1952 (holotype of *T. gallica* var. *tenuior* Zoh., HUJ); *Meyers* 2228, Jaffa 3.11.1905 (E, US). LEBANON: *Gaillardot* 543b, prope Saida (E, G, P, PRC, S, W).

Var. **abyssinica** Bge. Differs from var. *nilotica* in its longer racemes (up to 9 cm long). Sepals 1.25 mm long. Petals 2–2.25 mm long, elliptic-obovate. Androecium

16 Because of *T. gallica* L. var. *micrantha* Ledeb., 1843.

with 5 antesepalous stamens inserted as in var. *nilotica*, but occasionally also with 1 or 2 antepetalous ones.

Flowering: July to December.

Selected specimens: KENYA: *Elliot* 38, Sultan Hammud also on Jana River 3500 ft (EA). ETHIOPIA: *Schimper* 728, ad litus fluvii Tacaze infra Dscheladschegenne 4.11.1839 (lectotype of *T. nilotica* var. *abyssinica* Bge., P; isolectotypes B, G, K, OXF, P, S). SOMALIA: *Glover & Gilliland* 1054, from the tug (dry stream-bed) leading to Domei Doho (EA, FHO, K). SUDAN: *Ourem* 20035, Khartum Blue Nile, bank below university near the bridge 19.7.1961 (BG).

Observations: (a) Closely related to *T. mannifera* from which it differs mainly in its looser and divaricate leaves and in the hypodiscal configuration of the disk. (b) The holotypes of Ehrenberg were lost during World War II in Berlin. (c) The author has not seen the authentic specimen of *T. nilotica* var. *glaucescens* which was lost during World War II: however, he was able to examine many specimens from the locus classicus, Wadi Goaebe (Egypt), and they are identical with *T. mannifera*.

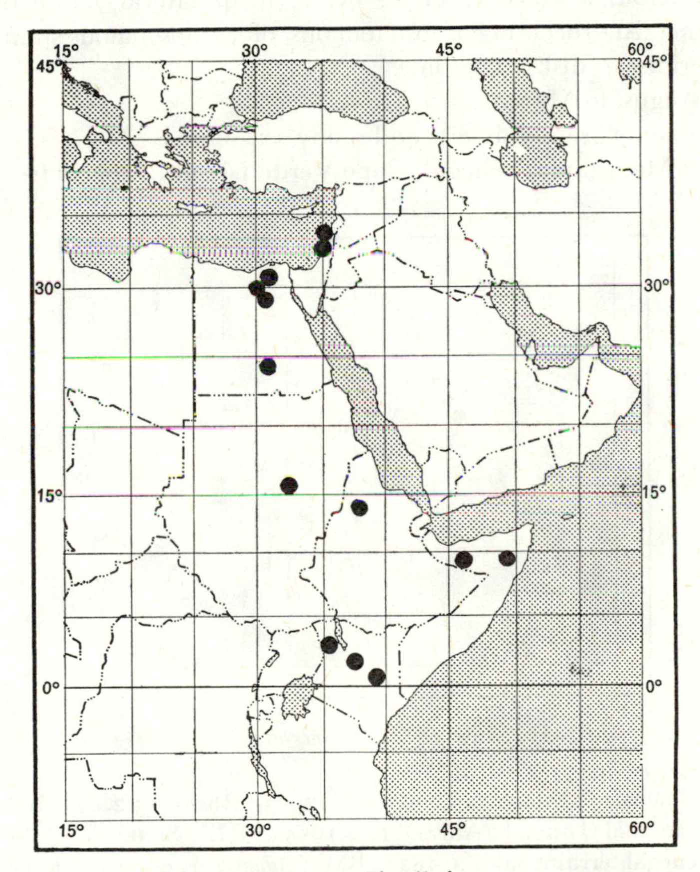

Map 17: *T. nilotica*

18. **T. senegalensis** DC., Prodr., 3:96 (1828) [Plate XVIII]

T. gallica L. subsp. *nilotica* (Ehrenb.) Maire var. *monodiana* Maire, Bull. Soc. Hist. Nat. Afr. N., 28:332 (1937).
T. gallica L. var. *monodiana* Maire, *op. cit.*, 30:327 (1938).

Type: SENEGAL: *Perrottet* 349, Senegal 20.1.1825 (holotype G-DC, isotypes BM, G, P, S)

Small tree or shrubby tree with reddish-brown to brownish-black bark, glabrous to entirely papillose, often only rachis of racemes papillose. Leaves (including the younger leaves) sessile with narrow base, sometimes subauriculate (never fully auriculate), 1.5–2.5 mm long. Inflorescences rarely vernal and simple, generally aestival and loosely or densely composed of racemes. Racemes 2–4 mm broad, 3–7 cm long. Bracts longer than pedicels, narrowly trullate, their lower parts with somewhat denticulate margins. Calyx pentamerous. Sepals trullate, deeply, irregularly and finely denticulate, especially towards apex, acute,[17] 1–1.5 mm long. Corolla pentamerous, caducous. Petals ovate, inequilateral, more or less retuse, 1.5–1.75 mm long. Androecium haplostemonous, of 5 antesepalous stamens; insertion of filaments peridiscal; disk hololophic.

Flowering: August to March.

Habitat: Saline soil, sandy deserts and sandy sea-shore.

Distribution: Mauritania, Senegal, Cape Verde Islands, Liberia (see Map 18).

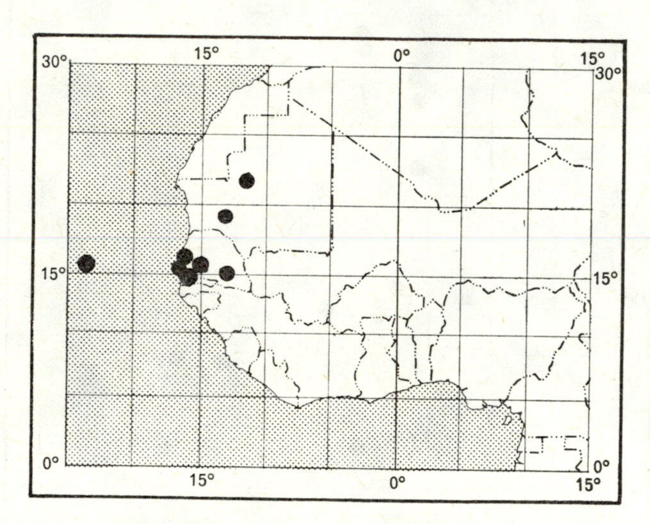

Map 18: *T. senegalensis*

Selected specimens: MAURITANIA: *Adam* 12973, Rasso 13.2.1957 (P); *Naegeli* 85, Sahara Sud Occidental Tongad (Adrar) 11.8.1954 (PRE). SENEGAL: *Perrottet*, Rivières inférieures en Senegal terrains salés .9.1825 (BM); *Adam* 301, environs de Dakar 14.1.1948

17 In *T. indica* obtuse to truncate.

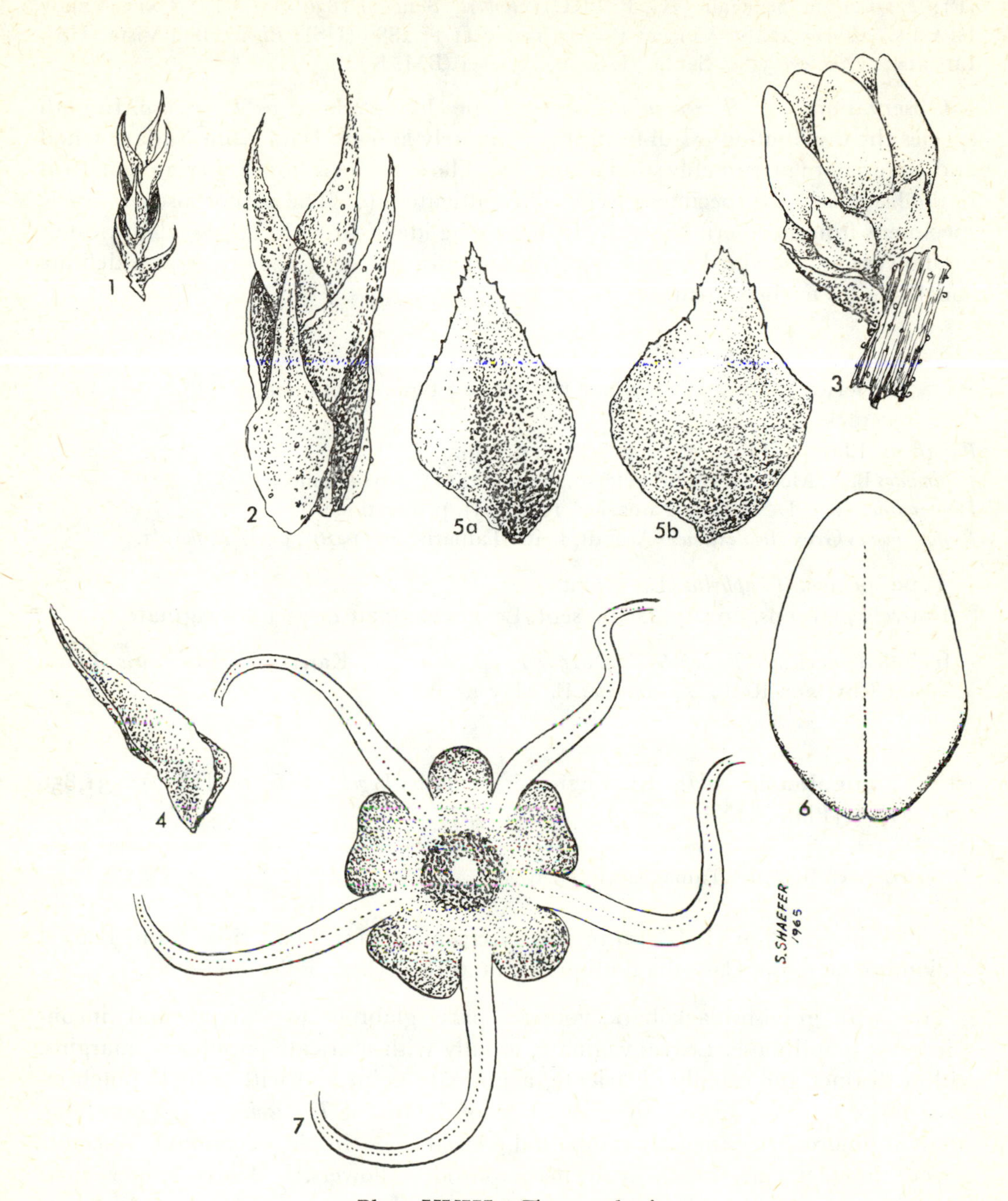

Plate XVIII *T. senegalensis*
1. Young twig (x 5); 2. id (x 20); 3. Flower (x 15); 4. Bract (x 30); 5 a. Outer sepal (x 30); 5b. Inner sepal (x 30); 6. Petal (x 30); 7. Androecium (x 30).

(P); *Leprieur*, in Senegalia (G, P, PRC); *Robert*, Senegal 1836 (GL, P). CAPE VERDE ISLANDS: *Brown* 22, St. Vincent Proto Grande 11.11.1889 (US); *Eights*, Boa Vista (US). LIBERIA: *Dinklage* 3180, Santa Maria 19.10.1934 (BM, K, P).

Observations: (a) *T. senegalensis* seems to be closely related to *T. arabica*. In both species the distribution is still far from adequately known. Data from Sudan, Chad and Nigeria are presumably still lacking. (b) The specimens from Liberia mentioned in the list of selected specimens seem to be cultivated. (c) The author also saw a few specimens from northern Nigeria which may be identical with *T. senegalensis*. Since the flowers he examined were not yet in anthesis, he could not arrive at a definite conclusion as to their identity.

Series 3. VAGINANTES (Bge.) Bge., Tentamen, 6 (1852)

Parviflorae Ehrenb., Linnaea, 2:257 (1827), p. min. p.
Vaginantes Bge., Mém. Acad. St. Pétersb., 7:292 (1851), pro sectione.
Haplandrae Ndz., De Genere Tamarice, 11 (1895), pro sectione.
Pentastemones Grex *Amplexicaules* Arendt, Beitr. Tamarix, 42 (1926), excl. *T. florida*.

Type species: *T. aphylla* (L.) Karst.
Entirely glabrous, no papillae present. Leaves vaginate or pseudo-vaginate.

Included species: *T. angolensis* Ndz., *T. aphylla* (L.) Karst., *T. bengalensis* Baum *T. dioica* Roxb. ex Roth, *T. usneoides* E. May ex Bge.

19. **T. angolensis** Ndz., in: Engler & Drude, *Veget. Erde*, 9 (III, 2): 531,850 (1921) [Plate XIX]

T. engleri Arendt, Beitr. Tamarix, 51 (1926).

Lectotype: ANGOLA: *Welwitsch* 1086, freq. in arenosis juxta ripas flum. Bero et Maiombo, saepiusa, Cassytha undique tecta .6.1859 (BM; isolectotype K).

Tree with grayish-black bark, younger parts glabrous to scarcely and inconspicuously papillulose. Leaves vaginate, usually with muricate-papillulose margins, with a distinct and usually divaricate, acuminate point and with a little notch on its opposite side (reminiscent to some extent of the leaf of *T. ericoides*), 1–3 mm long. Aestival inflorescences densely compound, vernal ones simple, occasional. Racemes 2–5 cm long, 5 mm broad, with hermaphrodite flowers.[18] Bracts longer than pedicels or of equal length, vaginate, as are leaves, but more acuminate, with more or less minutely muricate-papillose margins. Pedicels shorter than calyx. Calyx pentamerous. Sepals 1–1.25 mm long, minutely denticulate or subentire, the outer 2 ovate, acute, keeled, the inner broadly trullate-ovate, more obtuse, slightly larger

18 The nearest and closest related species, *T. usneoides*, is a dioecious tree.

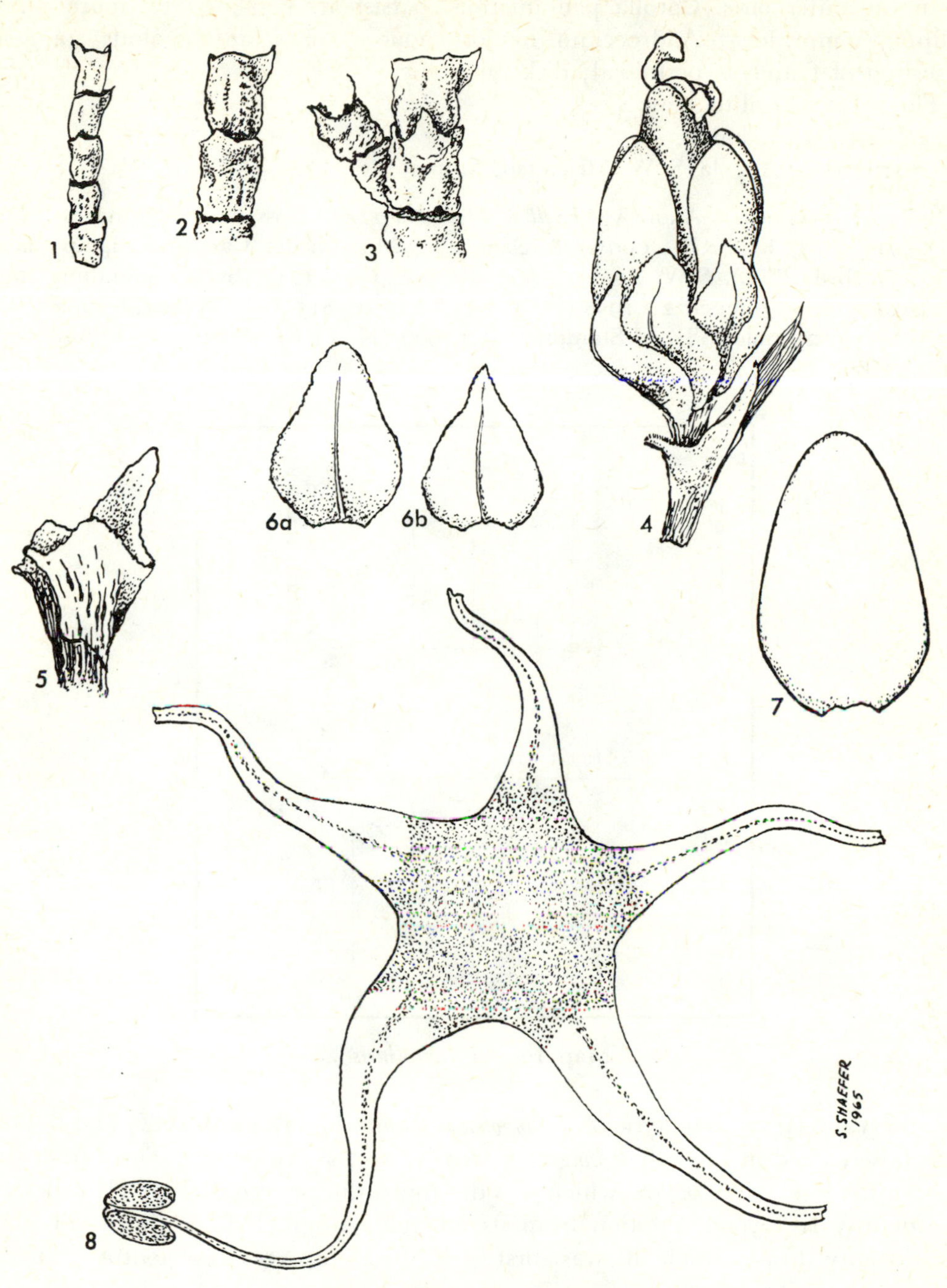

Plate XIX *T. angolensis*
1. Young twig (x 5); 2–3. id (x 10); 4. Flower (x 10); 5. Bract (x 20); 6a. Inner sepal (x 20); 6b. Outer sepal (x 20); 7. Petal (x 20); 7. Androecium (x 30).

than the outer ones. Corolla pentamerous, persistent. Petals ovate, more rarely elliptic, 2 mm long. Androecium haplostemonous, of 5 antesepalous stamens; insertion of filaments peridiscal; disk synlophic.

Flowering: April to May.

Habitat: River banks.

Distribution: Angola, S. W. Africa (see Map 19).

Selected specimens: ANGOLA: *Exell* & *Mendonca* 2144, Mossâmedes Giraul de Cinra 20.5.1933 (BM, K, PRE); *Carisso* & *Sousa* 252, Mossâmedes Carvalho, damp places 3.5.1937 (BM, PRE). S.W. AFRICA: *Merxmüller* & *Giess* 174b, Swakopmündung stark salzige Sände und Jumpel 22.2.1958 (PRE, US); *Leistner* 1815, Gibeon distr. compact clay in river bed 25 miles SSE of Stampnet 13.4.1960 (K, PRE); *Fisher* 120, Otjimbingue 1897 (HBG).

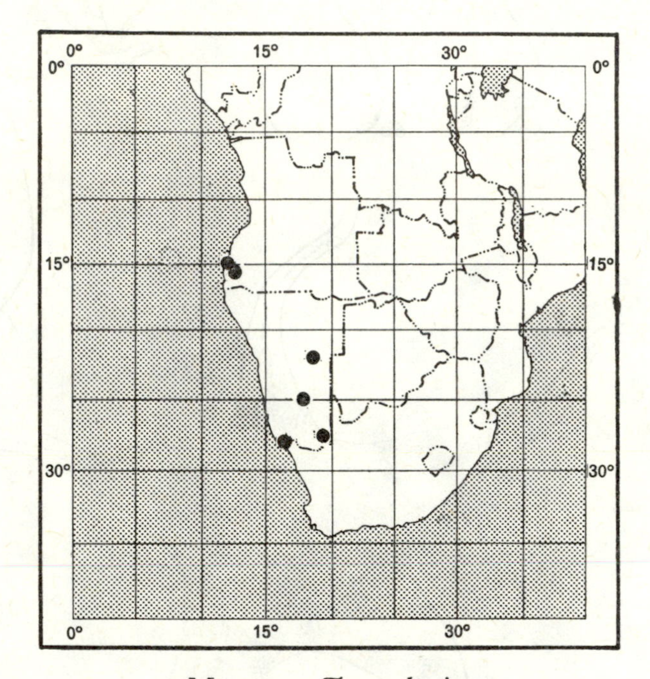

Map 19: *T. angolensis*

Observations: (a) The type of *T. angolensis* should be *Welwitsch* 1086, and not 86, as is falsely cited in Engler's *Pflanzenfam.*, ed. 2, 21:288 (1925). (b) The author did not see the type of *T. engleri*, which was destroyed during World War II in Berlin. Its identity is beyond question from its adequate original description and from the locality from which it was first described. (c) See also footnote under description of *T. usneoides*.

20. **T. aphylla** (L.) Karst., Deutsch. Fl., 641(1882) [Plate XX]

Thuja aphylla L., Cent. I. Plant., 32 (1755), p.p. excl. syn. Shaw (Cat. Pl. Afr., 188 f. 180 [188] 1838).
Tamarix orientalis Forsk., Fl. Aegypt.-Arab., 206 (1775).
T. articulata Vahl, Symb. Bot., 2:48 (1791), nom. illegit.
T. aphylla (L.) Lanza, Boll. R. Orto Bot. Palermo, 8:82 (1909), comb. illegit.
T. aphylla (L.) Warb., Beitr. Kent. Sinai, 139 (1929), comb. illegit.

Type: EGYPT: *Linnean Herb.* No. 1136.3 (holotype LINN; isotypes S, UPS).

Tree or high shrub with reddish-brown to grey bark, younger parts entirely glabrous. Leaves vaginate, abruptly and shortly pointed, about 2 mm long. Vernal inflorescences simple, aestival ones compound and more common. Raceme 3–6 cm long, 4–5 mm broad, with subsessile flowers. Bracts triangular to broadly triangular, acuminate, somewhat clasping, longer than pedicels. Pedicel much shorter than calyx. Calyx pentamerous. Sepals c. 1.5 mm long, entire, obtuse, the 2 outer slightly smaller, broadly ovate to broadly elliptic, slightly keeled, the inner slightly larger, broadly elliptic to suborbicular. Corolla pentamerous, subpersistent to caducous. Petals 2–2.25 mm long, elliptic-oblong to ovate-elliptic. Androecium haplostemonous, of 5 antesepalous stamens; insertion of filaments peridiscal; disk hololophic.

Flowering: August to November.

Habitat: Sandy soil and dunes, canal and river banks, salty deserts, fields.

Distribution: Morocco, Algeria, Libya, Egypt, Senegal, Sudan, Abyssinia, Eritrea, Somalia, Kenya, Israel, Saudi Arabia, Yemen, Iraq, Kuwait, Iran, Pakistan, Afghanistan (see Map 20).

Selected specimens: MOROCCO: *Sauvage* 8579, Moyen Oum-er-Rebia Oued bou Rhamoun entre Beni Mellal et Dar Ould Zidouk 2.1.1951 (RAB). ALGERIA: *Chevallier* 576, Sahara Inifel in arenis 'Oued Mya' frequentiss. et in omnibus vall. australior. 25.2.1904 (B, FI, K, PRC, US, WU); *Cosson* 39, in alveo exsiccato Oued en Nasr pr. Mguima, in parte australiore ditionis Mzab 19.5.1858 (G, P, S, US, W). LIBYA: *Bornmüller* 728, Tripolitania: Tripolis in palmetis prope Fadjura 18.4.1933 (B, S); *Keth*, 254 Giorgimpoli 5.8.1958 (L). EGYPT: *Letourneux* 32, ad fossas Ramle 25.8.1877 (B, E, FI, G, K, P, S, W); *Täckholm*, Kassas, Samy, Girgis, Zahran, Wadi Ghuwebba of Red Sea coast 9.6.1960 (CAI); *Täckholm* 264, Wadi Hammamet Faraon, Sinai 15.5.1956 (CAI); *Forskhål* ex oriente (isotype of *T. orientalis* Forsk., S, BM). SENEGAL: *Adam* 194, environs de Dakar 26.8.1948 (P). SUDAN: *Ouren* 20010, Khartoum prov. Khartoum Blue Nile bank below university (near the bridge) 18.7.1961 (BG). ABYSSINIA: *Ouren* 20268, Arussi prov. near warm spring by Awash river about 5 km from Koka 2.8.1961 (BG); *West* 5400, Belet Hen, river banks 9.3.1941 (EA, PRE). ERITREA: *Pappi* 2601 & 224, Assaorta Lungo il Torrente 14–15.8.1902 (BM, EA, G, HBG, K, P, S, U). SOMALIA: *Keller* 51, Webi-Habir 1891 (K). KENYA: *Gillet* 13301, between Yabichu and Mandera banks of water-course 23.5.1952 (B, EA, K, S); *Adamson* 83, between Mandera and Remu 10.10.1955 (EA, PRE, K). ISRAEL: *Waisel* 553, Negev, Gevulot sandy soil 30.8.1954 (B, DEL, E, EA, G, K, L, OXF, S, U, UC, US); *Dinsmore* 1317, environs of Gilgal in campestre arbor 8 m—250 m, 6.8.1908 (E, L). SAUDI ARABIA: *Wissman* 1119, Gebirge des Hinterlandes von Aden

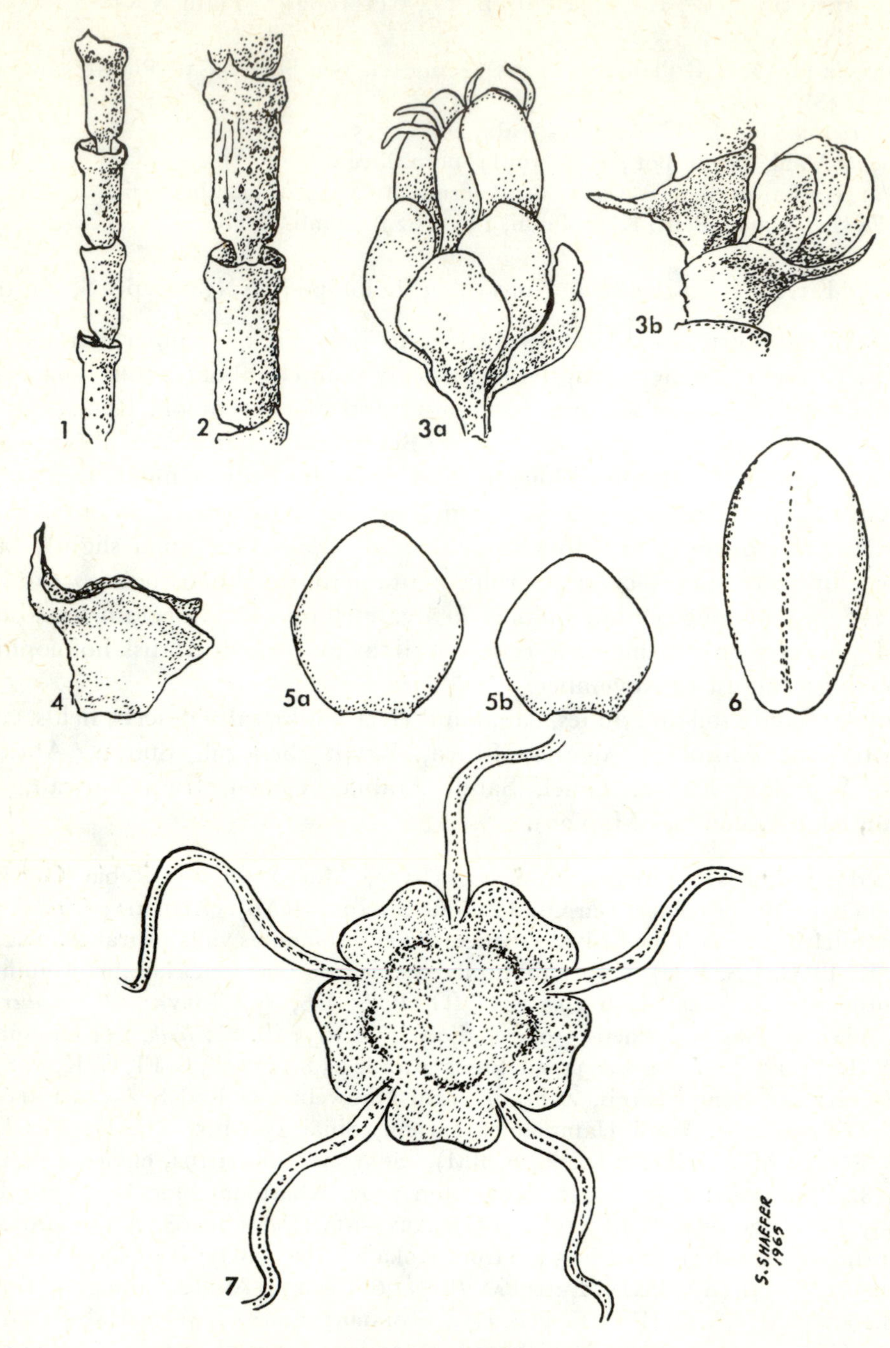

Plate XX *T. aphylla*

1. Young twig (x 5); 2. id (x 10); 3a. Flower (x 10); 3b. Flower bud (x 10); 4. Bract (x 20); 5a. Inner sepal (x 20); 5b. Outer sepal (x 20); 6. Petal (x 20); 7. Androecium (x 30).

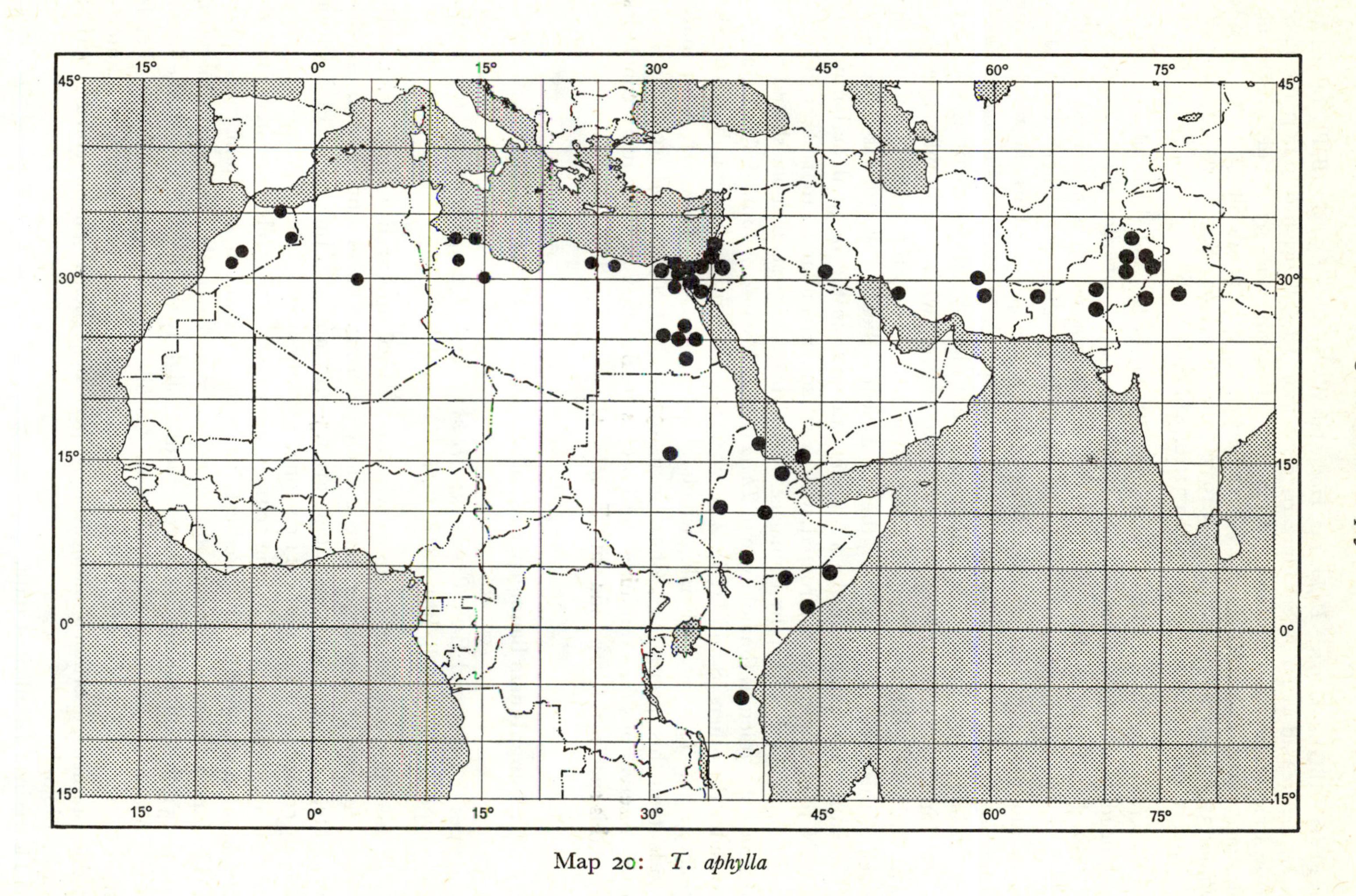

Map 20: *T. aphylla*

Ka'taba (HBG). YEMEN: *Deflers* 132, Hodjeilah Wadi Chaba 2.5.1887 (B, P). IRAQ: *Kotschy*, arborea ad ascem Nebokadnezaris districtum urbis Babylon 6.10.1841 (G, P, S). KUWAIT: *Willeson* 315, Kuwait 10.10.1935 (K). IRAN: *Alexeenko* 937, in ditione Daschistan prope pag. Daleki in planitie. PAKISTAN: *Lace*, Baluchistan Sibi 2.1.1899 (E, K); *Ritchie* 37, upper Scinde Khuclya .3.1937 (E); *Stocks* 400, Scinde 1880 (G, K, P). AFGHANISTAN: *Aitchison* 31, Omar-Sha 31.10.1884 (cult. 1., K).

Observations: (a) Cultivated as an ornamental hedge and shade plant. (b) Introduced into the New World. (c) Typification of *T. aphylla*: *Thuja aphylla* was collected in Egypt by Hasselquist. In his herbarium, which is preserved in UPS, there is a nice specimen numbered 544. In S there is a herbarium sheet with mounted fragments of branchlets of the same species. This sheet bears the following label: (1) '*Thuya aphylla* Linn. Hasselquist'; (2) '*Tamarix articulata Vahl*'; (3) '*Thuja aphylla* Linn.'. The author could not identify the handwriting of No. 1; only 'Linn.' was written by Vahl. No. 2 was entirely written by Vahl, while No. 3 was written by Linnaeus himself and Vahl added to it only the author 'Linn.' On the back of this sheet, in Stockholm, there is a clear indication that it was collected by Hasselquist and might thus be a fragment taken from the original herbarium. In LINN the author was confronted with a specimen similar to that of Stockholm, but with a label marked by Linnaeus. W.T. Stearn suggests that one should regard this last specimen as the holotype of *Thuja aphylla* because of Linnaeus' signs, and to consider the others (S and UPS) as isotypes. (d) The author was not able to see the holotype of *T. orientalis* Forsk., which is in Copenhagen. However, with many students of this species (e.g., Christensen, 1922; Hunt, 1963), and according to his own observations on examining the isotypes in BM and S, the author concludes that this species is conspecific with *T. aphylla*.

21. **T. bengalensis** Baum (sp. nov., see Appendix) [Plate XXI]

Type: INDIA and EAST PAKISTAN: *Hook. f.* & *Thomson*, Bengal or. reg. temp. (holotype W; isotypes B, BM, CGE, G, K, OXF, P, S, U, W).

Monoecious tree, often shrubby, with the general appearance of *T. dioica*, with brown to grey bark, younger parts entirely glabrous. Leaves amplexicaul, pseudovaginate, without conspicuous salt glands, 1.5–2.25 mm long. Inflorescences simple or loosely composed of racemes 4–11 cm long, 4–5 mm broad. Bracts simple, triangular-acuminate (with a relatively long acumen), longer than pedicels, sometimes more or less equal in length with calyx. Pedicel shorter than calyx. Calyx pentamerous. Sepals orbicular to broadly obovate, with very finely and densely denticulate margins, 1.25–1.5 mm long, the outer 2 more orbicular, slightly shorter and keeled, the inner more obovate, sometimes faintly keeled. Corolla pentamerous, persistent. Petals obovate to broadly obovate, 2 mm long. Androecium haplostemonous, of 5 antesepalous stamens; insertion of filaments hypo-peridiscal (i.e., 1–3 hypo- and 2–3 peri-); disk hololophic.

Flowering: Insufficient data.

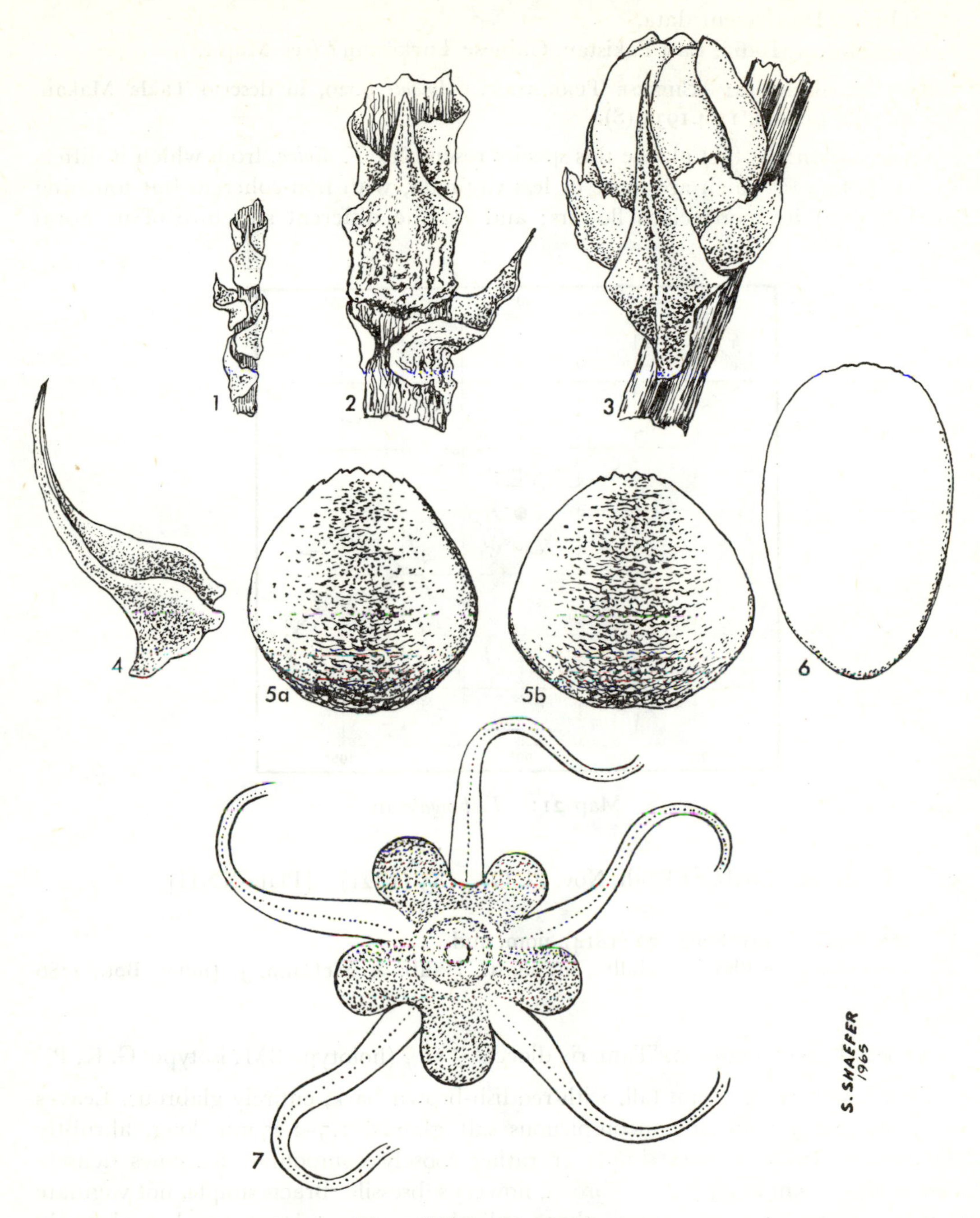

Plate XXI *T. bengalensis*
1. Young twig (x 5); 2. id (x 10); 3. Flower in bud (x 10);
4. Bract (x 20); 5a. Outer sepal (x 22); 5b. Inner sepal (x 22);
6. Petal (x 22); 7. Androecium (x 35).

Habitat: Insufficient data.

Distribution: India, East Pakistan, Chinese Turkestan? (see Map 21).

Selected specimens: CHINESE TURKESTAN: *Ambolt* 6029, in deserto Takla Makan, Yartongguz 1400 m, 12.4.1933 (S).

Observation: At first glance this species resembles *T. dioica*, from which it differs in (1) its leaves, which are more or less vaginate, with non-coherent but touching margins; (2) its monoecious flowers; and (3) the different structure of its floral parts.

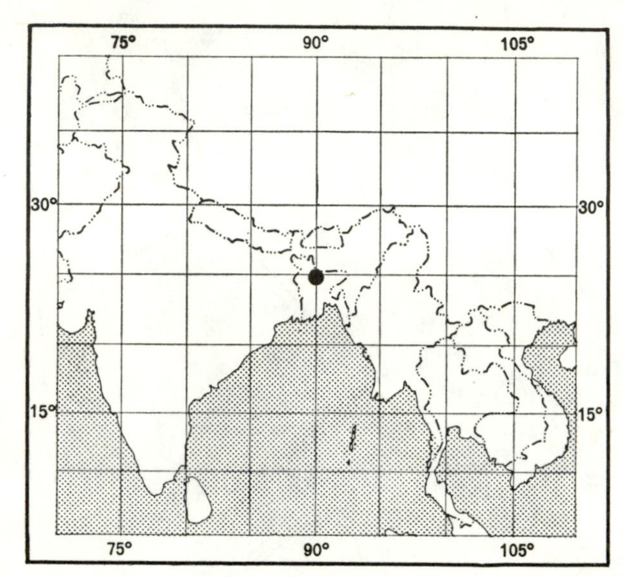

Map 21: *T. bengalensis*

22. **T. dioica** Roxb. ex Roth, Nov. Pl. Or., 185 (1821) [Plate XXII]

T. dioeca Roxb., Hort. Beng., 22 (1814), nom. nud.
T. longe-pedunculata Blatt. & Hallb., in: Blatt., Hallb. & McCann, J. Indian Bot., 1:86 (1919).

Type: INDIA: *Roxburgh*, 'Tamarix dioeca' ♂ et ♀ (holotype BM; isotypes G, K, P).

Dioecious tree, 2–3 mm tall, with reddish-brown bark, entirely glabrous. Leaves vaginate, with very tiny inconspicuous salt glands, 1.5–2.25 mm long, abruptly acuminate. Inflorescences simple or rather loosely compound. Racemes densely flowered, 3–8 cm long, 7–8 mm broad, flowers subsessile. Bracts simple, not vaginate but somewhat clasping, longer than pedicels, acuminate to somewhat abruptly acuminate. Pedicel shorter than calyx. Calyx pentamerous. Sepals more or less broadly ovate, 1 mm long, obtuse, entire or subentire in male flowers, very minutely denticulate, especially at apex of the female flowers, the outer 2 keeled. Corolla pentamerous, persistent. Petals more or less 2 mm long, obtriangular-obovate.

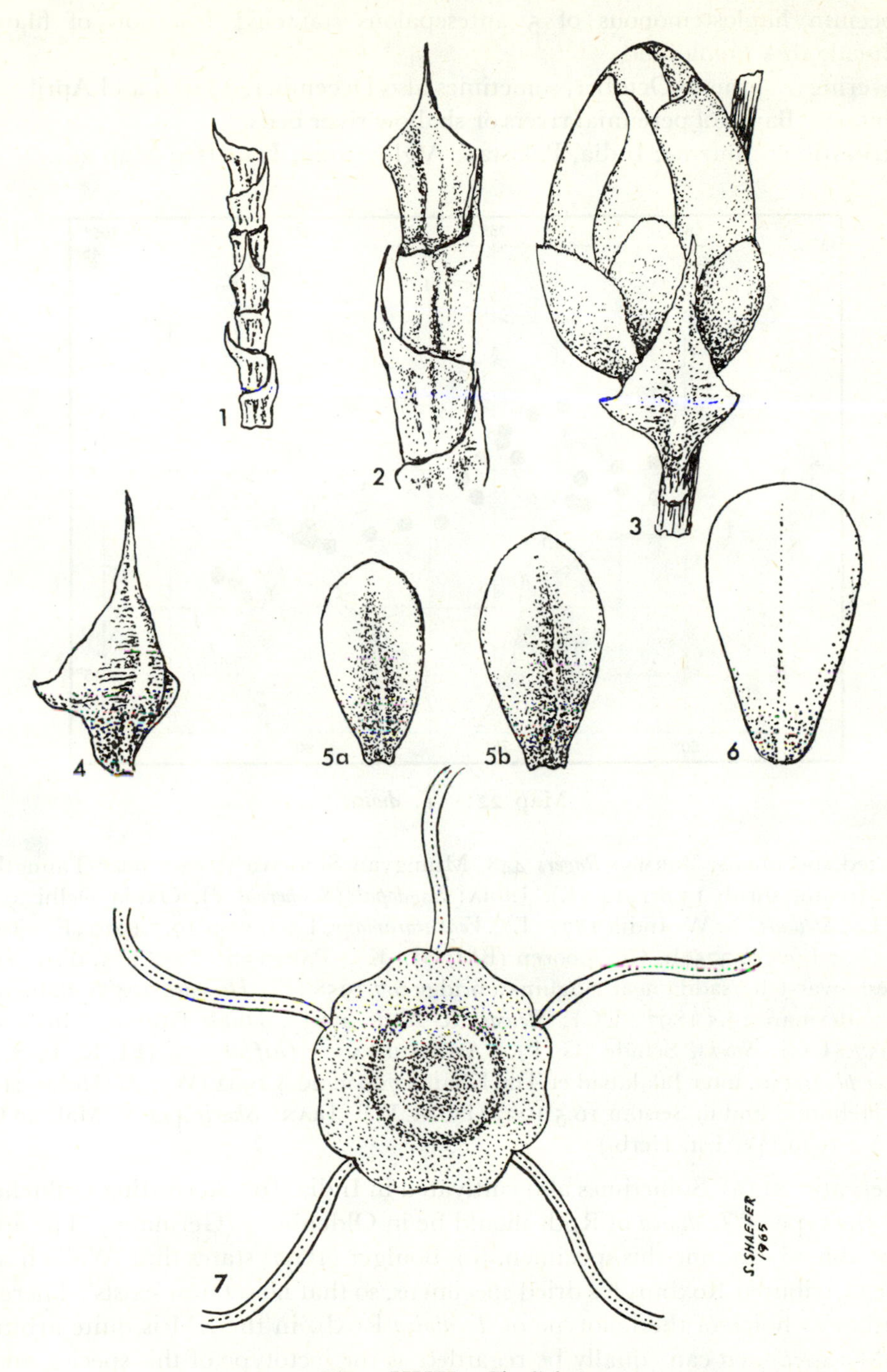

Plate XXII *T. dioica*
1. Young twig (x 5); 2. id (x 10); 3. Flower in bud (x 10); 4. Bract (x 20); 5a. Outer sepal (x 20); 5b. Inner sepal (x 20); 6. Petal (x. 20); 7. Androecium (x30).

Androecium haplostemonous of 5 antesepalous stamens; insertion of filaments hypodiscal; disk hololophic.

Flowering: June to October, sometimes also December, March and April.

Habitat: Banks of perennial rivers or shallow river beds.

Distribution: Burma, India, Pakistan, Afghanistan, Iran (see Map 22).

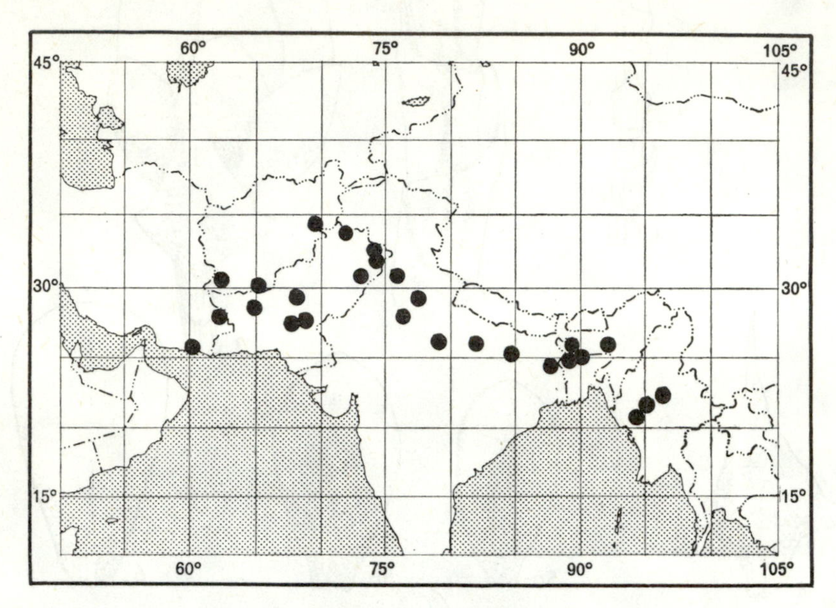

Map 22: *T. dioica*

Selected specimens: BURMA: *Rogers* 448, Muingyan Sindewa stream near Taungtha in bed of stream, shrub 15.8.1914 (E). INDIA: *Sagdopal* (*Sabherval* ?), Okhla Delhi .9.1958 (B, DEL); *Stewart*, N. W. India 1871 (E); *Venkataramany*, Lachivala 19.7.1930 (E); *Jenkins*, Assam (L); *Wallich* 3756b, Gualpooren (BM, GL, K). PAKISTAN: *Burtt* 975, distr. Peshawar, Peshawar-Charsadda near Naguman bridge 6.7.1958 (E); *Hersulch* 20486, Baluchistan Fort Landerman 13.5.1897 (UC); *Stewart* & *Nasir* 27855, Dhak Pathan, Altock distr. 22.3.1957 (UC); *Stocks*, Scinde (G, P). AFGHANISTAN. *Griffith* 959 (FI, K, P, S, W); *Rechinger fil.* 19356, inter Jalalabad et Tor Khama 700 m 20.8.1962 (W); *McMahon* 41/111, on the Helmund and in Seistan 16.5.1903 (POM W). IRAN: *Sharif* 444–E, Makran Chah Bahar 4.4.1949 (W, Esf. Herb.).

Observations: (a) Sometimes also cultivated in India. (b) According to Buchenau (1868) the type of *T. dioica* of Roth should be in Oldenburg (Germany). The author was not able to examine this specimen. (c) Boulger (1897) states that 'Wallich seems to have distributed Roxburgh's dried specimens, so that no set now exists'. Therefore, the author's choice of the holotype of *T. dioica* Roxb. in the BM is quite arbitrary. The BM's specimen can equally be regarded as the lectotype of this species, and the other ones mentioned, as isolectotypes. (d) *T. longe-pedunculata* Blatt. & Hallb.: the author was not able to see the type, however its description falls within the range of variability of *T. dioica*.

23. **T. usneoides** E. Mey. ex Bge., Tentamen, 74 (1852) [Plate XXIII]

T. usneoides E. Mey., in: Drege, Zwei Pflzn. Dokum., 225 (1844).
T. austro-africana Schinz, Bull. Herb. Boiss., 2:183 (1894).

Lectotype: S. AFRICA: *E. Meyer*, a, b, c, d, Cap de Bonne Espérance (P; isolectotypes CGE, G, GL, K, OXF, PRC, S).

Dioecious[19] tree, often shrubby, with brown, later grey, bark, entirely glabrous. Leaves vaginate with a short to inconspicuous herbaceous point,[20] 1.25 mm long. Inflorescences loosely to more or less densely composed of racemes. Racemes 2–6 cm long, 5–6 mm broad. Bracts shorter than to slightly exceeding pedicels, similar to leaves[21] (i.e., vaginate). Pedicels half as long as sepals. Calyx pentamerous. Sepals entire, 1–1.25 mm long, the outer 2 ovate-acute, the inner trullate-ovate. Corolla pentamerous, persistent. Petals 2.25 mm long, inequilateral, ovate to somewhat elliptic. Androecium haplostemonous, of 5 antesepalous stamens; insertion of filaments peridiscal; disk synlophic to para-synlophic in female flowers, and hololophic to paralophic in male flowers.

Flowering: July to October, sometimes January, February and May.

Habitat: Sandy and salty dunes and flats, rocky deserts, temporary river beds.

Distribution: Union of South Africa, S. W. Africa (see Map 23).

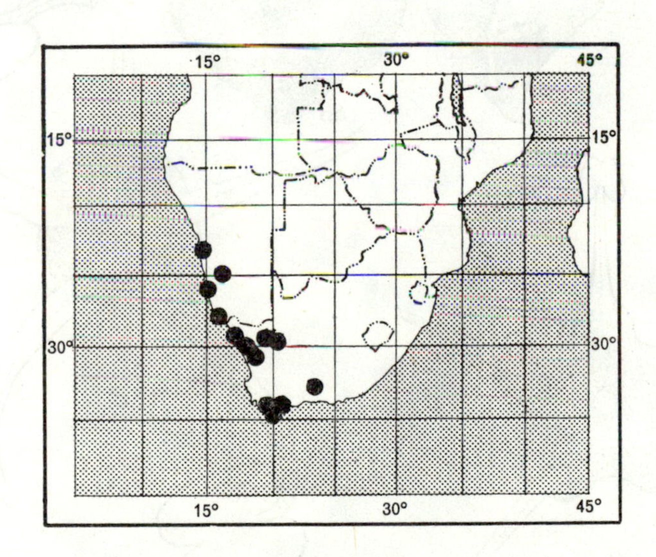

Map 23: *T. usneoides*

19 In *T. angolensis*, monoecious.
20 In *T. angolensis*, with a long acuminate and divaricate point.
21 In *T. angolensis*, longer than pedicels and notched opposite the point.

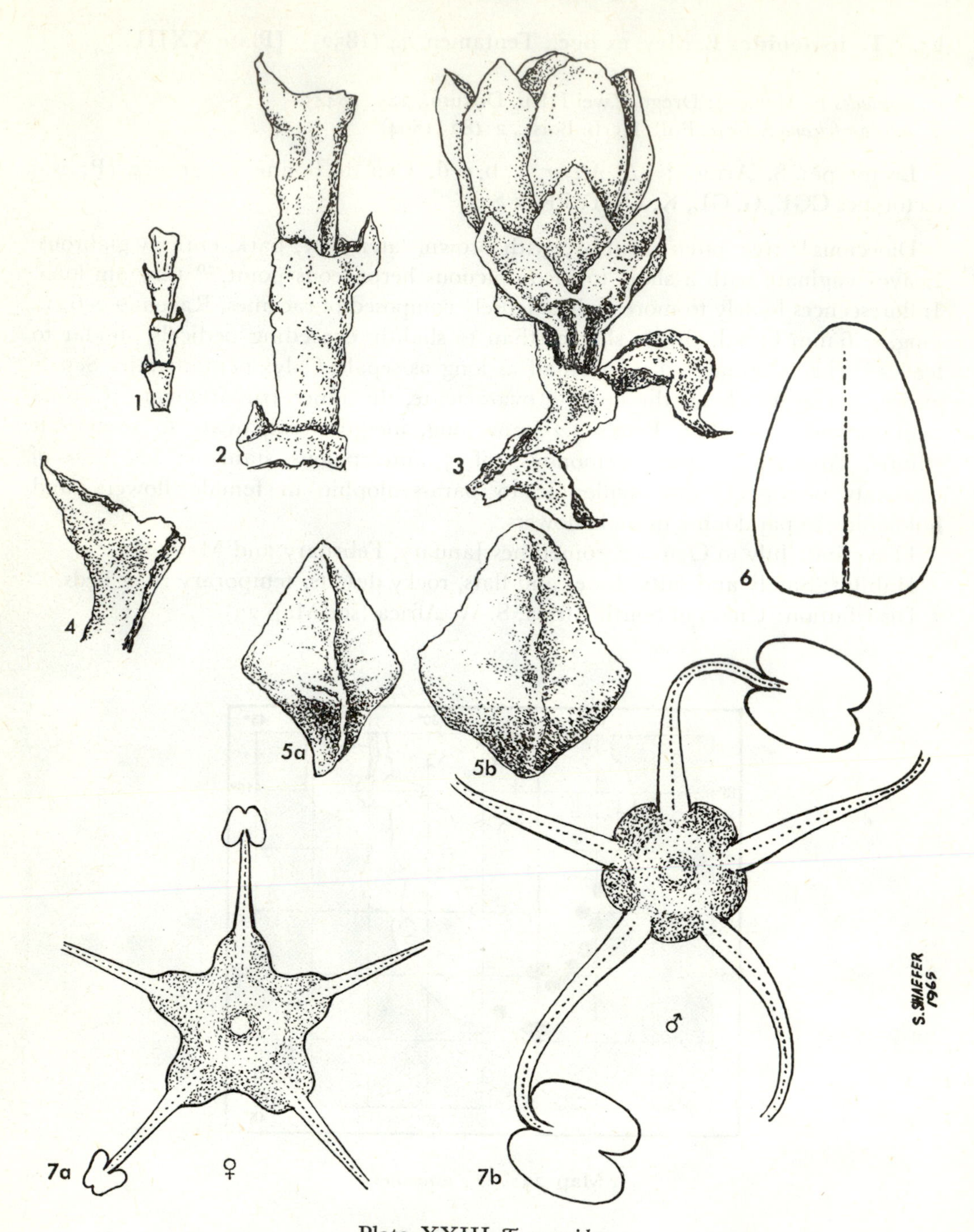

Plate XXIII *T. usneoides*
1. Young twig (x 5); 2. id (x 20); 3. Flower (x 20); 4. Bract (x 20);
5a. Outer sepal (x 30); 5b. Inner sepal (x 30); 6. Petal (x 17);
7a. Androecium of female flower (x 25); 7b. Androecium of male flower (x 25).

Selected specimens: UNION OF SOUTH AFRICA: *Drège*, in Zondagrivier 9.3.1833, im Fluss bei Kunkunnouioub 13.9.1830, zw. Blaauwekran und Bitterwater bei Gourka 20.3. 1827, bei Ausfluss der Gariep .9.1830, bei Natvaet im Fluss 14.9.1830 (paralectotypes of *T. usneoides* Mey., P); *Zeyher* 725, Boschemans-land in collibus campisque lapidosis ad Kamos et Springbokkeel .2.1830 (CGE, G, K, P, PRE, W, WU; also the isolectotypes of *T. austro-africana* Schinz), paralectotypes of *T. austro-africana* Schinz; *Ecklon* 2150, Cap Kolonie, in locis arenosis ad flumen T'Kausierivier (CGE, OXF, P, PRE, S, W); *Drège*, Cap Kolonie (G, GL, HBG, P, PRC, PRE, W); *Schrenk* 253, Gross Namaland (PRE); *Rodin* 1374, Namaqualand dry rocky and sandy desert country near Garies 26.9.1947 (K, PRE, UC). S.W. AFRICA: *Dinter* 5185, Vellardroft 10.6.1924 (B, HBG); *Landjouw* 52, Swakopmund Rabowsky 30.8.1938 (U).

Observation: Walter & Walter (1953) postulated that the distribution of *T. austro-africana* (*T. usneoides*) depends on groundwater conditions (not on climate).

Section Two. OLIGADENIA (Ehrenb.) Endl.,

Gen. Pl., 1039 (1840)

Oligadenia Ehrenb., Linnaea, 2:253 (1827), pro subgenere.
Decadenia Ehrenb., *loc. cit.*, pro subgenere, p.p.
Vernales Bge., Tentamen, 5 (1852), p.p.
Aestivales Bge., *op. cit.*, 6, p. min. p.
Sessiles Ndz., De Genere Tamarice, 4 (1895), pro subgenere, p.p.
Primitivae Arendt, Beitr. Tamarix, 33 (1926), p. maj. p.
Pentamerae Arendt, *op. cit.*, 36, p.p.
Tetramerae Arendt, *op. cit.*, 46.
Eutamarix Gorschk., Not. Syst. Leningrad, 7:81 (1937), pro subgenere, p.p.

Type species (lectotype): *T. laxa* Willd.

Leaves usually sessile with narrow bases, rarely amplexicaul at their decurrent base (in *T. kotschy*). Racemes 3–12 mm broad, or 3–5 mm broad and then flowers purely tetrandrous. Bracts shorter to longer than pedicels. Flowers tetramerous, tetra-pentamerous, pentamerous or all intermixed in one raceme. Petals 2–6 mm long. Androecium haplostemonous, partially diplostemonous or diplostemonous, of 4–5 antesepalous stamens and 0–4 shorter antepetalous ones, and of various discal structures.

Series 4. LAXAE Gorschk., in: Komarov, Fl. URSS, 15:302 (1949)

Grandiflorae Ehrenb., Linnaea, 2:257 (1827), p.p.
Paniculatae Bge., Mém. Acad. St. Pétersb., 7:294 (1851), pro sectione, p.p.
Pachybotryae Bge., Tentamen, 5 (1852), p.p.
Leptobotryae Bge., *loc. cit.*, p.p.
Pycnocarpae Bge., *op. cit.*, 6, p.p.
Xeropetalae Bge., *loc. cit.*, p. min. p.
Haplostemones Ndz., De Genere Tamarice, 5 (1895), pro subsectione, p.p.
Epilophus Ndz., *op. cit.*, 8, pro subsectione, p. min. p.
Mesodiscus Ndz., *loc. cit.*, p. min. p.
Tetrascopae Arendt, Beitr. Tamarix, 35 (1926), pro subsectione, p.p.
Pentastemones Grex *Sessiles* Arendt, *op. cit.*, 36, p. min. p.
Tetrastemones Grex *Sessiles* Arendt, *op. cit.*, 47, p.p.
Graciles Gorschk., *op. cit.*, 307, p.p.

Type species: *T. laxa* Willd.

Bracts, at least the lower ones of the vernal racemes, shorter to about as long as pedicels. Flowers tetra-pentamerous; androecium haplo- to partially diplostemonous, with not more than 1 additional antepetalous stamen.

Included species: *T. chinensis* Lour., *T. gracilis* Willd., *T. laxa* Willd., *T. szowitsiana* Bge.

24. **T. chinensis** Lour., Fl. Cochinch., 1:182 (1790) [Plate 24]

T. gallica L. var. *chinensis* (Lour.) Ehrenb., Linnaea, 2:267 (1827).
T. gallica L. var. *narbonensis* Ehrenb. f. *virgata* Ehrenb., *loc. cit.*
T. gallica L. var. *subtilis* Ehrenb., *loc. cit.*
T. juniperina Bge., Mém. Acad. St. Pétersb. Sav. Str., 2:103 (1833).
T. elegans Spach, Hist. Nat. Veget. Phan., 5:481 (1836).
T. plumosa Hort. ex Carr., Revue Hort., 40:358 (1868), nom. nud.
T. libanotica Hort. ex Koch, Dendrol., 1:454 (1869).
T. plumosa Hort. ex Lavalle, Arboretum Segrezianum, 113 (1877).
T. caspica Hort. ex Dippel, Handb. Laubholzk., 3:9 (1893), nom. nud.
T. japonica Hort. ex Dippel, *loc. cit.*, pro syn.
T. amurensis Hort. ex Chow, Familiar Trees Hopei, 329 (1934), pro syn. *T. pentandrae* Pall.

Type: China: *Loureiro*, '*Tamarix sinica*' (holotype P; isotype 'prope Canton' coll. *Louriero* [B, Willd. Herb. No. 6064]).

Tree with brown to black bark, entirely glabrous. Leaves sessile with narrow base, 1.5–3 mm long. Vernal inflorescences pyramidal, of many dense racemes, aestival inflorescences loose, of slender racemes. Racemes 2–6 cm long, 5–7 mm broad. Pedicel about as long as calyx. Bracts equalling pedicels to slightly longer, linear to linear-oblong; the lower bracts of the vernal racemes oblong and equalling pedicels, the upper bracts and those of the aestival racemes longer, narrowly triangular, acuminate, entire, herbaceous. Calyx pentamerous. Sepals 0.75–1.25 mm long, subentire, trullate-ovate to narrowly trullate-ovate, acute, the outer 2 keeled; sepals somewhat connate at base in aestival inflorescences. Corolla pentamerous, persistent. Petals elliptic to ovate, usually ovate-elliptic, rarely obovate and keeled at base, 1.5–2.25 mm long. Androecium haplostemonous, of 5 antesepalous stamens; insertion of filaments hypodiscal in vernal flowers, hypo-peridiscal (1–3 hypo-, 2–4 peridiscal) in aestival flowers; disk hololophic.

Flowering: March to November.

Habitat: River banks, humid plains, mountain slopes.

Distribution: Mongolia, China, Japan (see Map 24).

Selected specimens: Mongolia: *David* 2721, Mongolie orientale, abonde les plaines stériles des Ortous 1863 (P). China: *Chien* 5261, Szechuan Chien-Yang-Hsiou 24.4.1936 (E); *Willdenow* Herb. No. 6059, 1a (holotype of *T. gallica* var. *subtilis* Ehrenb. B); *Willdenow* Herb. No. 6059, 3 (holotype of *T. gallica* var. *narbonensis* Ehrenb., B); *Bunge*, China Borealis 1831 (holotype of *T. juniperina* Bge., P; isotypes G, K, P); *Rock* 4659, Yunnan Yangtze

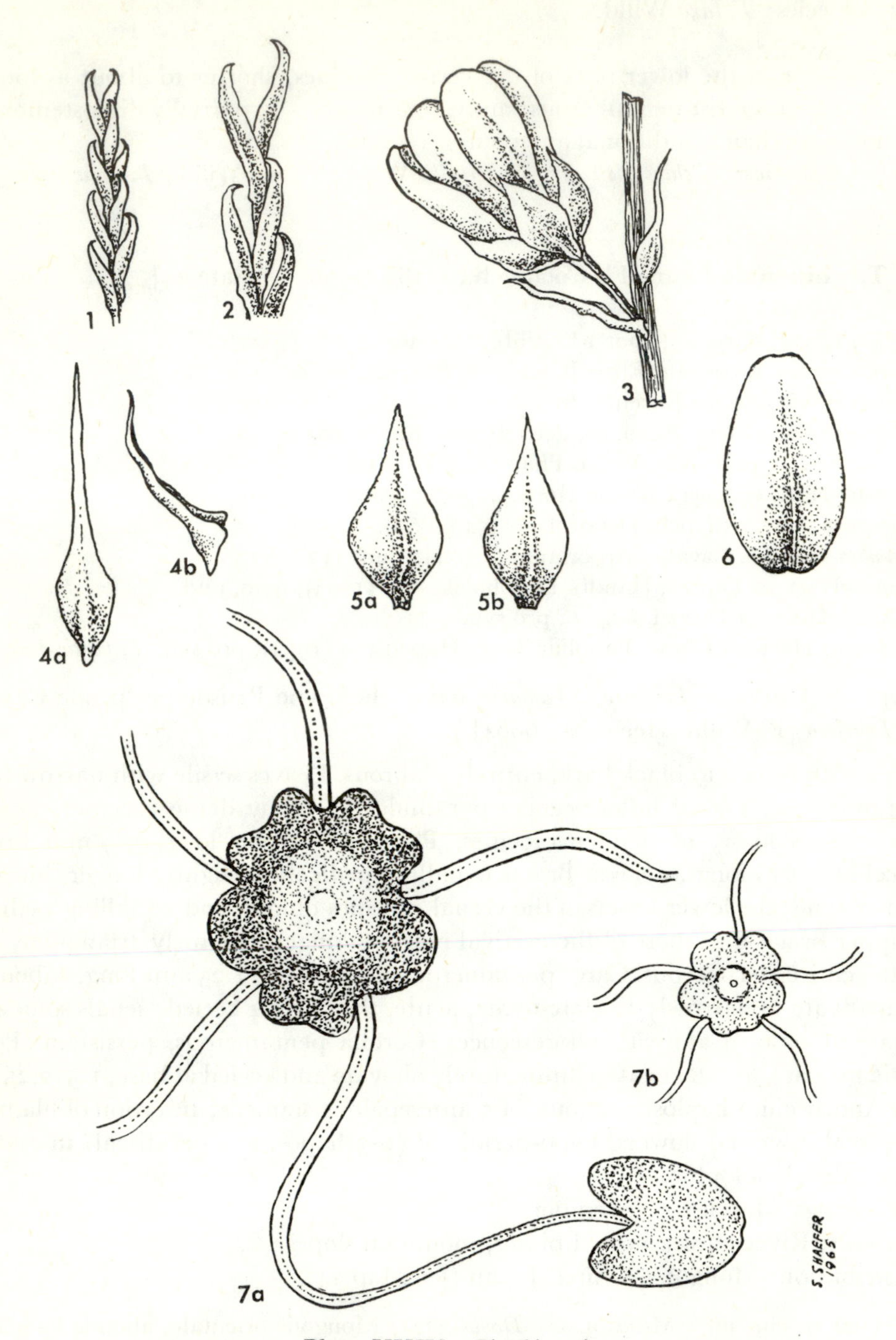

Plate XXIV *T. chinensis*

1. Young twig (x 5); 2. id (x 10); 3. Flower (x 10); 4a.–b. Bracts (x 20); 5a. Inner sepal (x 20); 5b. Outer sepal (x 20); 6. Petal (x 20); 7a. Androecium of vernal flower (x 30); 7b. Androecium of aestival flower (x 15).

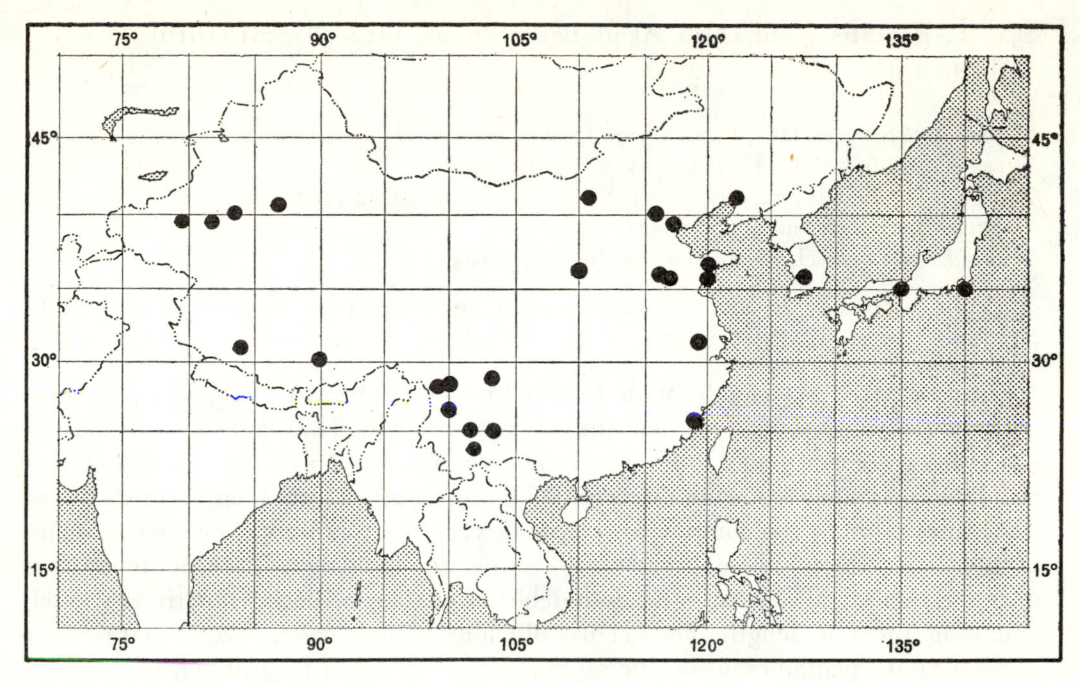

Map 24: *T. chinensis*

watershed in the prefectural district of Likiang eastern slopes of Likiang snow range .6.1922 (E, S, UC); *Chung* 2787, Fukien Foochow Electric Light Company, growing on the plain 25.7.1927 (E, K, UC, W); *Deasy*, Chinese Turkestan Takla-Makan Desert, Mt. Kara Jargaz, on sand dunes (BM); *Farges* 2121, Su-Tchuen oriental, distr. de Tchen-Keou Lin (B, P, S, UC). JAPAN: *Tagawa* 5307, Kyoto bor., Minatomyia Minato-Mura Kumano-Gun (cultivated) 31.8.1953 (DEL); *Siebold*, Japonia (E, FI, L, P, U, W); cultivated types: *Spach*, '*Tamarix indica* h. par. 1831–1836' (holotype of *T. elegans* Spach, PCU; isotype '*T. indica* h. par. 7.10.1835' FI); *Herb. Presl*, 'Cult. in Hort. Bot. Vind.' (holotype of *T. libanotica* Hort. ex Koch, PRC).

Observations: (a) *T. chinensis* is extensively cultivated in gardens as a successful ornamental plant and also as a hedge plant along the sea-shore. (b) Merrill (1935) states that Loureiro undoubtedly had specimens from cultivated plants. Hemsley (1888) states that there is no evidence that the tamarisk is wild anywhere in China. This does not seem to be true according to collected specimens which the author was able to see; according to Debeaux (1879) it grows on alluvial sand in Prov. Pe-tche-ly; Teng (1947) reports: 'The steppe and desert of Hosi, in the N.W. arm, is practically treeless, having occasional xerophytic species such as *Populus euphratica*, *T. chinensis* and *Ulmus pumila*.' (c) following Gomes (1868) and Merrill (1935) the author found the holotype of *T. chinensis* Lour. in Paris (Herbiers Historiques) in a package of Loureiro's material, to which Desvaux' manuscript enumeration of species is attached. An isotype (or a fragment of this holotype) is in Berlin: Willd. Herb. No. 6064.

25. **T. gracilis** Willd., Abh. Akad. Berlin Physik, 1812–1813:81 (1816) [Plate 25]

T. paniculata Stev. ex DC., Prodr., 3:96 (1828), pro syn. *T. pallasii* Desv.
T. cupressiformis Ledeb., Fl. Alt., 1:423 (1829).
T. angustifolia Ledeb., in: Eichw., Pl. Nov. Casp.-Cauc., 1:12 (1831).
T. affinis Bge., Tentamen, 36 (1852).
T. spiridonowii Fedtsch., Not. Syst. Leningrad, 3:183 (1922).

Type: Russian SFSR: *Pallas*, habitat in Siberia (holotype, Herb. Willd. No. 6066, B; isotype PRC).

Shrubby tree, 2–4 m tall, with brown to blackish-brown bark, younger parts entirely glabrous, or with papillae (in var. *cupressiformis*). Leaves sessile with narrow base, much longer than broad, 1–3 mm long. Inflorescences simple, the aestival usually immediately following the vernal.[22] Racemes 2.5–6.5 cm long, 6–7 mm broad, sometimes more or less compound, i.e., with very short axillary flowering branches at their base. Bracts in vernal racemes spoon-like to narrowly trullate, acute in upper part and in aestival racemes, usually much shorter than pedicels, sometimes scarcely equalling them in length. Pedicel usually longer than calyx. Calyx tetra-pentamerous (tetra-pentamerous on the same raceme.)[23] Sepals 1 mm long, more or less connate at their base, all obtuse, subentire or irregularly denticulate, ovate, the outer 2 broader, more acute, more or less keeled. Corolla tetra-pentamerous, caducous. Petals elliptic to ovate, 1.75–2.5 mm long. Androecium haplostemonous to partially diplostemonous, of 4–5 antesepalous stamens, or 4 antesepalous and 1 antepetalous stamens; insertion of filaments peridiscal; fleshy disk hololophic to paralophic.

Flowering: April to August.

Habitat: Steppes, salty loam; var. *angustifolia* on somewhat salty banks of rivers and shores of lakes; var. *gracilis* on solonchak.

Distribution: Russian SFSR, Kazakh SSR, Tadzhik SSR, Turkmen SSR, Turkey (see Map 25).

Var. **gracilis.** Flowers regularly pentamerous, pentandrous; younger parts entirely glabrous.

Selected specimens: Russian SFSR: *Ledebour*, Altai (P, PRC, K); *Iljin & Heinrichson* 642, Palatinsk region Mai-Tshak on solontshak 18.8.1928 (LE); *Blumenthal* 324, Iter Semipalatense 1928 (LE); *Dubianski* 604, Uralskaya Oblast 3.6.1904 (LE); *Bunge*, Sibir. Altaicae, in salsis deserti p. fodinam Loktewsk (P); *Steven*, prope Kisliar 1825 (holotype of *T. paniculata* Stev., G-DC; isotype G-Boiss.).

Var. **angustifolia** (Ledeb.) Baum. Flowers tetra-pentanisomerous, tetra-pentandrous; younger parts entirely glabrous.

22 Both kinds are usually found on the same specimen.
23 Different states of meiomery of sepals and/or petals are observed.

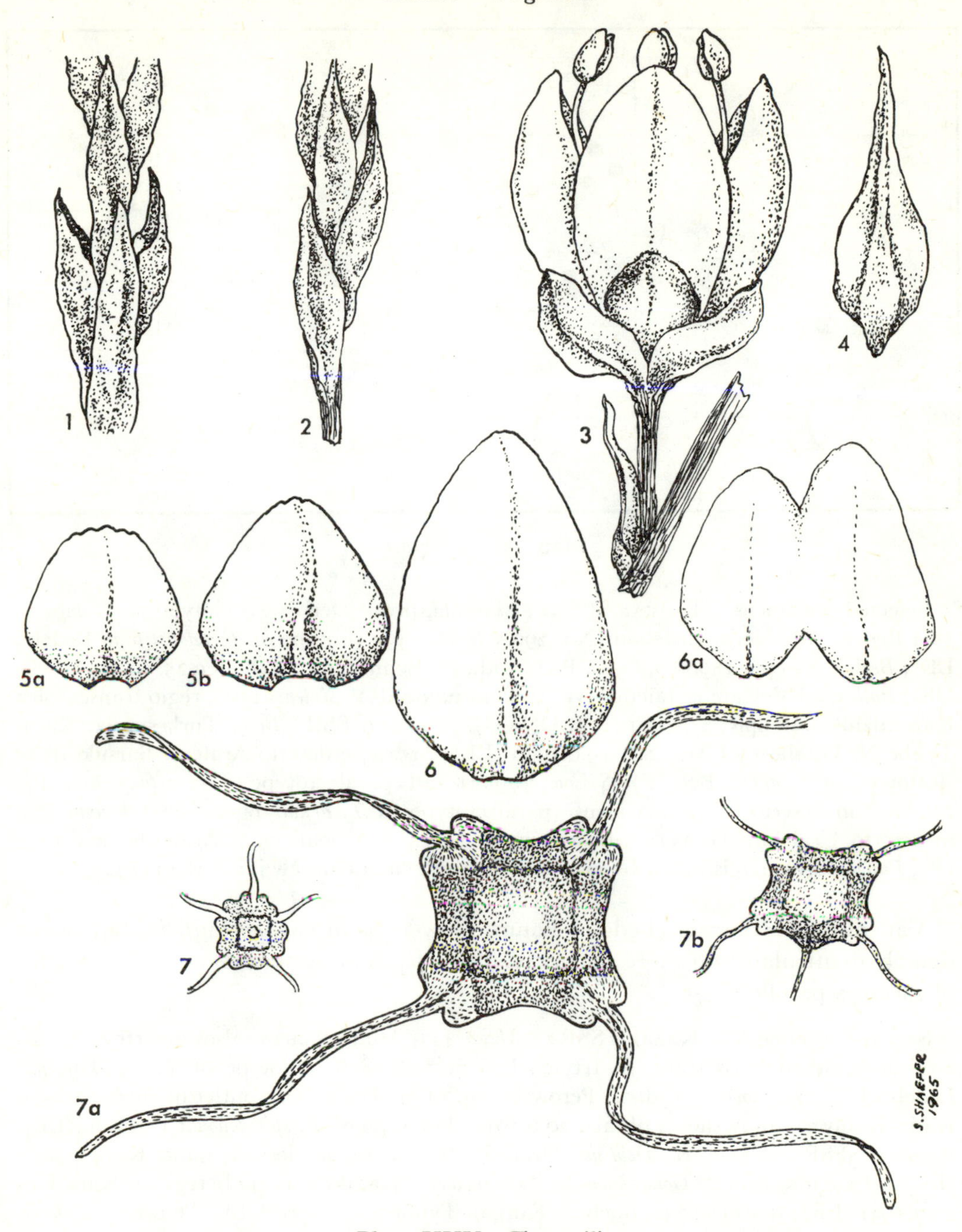

Plate XXV *T. gracilis*

1–2. Young twigs (x 20); 3. Flower (x 20); 4. Upper vernal or aestival bract (x 20); 5a. Inner sepal (x 30); 5b. Outer sepal (x 30); 6. Petal (x 30); 6a. Meiomeric petal (2 petals fused) (x 15); 7. Regular pentamerous androecium (x 10); 7a. Regular tetramerous androecium (x 40); 7b. Androecium with 4 antesepalous and 1 antepetalous stamens (x 20).

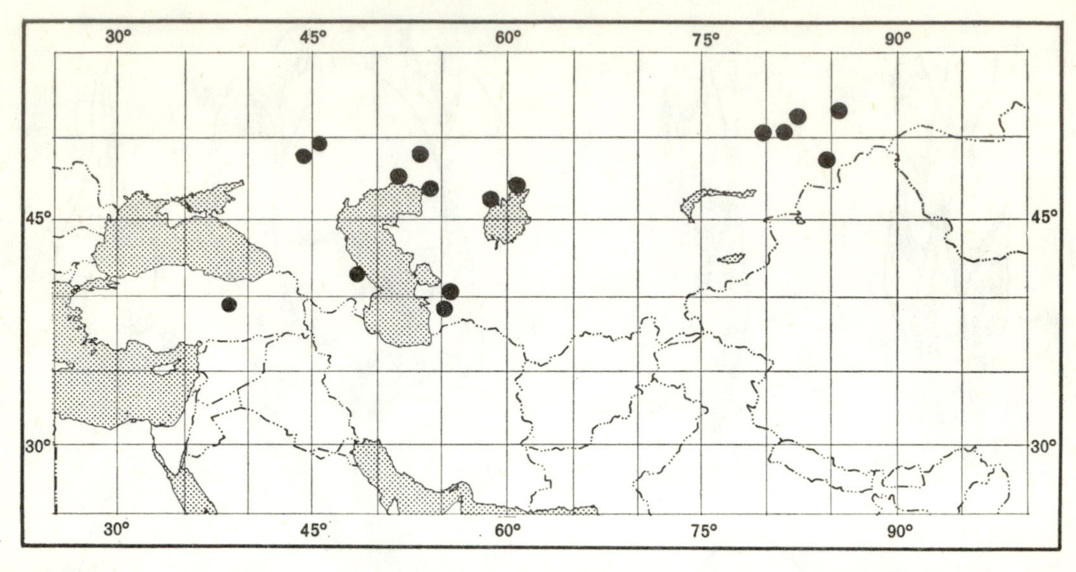

Map 25: *T. gracilis*

Selected specimens: Russian SFSR: *Eichwald*, prope Derbent (holotype of *T. angustifolia* Ledeb., P); Herb. Ledebour No. 299.5 & No. 394 (isotype of *T. angustifolia* Ledeb., LE); *Becker*, Sarepta 1896 (B, G, L, PRC); id., im Lehm- und Salzboden 25.5.1884 (PRC, US); *Pallas*, ad Volgam et Jaicum (W). Turkmen SSR: *Sintenis* 1592, regio transcaspica Kazandzhik in steppis 28.4.1901 (B, E, G, WU). Kazakh SSR: *Berg*, Turkestania, Kara-Tyube N. W. shore of Aral Sea 19.6.1900 (LE); *Syssow*, e deserto Aralensi Barsuki dicto (lectotype of *T. affinis* Bge., P); *Syssow*, Jumart-Kul (paralectotype of *T. affinis* Bge., P); *Lehmann*, in deserto Caspico-Aralensi (paralectotype of *T. affinis* Bge., G); *Lehmann* 493, in deserto Ural inf. (G, W); *Lehmann*, in littore septentrionali lacus Aralensis 30.6.1841 (P); *Bunge*, legi copiosissimam versus mare Caspium circa Guryew junio 1840 (P).

Var. **cupressiformis** (Ledeb.) Baum. Flowers as in var. *angustifolia*, but sepals densely denticulate and more obtuse; younger parts of branchlets or at least rachis of racemes papillose.

Selected specimens: Kazakh SSR: *Meyer* 443, Altai ad lacos salsos deserti Songoro-kirgisici occidentalioris trans. fl. Irtysch ledi d. 19.8.1826 (holotype of *T. cupressiformis* Ledeb., LE); *Spiridonow* 73, distr. Perowski reg. of Sir-Daria near outlet of river Sara-Su Katin-Kamyz., sandy shores of lake 20.6.1914 (holotype of *T. spiridonowii* Fedtsch., LE). Russian SFSR: *Nikitin* & *Deulina*, Uralo-Embenski rayon leskaja dana Kara-Agach (LE). Tadzhik SSR: *Gorschkova* & *Tschernov* 56, Kasakstania po beregi oz. Kara Kul 22.7.1931 (LE); *Roshewitz* 332, Buchara Kurgan Tyube 20.4.1906 (US). Turkey: *Zohary*, env. of Malatya, great salt valley 27.7.1962 (HUJ).

Observation: The whereabouts of the type of *T. angustifolia* Ledeb. is unknown; a fragment of the type is in Paris.

26. **T. laxa** Willd., Abh. Akad. Berlin Physik, 1812–1813: 82 (1816) [Plate XXVI]

T. pallasii Desv., Ann. Sci. Nat. Bot., I, 4:349 (1824).
T. laxa Willd. var. *racemosa* Ehrenb., Linnaea, 2:254 (1827).
T. laxa Willd. var. *occidentali-caspica* Bge., Tentamen, 35 (1852).
T. laxa Willd. var. *vulgaris* Bge., *loc. cit.*
T. astrachanica Gand., Nov. Consp. Fl. Eur., 191 (1910), pro syn.
T. chersonensis Gand., *loc. cit.*, pro syn.

Lectotype: Russian SFSR: *Baron Marschall*, ad Wolgam legit et perhibet esse *tetrandra*, certe a *taurica* differt! Willd. Herb. No. 6068 fol. 1 (B; isolectotypes P, *Marschall de Bieberstein*, in Rossia ad Wolgam, PRC).

Low tree or shrub, 2–3 m high, with brown bark, entirely glabrous. Leaves sessile with narrow base, 1–4 mm long. Vernal inflorescences simple, aestival ones loosely compound. Racemes 1–7 cm long, 6–8 mm broad. Bracts much shorter than pedicels, the uppermost in each raceme sometimes equalling their length, spoon-shaped, blunt, with diaphanous upper parts. Pedicel longer than calyx. Calyx tetramerous. Sepals 1–1.5 mm long, connate at base, the outer 2 trullate-ovate, acute, keeled, subentire, the inner more or less ovate, obtuse, erose-denticulate. Corolla tetramerous, caducous. Petals more or less broadly elliptic to obovate, 2.25–3 mm long. Androecium haplostemonous, of 4 antesepalous stamens; insertion of filaments peridiscal; fleshy disk hololophic.

Flowering: March to June.

Habitat: Salty river beds, sand dunes.

Distribution: Russian SFSR, Kazakh SSR, Turkmen SSR, Mongolia (see Map 26).

Selected specimens: Russian SFSR: paralectotypes of *T. laxa* Willd.: '*Pallas*, ad mare Caspium' Willd. Herb. No. 6068 fol. 1, also holotype of *T. laxa* var. *racemosa* Ehrenb.; Fol. 2, 3 et 5 sine coll. (B); *Patrin*, Sibiria ad lacus salsos inter Obum et Irtim 30.6.1781 (holotype of *T. pallasii* Desv., P-Jussieu, a fragment in K; isotypes G, *Tamarix pentandra* a Sibiria *D. Patrin*, P-Lamarck); *Steven, Kisliar* 1850 (syntypes of *T. laxa* var. *occidentali-caspica* Bge., G, P, a fragment in K); *Eversmann*, ad Algelme Siberia (PRC); *Politow*, Irtysch (WU). Mongolia: *Chaney* 78, Shabarakh Usu, Outer Mongolia 1925 (US). Kazakh SSR: *Ledebour*, Altai 1834 (P, PRC). Turkmen SSR: *Nietschasva* & *Prikhodo* 416, Turkmenia Parapamizk distr. basin of Kushki River 5.5.1951.

Observations: (a) Willd. Herb. No. 6068, fol. 1, has 3 specimens which are mounted on one sheet. Two on the left side bear a clear indication that they were collected by Baron Marschall (Marschall von Bieberstein); the other is a specimen collected by Pallas, according to the note below it. This is also in full agreement with Ledebour's cited specimens of *T. laxa* (in *Fl. Ross.*, 2:133). The first-mentioned specimen is the author's choice as lectotype, and the last-mentioned as paralectotype. Fol. 4 of the same number is identical with *T. polystachya* Ledeb. (b) In Stockholm there is one specimen of *T. laxa* which was collected by Pallas. It could be another paralectotype or an iso-paralectotype. (c) J. Gay was the first to remark (in Blanche & Gaillardot, 1854, p. 10) that *T. pallasii* Desv. is synonymous with *T. laxa* Willd.

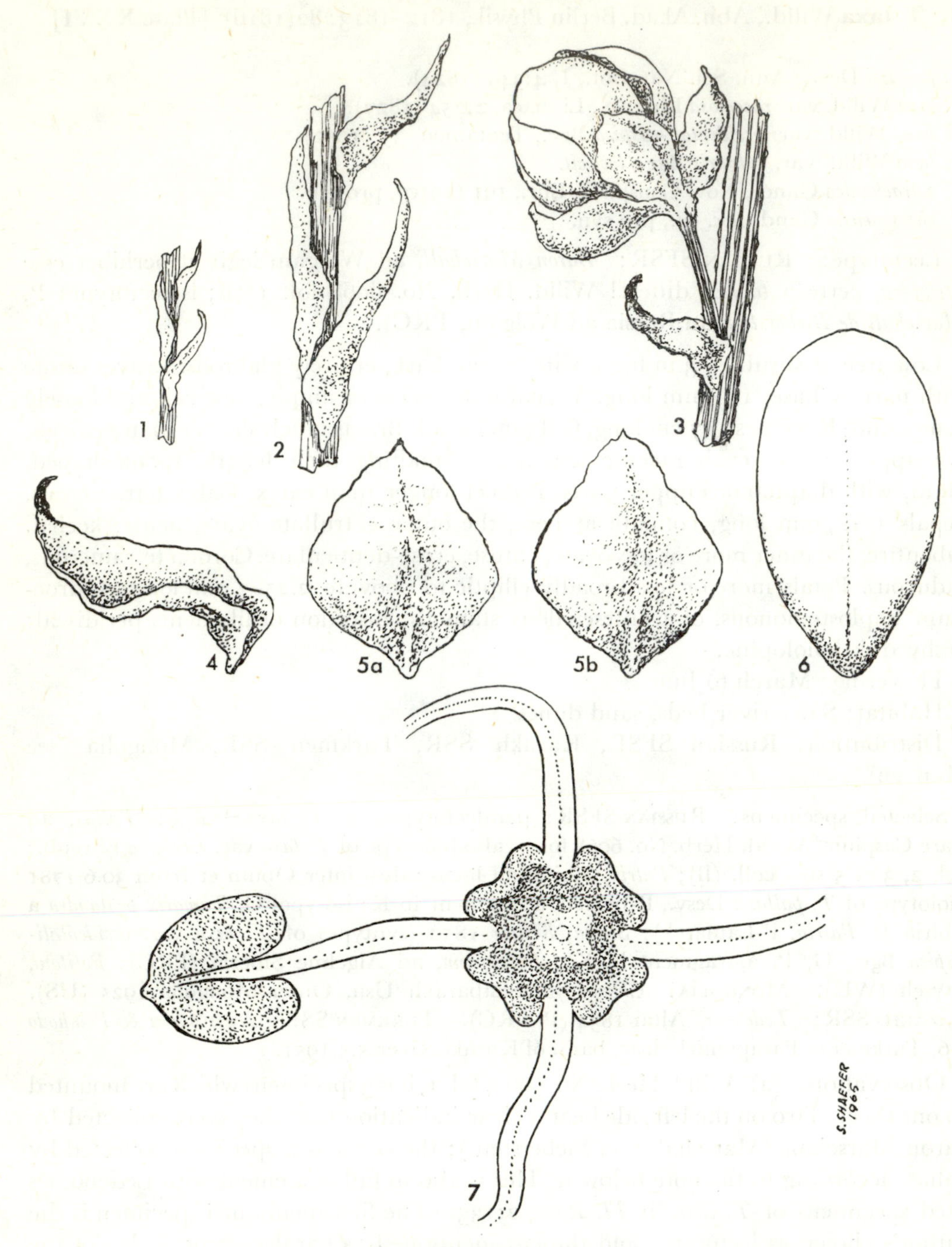

Plate XXVI *T. laxa*

1. Young twig (x 5); 2. id (x 10); 3. Flower (x 10); 4. Bract (x 20);
5a. Outer sepal (x 20); 5b. Inner sepal (x 20); 6. Petal (x 20);
7. Androecium (x 20).

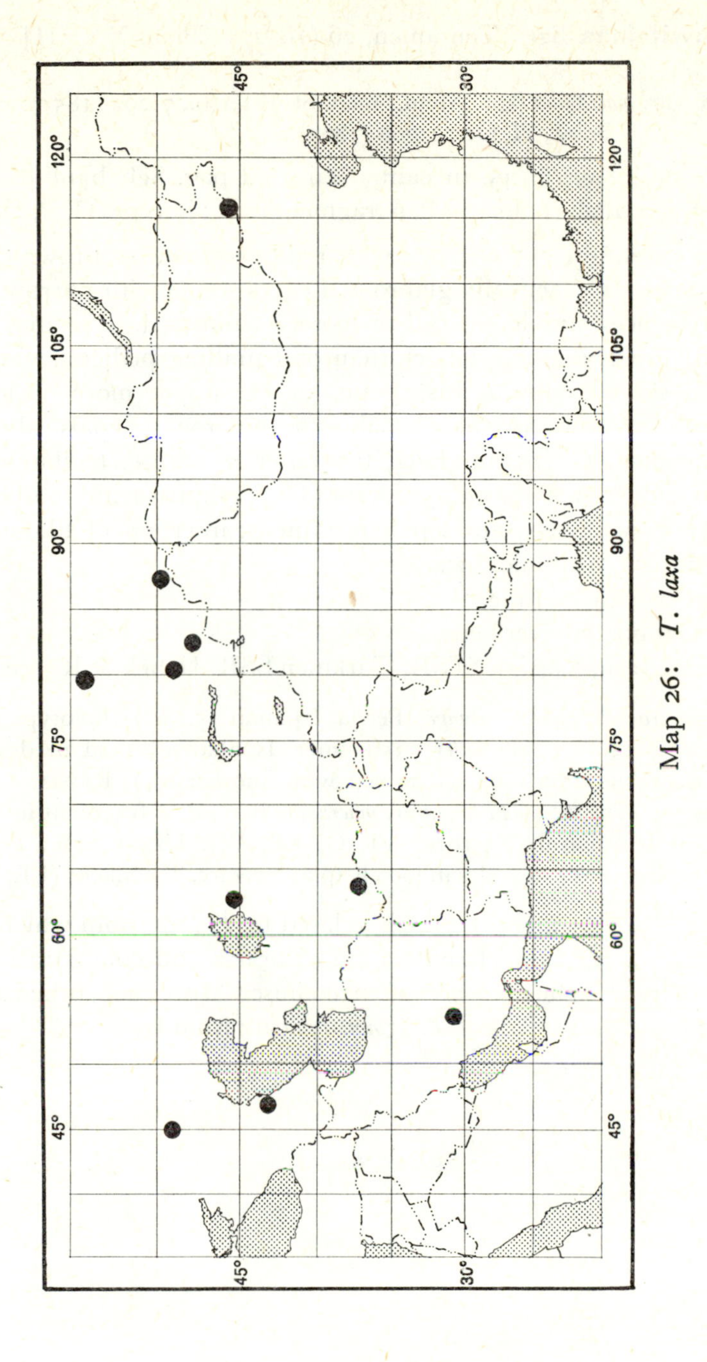

Map 26: *T. laxa*

27. **T. szowitsiana** Bge., Tentamen, 26 (1852) [Plate XXVII]

T. laxa Willd. var. *macrantha* Bge., Mém. Acad. St. Pétersb., 7:292 (1851).
T. ispahanica Bge. ex Boiss., Fl. Or., 1:768 (1867).

Type: IRAN: *Szovits* 135, in campo salso ad pag. Schabanli distr. Khoi prov. Aderbeidzan 27.4.1826 (holotype P, a fragment in K; isotypes G, P, US).

Small tree or shrub, 3–7 m high, with reddish-brown to brown bark, minutely papillose, sometimes practically glabrous. Leaves sessile with narrow base, 2–3 mm long. Inflorescences simple or rather loosely composed. Racemes 2–4 cm long, 6 mm broad. Bracts scarcely longer than or equalling pedicels, narrowly oblong, blunt with a short narrow obtuse point, or the upper more or less acuminate. Pedicel about as long as calyx. Calyx tetramerous. Sepals almost entire or slightly denticulate, 1–1.25 mm long, the outer 2 obtuse, trullate-ovate. Corolla tetramerous, caducous. Petals 2.25–2.5 mm long, elliptic-ovate to obovate. Androecium haplostemonous, of 4 antesepalous stamens; insertion of filaments peridiscal; disk synlophic or para-synlophic.

Flowering: March to June.

Habitat: Temporary river beds.

Distribution: Iran, Russian SFSR, Turkmen SSR, Uzbek SSR (see Map 27).

Selected specimens: IRAN: *Bunge*, Persia Ispahan .5.1859 (holotype of *T. ispahanica* Bge., G; isotypes L, P); *Bornmüller* 3359, inter Ispahan et Yesd in desertis salis inter Banbis et Hasserabad 27.3.1892 (B, mixed with another sp.). RUSSIAN SFSR: *Pallas*, prope Loktensem (syntype of *T. laxa* var. *macrantha* Bge., P, a fragment in K). TURKMEN SSR: *Michelson* 101, Farab 9.4.1910 (FI, G, LE, W). UZBEK SSR: *Popov* 313, Tian-Schan occidentalis in valle fl. Keles in loco Kaplanbek circ. Taschkent (LE, P).

Observation: *T. szowitsiana* is closely related to *T. laxa*, from which it differs only in minute characters, e.g., relative length of bracts and configuration of the disk. In different herbaria the two are often confused with each other and also with *T. gracilis*. Most of the records of *T. szowitsiana* are from Iran.

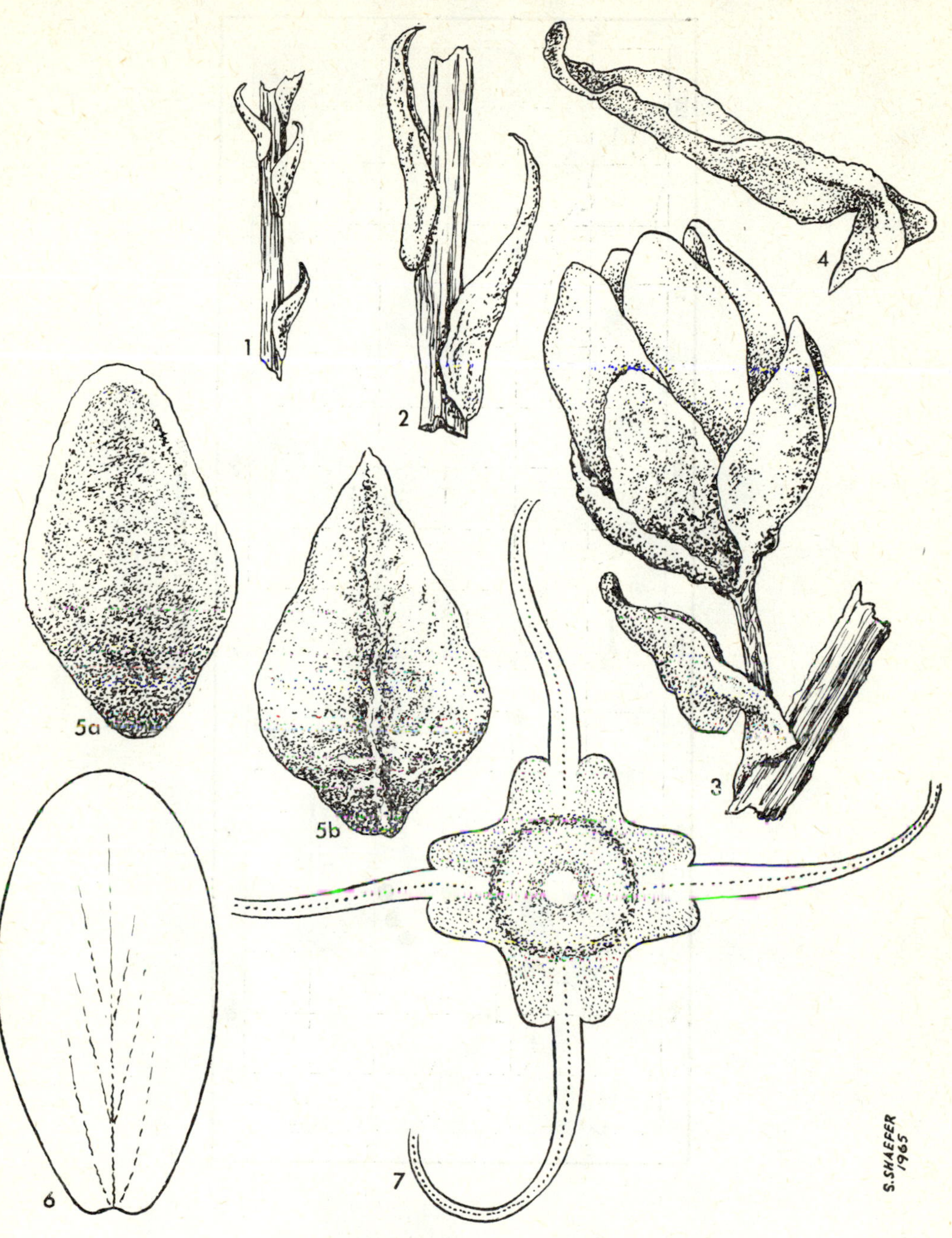

Plate XXVII *T. szowitsiana*
1. Young twig (x 5); 2. id (x 10); 3. Flower (x 10); 4. Bract (x 30); 5a. Inner sepal (x 30); 5b. Outer sepal (x 30); 6. Petal (x 25); 7. Androecium (x 30).

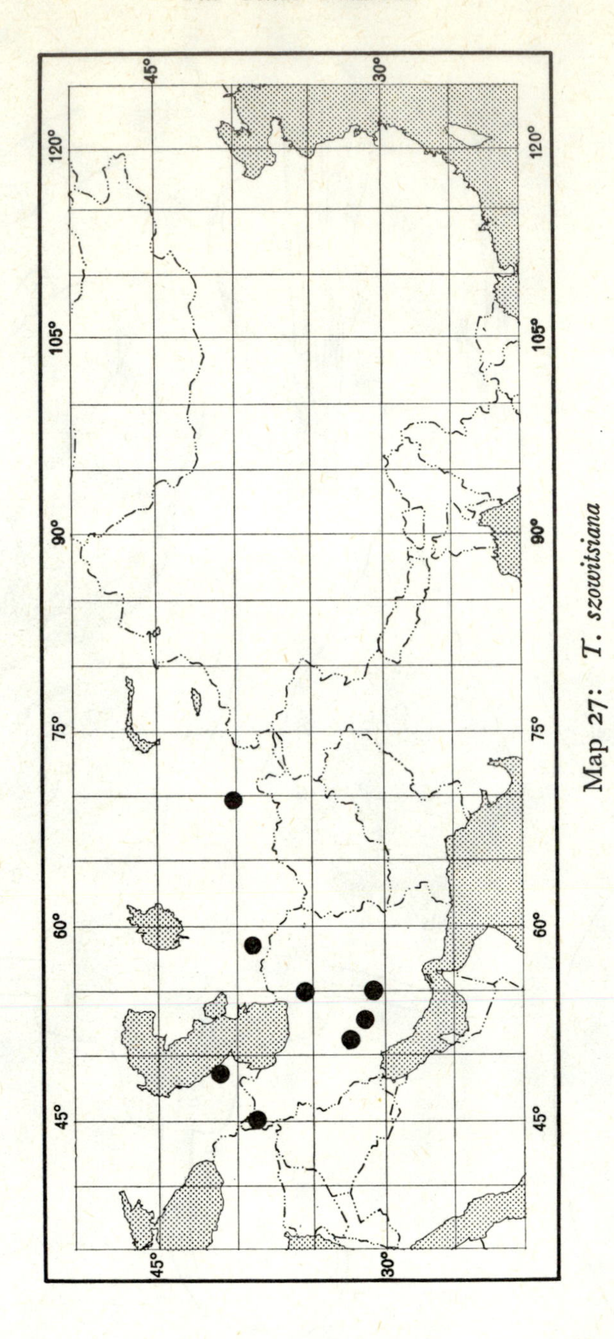

Map 27: *T. szowitsiana*

Series 5. ANISANDRAE Bge., Tentamen, 5 (1852), p. maj. p.

Grandiflorae Ehrenb., Linnaea, 2:257 (1827), p.p.
Parviflorae Ehrenb., *loc. cit.*, p.p.
Pachybotryae Bge., *loc. cit.*, p. maj. p.
Macrobotryae Bge., *loc. cit.*
Leptobotryae Bge., *op. cit.*, 6, p. min. p.
Macrostylae Bge., *loc. cit.*
Anisandrae (Bge.) Ndz., De Genere Tamarice, 4 (1895), pro subsectione, p. maj. p.
Haplostemones Ndz., *op. cit.*, 5, pro subsectione, p.p.
Pentascopae Arendt, Beitr. Tamarix, 34 (1926), pro subsectione, p.p.
Tetrascopae Arendt, *op. cit.*, 35, pro subsectione, p.p.
Pentastemones Grex *Sessiles* Arendt, *op. cit.*, 36, p. min. p.
Tetrastemones Arendt, *op. cit.*, 46, pro subsectione, p.p.
Tetrastemones Grex *Sessiles* Arendt, *op. cit.*, 47, p.p.
Tetrastemones Grex *Semiamplexicaules* Arendt, *op. cit.*, 48, p.p.
Octostemones Arendt, *op. cit.*, 49, pro subsectione.
Elongatae Gorschk., in: Komarov, Fl. URSS, 15: 295 (1949).
Tetrandrae Gorschk., *op. cit.*, 298.

Type species: *T. rosea* Bge.

Bracts exceeding calyx, or where shorter than pedicels to about equal in length with them, petals are parabolic (i.e., trullate-ovate). Flowers tetra-penta-(hexa)-merous. Petals more than 2.25 mm long. Stamens 4–8, of which 4–5 are antesepalous and 0–4 antepetalous. Androecium haplo- to diplostemonous.

Included species: *T. africana* Poir., *T. boveana* Bge., *T. brachystachys* Bge., *T. dalmatica* Baum, *T. elongata* Ledeb., *T. hampeana* Boiss. & Heldr., *T. meyeri* Boiss., *T. octandra* Bge., *T. rosea* Bge., *T. tetragyna Ehrenb.*, *T. tetrandra* Pall. ex M.B.

28. **T. africana** Poir., Voy. Barb., 2:139(1789) [Plate XXVIII]

T. gallica L., Sp. Pl., 1:270 (1753), p. min. p. quoad syn. Bauh., Pin., 485 (1623).
T. narbonensis Gars., Descr. Pl. Anim., 337 et t. 577 (1767), nom. illegit.
T. africana Desf., Fl. Atl., 1:269 (1798), nom. illegit.
T. gallica L. var. *hispanica* Bge., Tentamen, 62 (1852).
T. africana Poir. var. *saharae* J. Gay ex Coss., Ann. Sci. Nat. Bot., IV, 4:283 (1855), nom. nud.
T. hispanica Boiss., Diagn. Pl. Or. Nov., II, 2:56 (1856).
T. speciosa Hort. ex Koch, Dendrol., 1:453 (1869).
T. speciosa Ball, J. Bot., 1873:301 (1873), nom. illegit.
T. africana Poir. var. *saharae* J. Gay ex Batt. & Trab., Fl. Alg., 1(2):322 (1889).
T. segobricensis Pau, Butl. Inst. Catal. Hist. Nat., 18:160 (1918).
T. calarantha Pau, Mem. Mus. Ci. Nat. Barcelona Bot., 1:43 (1922).
T. tingitana Pau, Mem. Soc. Esp. Hist. Nat., 12:293 (1924).
T. castellana Pau & Villar, Brot. Bot., 23:107 (1927).
T. castellana Pau & Villar f. *rosea* Pau & Villar, *loc. cit.*
T. viciosoi Pau & Villar, *op. cit.*, 109.
T. celtiberica Sennen & Elias, Bot. Soc. Iber., 27:66 (1928).
T. uncinatifolia Sennen, Ann. Soc. Linn. Lyon, 73:9 (1928).

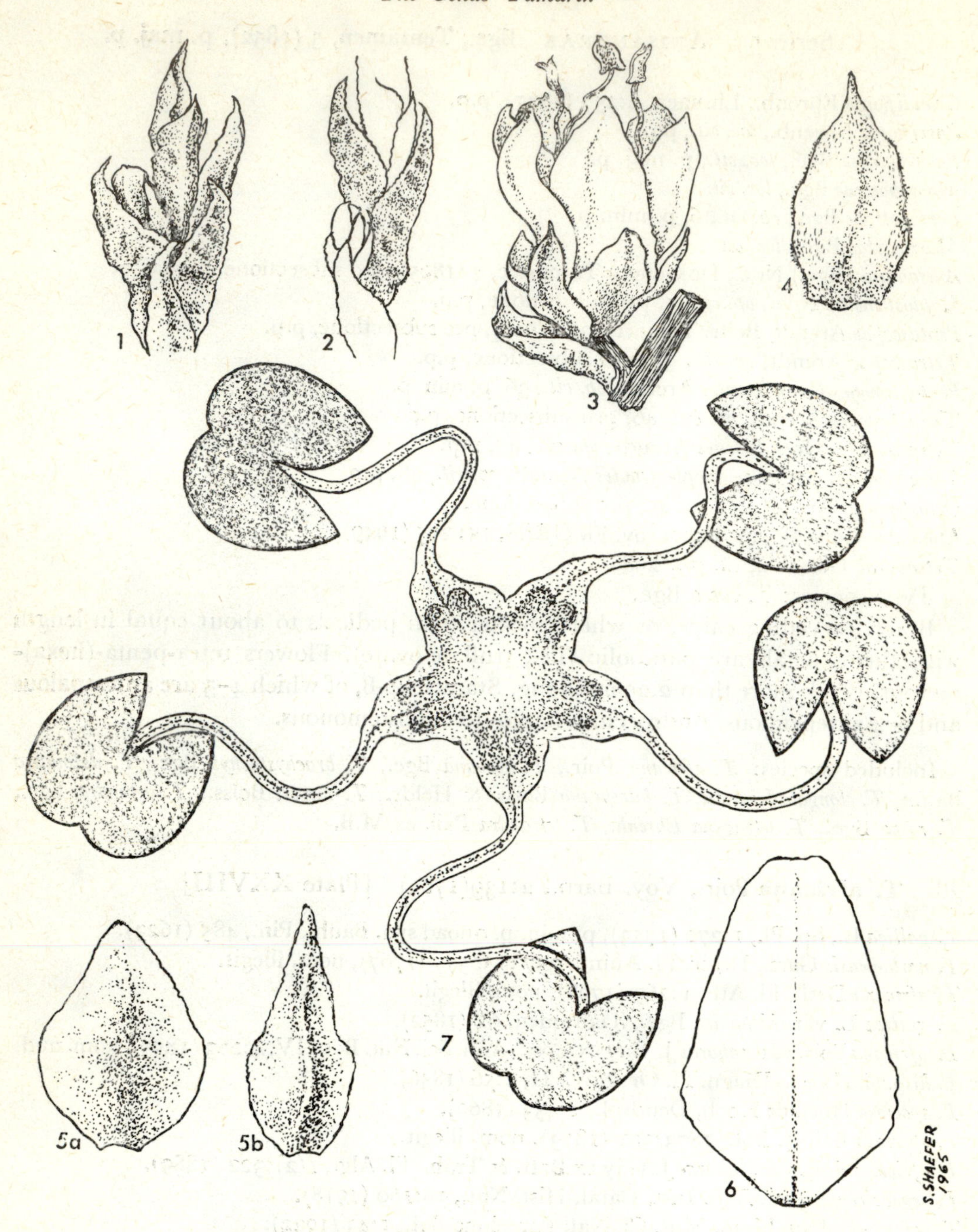

Plate XXVIII *T. africana*

1–2. Leaves with young axillary primordial branches (x 10); 3. Flower (x 10); 4. Bract of vernal racemes (x 20); 5a. Inner sepal (x 20); 5b. Outer sepal (x 20); 6. Petal (x 20); 7. Androecium (x 30).

T. longicollis Trab. ex Maire, in: Jahand. & Maire, Cat. Pl. Maroc., 2:488 (1932).
T. speciosa Ball emend. Maire var. *longicollis* Maire, in: Jahand. & Maire, *loc. cit.*
T. speciosa Ball emend. Maire var. *tingitana* (Pau) Maire, in: Jahand. & Maire, *loc. cit.*
T. atlantis Trab. ex Jahand. & Maire, *op. cit.*, 489.
T. africana Poir. var. *brevistyla* Trab. ex Jahand. & Maire, *loc cit.*, nom. nud.
T. africana Poir. var. *ducellieri* Maire & Trab. ex Jahand. & Maire, *loc. cit.*, nom. nud.
T. allorgei Sennen & Maur., Cat. Fl. Rif. Or., 147 (1933), nom. nud.
T. mauritii Sennen in: Sennen & Maur. *loc. cit.*, nom. nud.
T. lundibunda Maire, Bull. Soc. Sci. Hist. Nat. Maroc., 13:265 (1933).
T. brachystylis J. Gay var. *fluminensis* Maire, Bull. Soc. Hist. Nat. Afr. N., 26:184 (1935).
T. gallica L. subsp. *nilotica* (Ehrenb.) Maire var. *pleiandra* Maire, *loc. cit.*
T. africana Poir. var. *brevistyla* Trab. ex Maire, *loc cit.*,
T. africana Poir. var. *ducellieri* Maire & Trab. ex Maire, *loc. cit.*
T. africana Poir. var. *weilleri* Maire, *loc. cit.*
T. speciosa Ball var. *acutibracteata* Maire, *op. cit.* 29:403 (1938).

Type: ALGERIA: *Poiret*, 'Tamarix africana cote de barbarie' (holotype P, Herb. Lamarck; isotypes Herb. Cosson et Herb. Moq. Tand., P).

Tree or bushy tree with black or dark purple bark; younger parts glabrous except for some papillose parts of racemes. Leaves sessile with narrow base, 1.5–2.5 mm long, with narrow scarious margins, occasionally minutely papillose on both sides or underneath; margins often muricate-papillate. Inflorescences usually simple. Racemes 3–7 cm long, 6–9 mm broad, the aestival inflorescences somewhat smaller. Bract longer than pedicel (rarely exceeding calyx in subsessile flowers), narrowly oblong and shortly acute to triangular ovate and acuminate, entire, usually with dense minute tooth-like papillae on margins. Pedicel shorter than calyx or occasionally almost lacking. Calyx pentamerous. Sepals trullate-ovate, acute, subentire, 1.5 mm long, the outer 2 slightly keeled and usually somewhat longer than the more obtuse inner ones. Corolla pentamerous, subpersistent. Petals ovate to broadly trullate-ovate, 2.5–3 mm long in vernal flowers, 2 mm or more in aestival ones. Androecium of 5 antesepalous stamens; insertion of filaments peridiscal; disk synlophic.

Flowering: March to May, often February to August.

Habitat: Canal and river banks, ditches or sands near the sea-shore; also on humid calcareous soil.

Distribution: Italy, Sardinia, Sicily, France, Corsica, England (introduced), Spain, Portugal, Canary Islands, Morocco, Algeria, Tunisia (see Map 28).

Var. **africana.** Racemes 3–7 cm long, 6–9 mm broad; petals ovate to broadly trullate-ovate; bracts about as long as calyx.

Selected specimens: ITALY: *Bicknell*, Liguria Bordighera 21.4.1890 (B); *Porta & Rigo* 74, Calabria occid. p. Rhegium Julium 14.4.1877 (B, FI, K, P). SARDINIA: *Müller*, in humidis prope Cagliari 1828–1829 ? (GL, K, OXF, P, W); *Moris*, Sardaigne 1839 ? (G, P). SICILY: *Gussone*, Sicilia 1831 ? (G, K, P, S); *Todaro* 497, ad torrentium alveos Misilmeri (G, HBG, K, S, U). FRANCE: *Penchinat* 1874, alluvions de la vallée d'Argèles/ Mer le long de ruisseaux (Pyrénées Orientales) 1.5.1855 (B, P); *Steinheil*, Toulon (syntype of *T. gallica* var. *narbonensis* [Ehrenb.] Bge., B); *Bertrand*, Var, Roquebrune 18.5.1900

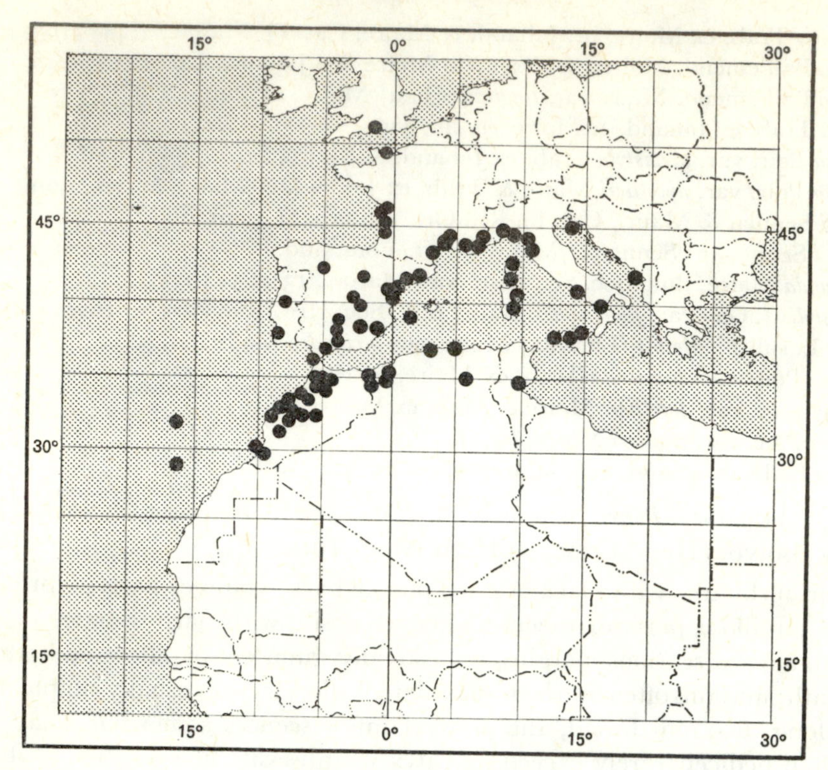

Map 28: *T. africana*

(B, P, W); *Reboud* 978, sur les bords du canal entre la porte Notre-Dame et la porte St. Nicolas près de la Rochelle (Charente Inférieure) 2.8.1853 (B); *Bauhin* 38, Monspelii, Arelati et in Italia (paralectotype of *T. gallica* L., UPS). CORSICA: *Spencer*, ad sepes Bastia 27.4.1906 (B). ENGLAND: *White* 245, in littoribus maritimis prope Weymouth commitatus Dorsetiensis oppidum habitat 20.7.1884 (B, HBG, S). SPAIN: *Sennen* 3066, Catalogne, Cambrils torrent de Janer 8.6.1917 (holotype of *T. uncinatifolia* Sennen, MA; isotypes G, HUJ, P, W); *Elias* 4988, Logrono; Recajo, sables de l'Ebre 10.5.1923 (holotype of *T. celtiberica* Sennen & Elias, MA; isotypes G, S); *Beltran* 78874, Vaciamadrid (holotype of *T. castellana* Pau & Villar, MA); *Villar* 78871, Vaciamadrid (Madrid) Tamaricetum in vega juxta vicum 18.6.1926 (holotype of *T. castellana* f. *rosea* Pau & Villar, MA; isotype No. 78873, MA, and syntypes); *Vicioso* 78880, Rivas de Javarna 9.6.1918 (holotype of *T. viciosoi* Pau & Villar, MA); *Pau* 203, de Torre del Mar a la Cala 3.5.1919 (holotype of *T. calarantha* Pau, MA); *Pau* (MA); *Pau* 79030, las orillas del Jucar .5.1918 (holotype of *T. segobricensis* Pau, MA); *Blanco* 340, Puerta Guadalmessa 1849 (holotype of *T. hispanica* Boiss., G; isotypes FI, G, K, P); *Porta* & *Rigo* 683, Prov. Gaditana, ad rivulos p. Algeiras, sol schiztoso .4.1895 (B, FI, HBG, PRC, WU); *Reverchon* 1198, prov. Grenada, La Puebla de Don Fadrique, lieux frais et ombrageux, sur le calcaire 1200 m .5.1900 (B, P, PRC); *Willkomm* 22 (52), in valle Loyola pr. St. Sebastian .5.1850 (holotype of *T. gallica* var. *hispanica* Bge., W; isotypes G, P, PRC). PORTUGAL: *Bornmüller* 614, Funchal, Madeira Islands, Gurgulho 22.3.1900 (P, PRC); *Rainha* 1013, Baixo Alentejo sinus na margem de uma ribuira 13.4.1946 (RAB, S); *Rainha* 3842, Beira-Baixa Portas de Radao na margens da ribuira do Acafol 23.9.1958 (G, U). CANARY ISLANDS: *Perrardière*, Teneriffe Santa Cruz in arenosis ad mare Azores 15.4.1855 (B). MOROCCO: *Maire, Weiller & Wilczek*, ad ripas fluminis Oum-el-Rebia prope Mechra ben Abbou 31.3.1934 (holotype of *T. africana* var. *weilleri* Maire, P; isotypes RAB, S); *Maire* 2595, secus torrentes prope Agadir et Imoucha, solo calcareo 4.4.1937 (holotype of *T. speciosa* Ball var. *acutibracteata* Maire, P); *Malençon*, ad ripas fluminis Draa prope Zagora 23.2.1934 (holotype of *T. malenconiana*, P); *Trabut* 2524, in ditione Ida-ou-Tanan; ad ripas amnis infra cataractem Immouzir, solo calcareo 18.3.1932 (holotype of *T. ludibunda* Maire, P; isotype RAB); *Sennen* 9370, Beni-Tuzin, bords du rio Kert 6.5.1932 (holotype of *T. mauritii* Sennen, MA; isotypes BM, G, RAB); *Pau* 78992, de Tanger al Fondak 2.5.1921 (holotype of *T. tingitana* Pau, MA); *Ball*, South Morocco Greater Atlas in clivo septentr. Atlantis majoris distr. Mesfioua 1871 (holotype of *T. speciosa* Ball, K; isotypes G, P, WU); *Jahandiez* 120, Hte Moulouya, Ksabi bords de oueds 28.4.1925 (E); *Sauvage* 2043, Tabahart, à la traversée de la piste Harchi-Tiliouine 23.5.1950 (RAB); *Maire* 254, Sahara in montibus Tefedest secus amnem Agelil 1200 m 11.4.1918 (RAB); *Maire* 249, Sahara in montibus Emmidir (Moudyr) Tigelgemin ad ripas lacus inferioris 400–500 m 29.2.1928 (isotype of *T. loecocharis* Maire, RAB); *Maire*, ad ripas fluminis Oum-er-Rebia prope Mechra-ben-Abbou 26.3.1922 (holotype of *T. africana* var. *brevistyla* Trab. ex Maire, P); *Ducellier*, in humidis ad ripas amnium prope urbem Taza, ad fontem Ain-bou-Khelel (holotype of *T. africana* var. *ducellieri* Maire & Trab., P). ALGERIA: *Desfontaines* 631, 'habitat algeria ad maris littora' (holotype of *T. africana* Desf., P; isotypes Willd. Herb. No. 6061, B, G, P); *Balansa* 673, bords du grand lac de Miserghin 1852 (B, FI, G, GL, K), et 672 (FI, G, GL, K, P); *Reverchon* 106, Kabylie Bougie sur les sables de la plage .5.1896 (E, G, P, WU); *Faure*, Oran Oued-Imbert bords d'un canal 18.5.1916 (BG, CAI, S, UC); *Balansa* 991, bords de l'Oued de Biskra à Biskra 28.6.1853 (BM, FI, G, P, US, W); *Bové*, Alger dans les marais .4.1837 (G, L, P); *Maire & Wilczek*, ad ripas fluminis Massa 3.4.1934 (holotype of *T. gallica* subsp. *nilotica* var. *pleiandra* Maire, P). TUNISIA: *Kralik* 58, in palmetis Oued Gabes 10.5.1895 (B, P).

Var. **fluminensis** (Maire) Baum.

T. africana Poir. var. *rungsii* Vill., Bull. Soc. Nat. Maroc., 28:36 (1948).

Racemes narrower and longer than in var. *africana*, 6–8 cm long, 5 mm broad, densely flowered; petals obovate; bracts usually exceeding calyces.

Selected specimens: MOROCCO: *Rungs* 2045, M'tal à 21 km S. de Sidi Bennour 21.3.1948 (holotype of *T. africana* var. *rungsii* Vill., RAB; isotype K); *Maire*, Mechra ben Abbou ad ripas fluminis 31.3.1934 (holotype of *T. africana* var. *weilleri* Maire, P; isotype RAB); *Maire*, in ditione Ida-ou-Tanan, Immouzer ad ripas amnis infra cataractam 18.5.1932 (P, RAB); *Faure*, Berkane bords de la Moulouya; *Maire*, secus amnis vallis Icafen infra Igherm Anti-Atlantis 1100 m 9.4.1934 (holotype of *T. brachystylis* J. Gay var. *fluminensis* Maire, P; isotype RAB).

Observations: (a) *T. narbonensis* Garsault is not a binomial. The word 'narbonensis' is not a specific epithet. Garsault did not follow the binomial nomenclature. The author has listed the name *T. narbonensis* here because it is cited in several works (e.g., *Index Kewensis*, Thellung, 1908). (b) A few intermediate forms, such as *Sauvage* 3687 (RAB), are apparently of hybrid origin between *T. africana* and *T. boveana*. (c) A few intermediate forms between *T. africana* and *T. tetragyna* (e.g., *Balansa* 673)

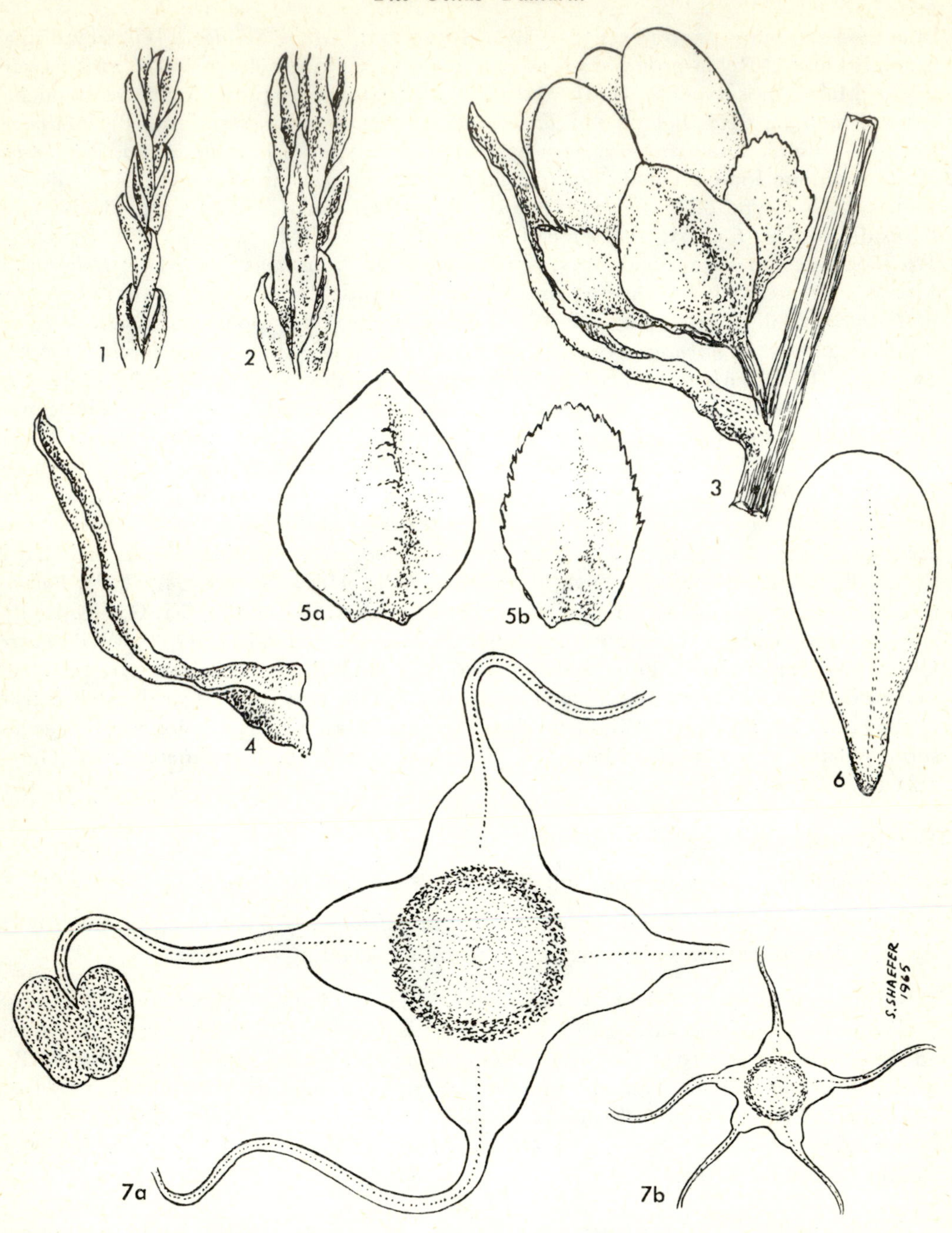

Plate XXIX *T. boveana*

1. Young twig (x 5); 2. id (x 10); 3. Flower (x 15); 4. Bract (x 17); 5a. Outer sepal (x 17); 5b. Inner sepal (x 17); 6. Petal (x 17); 7a. Regular androecium (x 30); 7b. Androecium of 4 antesepalous and 1 antepetalous stamens (x 15).

either indicate hybrid origin or are perhaps remnants of an ancient intermediate form linking *T. tetragyna* to *T. africana*. The specimen mentioned has the bark colour, leaves and petals typical of *T. africana*, but the androecium is typical of *T. tetragyna*. (d) Var. *fluminensis* seems to be closely related to *T. dalmatica*.

29. **T. boveana** Bge., Mém. Acad. St. Pétersb., 7:291 (1851) [Plate XXIX]

T. bounopea J. Gay ex Coss., Ann. Sci. Nat. Bot., IV, 1:239 (1854), nom. nud.
T. bounopea J. Gay ex Batt. & Trab., Fl. Alg., 1(2):321 (1889).
T. jimenezii Pau, Bull. Acad. Intern. Geogr. Bot., 16:75 (1906).

Type: Algeria: *Bové*, près de la Macta .4.1830 (holotype P; isotypes B, FI, G, K, W, WU).

Shrubby tree with reddish-brown bark, younger parts papillose to sparsely papillulose. Leaves sessile with narrow base, 2–4 mm long. Inflorescences usually simple and mostly vernal. Racemes 5–15 cm long, 8–9 mm broad. Bracts linear, sometimes slightly cuneate, exceeding flowers, acute, densely papillose. Pedicel mostly shorter than calyx. Calyx tetramerous. Sepals 1.5–2 mm long, the 2 outer broadly ovate-trullate, entire, acute, the inner ovate, obtuse, more or less denticulate and slightly shorter. Corolla tetramerous, caducous. Petals narrowly obovate, unguiculate, 3–4 mm long. Androecium haplostemonous, of 4–(5) antesepalous stamens, rarely upper flowers with 1 additional antepetalous stamen; insertion of filaments peridiscal; disk synlophic to para-synlophic.
Flowering: March to May.
Habitat: Wadis, banks of rivers, borders of swamps.
Distribution: Spain, Morocco, Algeria, Libya, Tunisia (see Map 29).

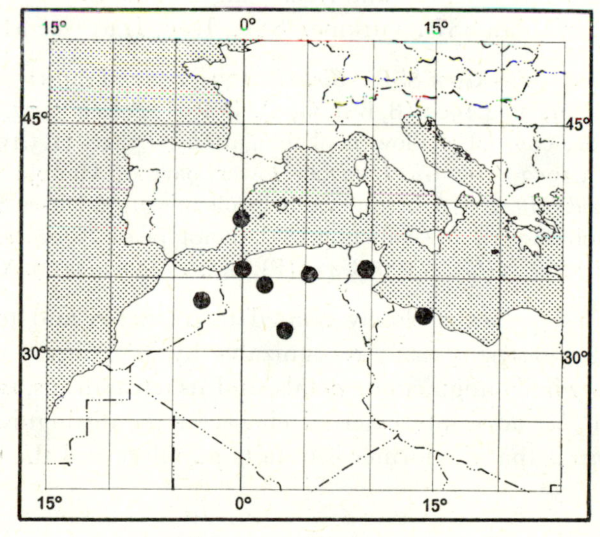

Map 29: *T. boveana*

Selected specimens: SPAIN: *Jimenez* 379004, Cartagena Pentamesas-Lopollo 31.5.1903 (holotype of *T. jimenezii* Pau, MA; isotype K); *Jimenez* & *Ibanez* 678, Murcie, Carthagens, Marais de Lopollo, .4.1908 (FI, MA, W). MOROCCO: *Debeaux*, ad ripas Oued-Auli prope Bognan 30.4.1857 (FI); *Maire*, ad ripas amnis Cheris 11.4.1933 (UC, RAB). ALGERIA: *Chevalier* 303, Sahara, El Golea, ad ripas amnis 'O. seggueur' prope Hassi-el-Gara 15.3.1899 (B, FI, G, P, PRC, US); *Balansa* 989, environs de Biskra .4.1853 (BM, E, FI, G, P, W); *Balansa* 671, bords du Chott-el-Chergui près de Khrider cercle Saida 30.5.1852 (holotype of *T. bounopoea* J. Gay, K; isotypes E, G, P, W); *Maire*, Sahara septentr. ditione Oued Rhir (FI, RAB). LIBYA: *Vaccari* 65, Tripolitania, Homs, Wadi Lebda 26.3.1914 (E, HUJ). TUNISIA: *Pitard* 1820, Gabes in aridis deserti .2.1900 (G, P).

30. **T. brachystachys** Bge., Tentamen, 26 (1852) [Plate XXX]

T. noëana Boiss., Diagn. Pl. Or. Nov., II, 2:56 (1856).

Lectotype: AZERBAIJAN SSR: *Abich*, Caucasus in campis ad mare Caspium (P, a fragment in K).

Low tree or shrub with reddish-brown to brown bark, usually glabrous. Leaves sessile with narrow base, long acuminate, 1.5–7 mm long. Inflorescences usually vernal, simple. Racemes thick and long, 4–15 cm long, 9 mm broad, densely flowered. Bracts more or less equalling calyx, occasionally exceeding it, with somewhat papillose margins, the middle and upper ones shortly acute.[24] Pedicel one-fourth as long as calyx or shorter. Calyx tetramerous, urceolate. Sepals more or less alike, without keel or the outer 2 only faintly keeled, entire to subentire, trullate-ovate to ovate, 2.5–2.75 mm long. Corolla tetramerous, caducous. Petals narrowly obovate to elliptic-obovate, 4.5 mm long. Androecium haplostemonous, of 4 antesepalous stamens, the upper flowers only with one entire or abortive antepetalous stamen; insertion of filament peridiscal; disk synlophic.

Flowering: March to May.

Habitat: Wadis, banks of canals and rivers, edges of pools.

Distribution: Azerbaijan SSR, Turkmen SSR, Iran, Iraq (see Map 30).

Selected specimens: TURKMEN SSR: *Sintenis* 139, regio Transcaspica Aschabad ad ripas fluvii prope Gjaurs 27.4.1900 (B, BM, G, S). IRAQ: *Polunin* 5007, Kurdistan Kut rd. opposite Salman Pak, edge of shallow pool, irrigation seepage 31.3.1958 (E, K); *Regel*, Baquba Wüste 27.3.1953 (B); *Gillet* 10706, lower part of Hamija irrigation scheme 30.4.1948 (US); *Hand.-Mazz.* 516, in limo vallis Euphratis supra Dir-es-Sar 28.3.1910 (W); *Noë* 907, Bassora Mohammera an Graben .3.1851 (holotype of *T. noëana* Boiss., G; isotypes K, P, US); paratypes: *Noë* 73 (G), *Noë* 76 & 435 (P). IRAN: *Macmillan* 69, Abadan .4.1927 (K).

Observation: This species is closely related to *T. meyeri* and to some extent to *T. tetragyna*. *T. brachystachys* is easily recognizable by its typically equal and obtuse sepals, its nearly or wholly unguiculate petals, and its tetrandrous, typically synlophic disk configuration. An adequate external character for distinguishing *T. tetragyna* from *T. brachystachys* is that the former is usually papillose and the latter is generally glabrous.

24 In *T. meyeri* and *T. elongata* the acumen is much longer.

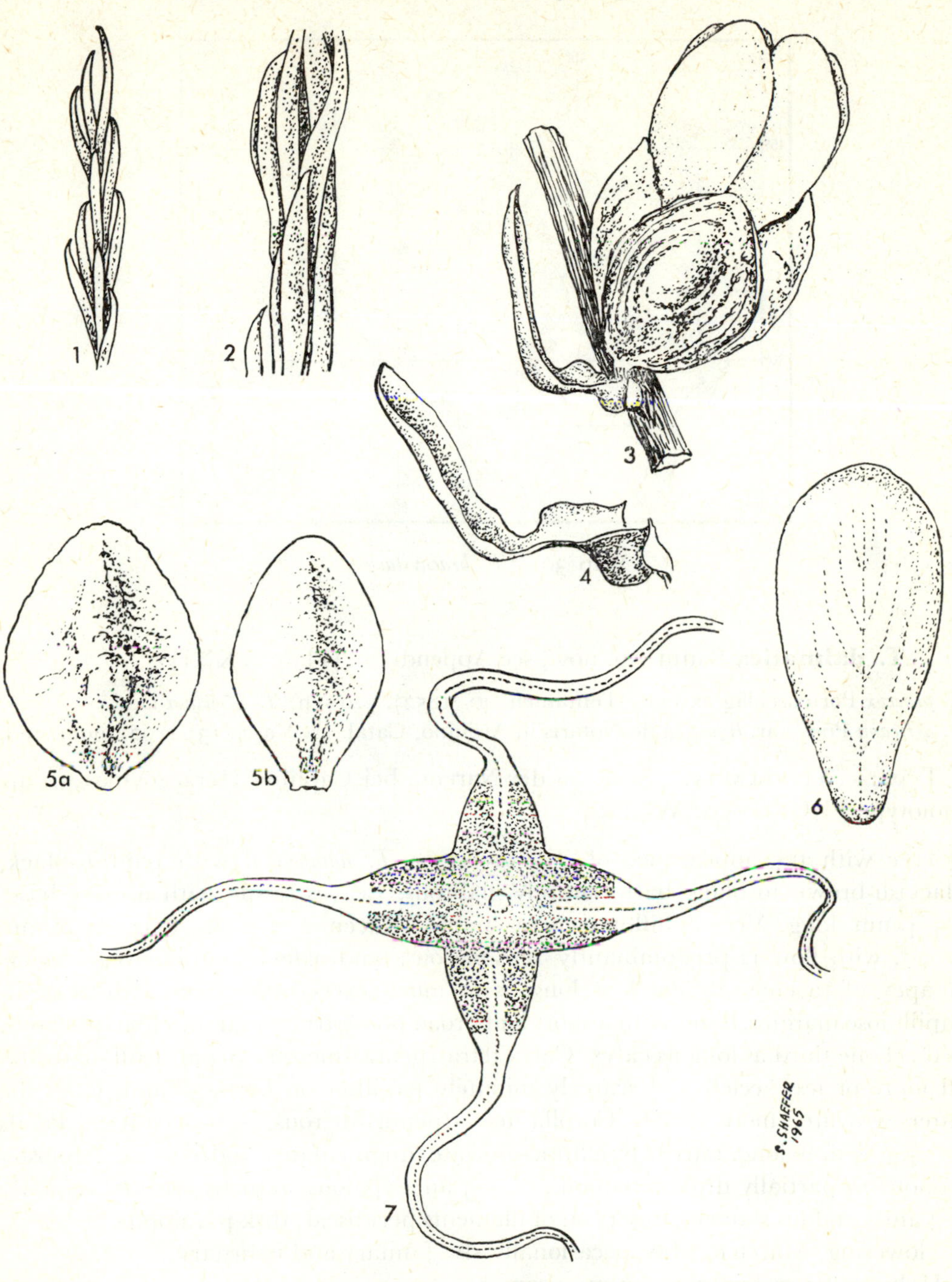

Plate XXX *T. brachystachys*

1. Young twig (x 5); 2. id (x 10); 3. Flower (x 10); 4. Bract (x 15); 5a. Outer sepal (x 15); 5b. Inner sepal (x 15); 6. Petal (x 15); 7. Androecium (x 25).

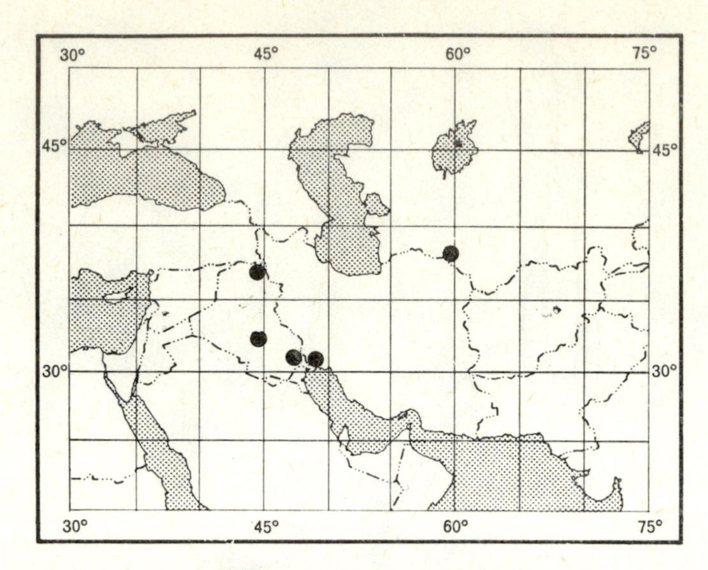

Map 30: *T. brachystachys*

31. **T. dalmatica** Baum (sp. nov., see Appendix) [Plate XXXI]

T. praecox Portenschlag ex Bge., Tentamen, 40 (1852), pro syn. *T. africana* Desf.
T. africana Poir. var. *ligustica* de Notaris in Ardoino, Catal. Pl. Vasc., 13 (1862), nom. nud.

Type: YUGOSLAVIA: *Fiala*, an der Narenta bei Capljina Herzegovina .5.1892 (holotype PRC; isotype WU).

Tree with the appearance of *T. africana* or *T. hampeana*, with reddish-black, blackish-brown to black bark, entirely glabrous. Leaves sessile with narrow base, 2.5–4 mm long. Vernal inflorescences simple. Racemes 2–6 cm long, 8–10 mm broad, with flowers predominantly tetramerous, and a few pentamerous flowers at apex of raceme. Bracts [25] as long as to much exceeding calyx, with scabrid-papillulose margins, blunt with a short and broad point, triangular, often diaphanous. Pedicel one third as long as calyx. Calyx tetra-(penta)-merous. Sepals trullate-ovate, all more or less keeled and scarcely minutely papillose on keels, 3.5 mm long, the outer 2 ovate, more acute. Corolla tetra-(penta)-merous, sub-persistent. Petals 2.5–4.5(5) mm long, narrowly elliptic-obovate, unguiculate. Androecium haplostemonous to partially diplostemonous, of 4–5 antesepalous stamens and occasionally 1–3 antepetalous stamens; insertion of filaments peridiscal; disk paralophic.

Flowering: March to May, occasionally also January and February.

Habitat: River banks, maritime swamps.

Distribution: Yugoslavia, Italy, France (see Map 31).

25 Broader than those of *T. africana*.

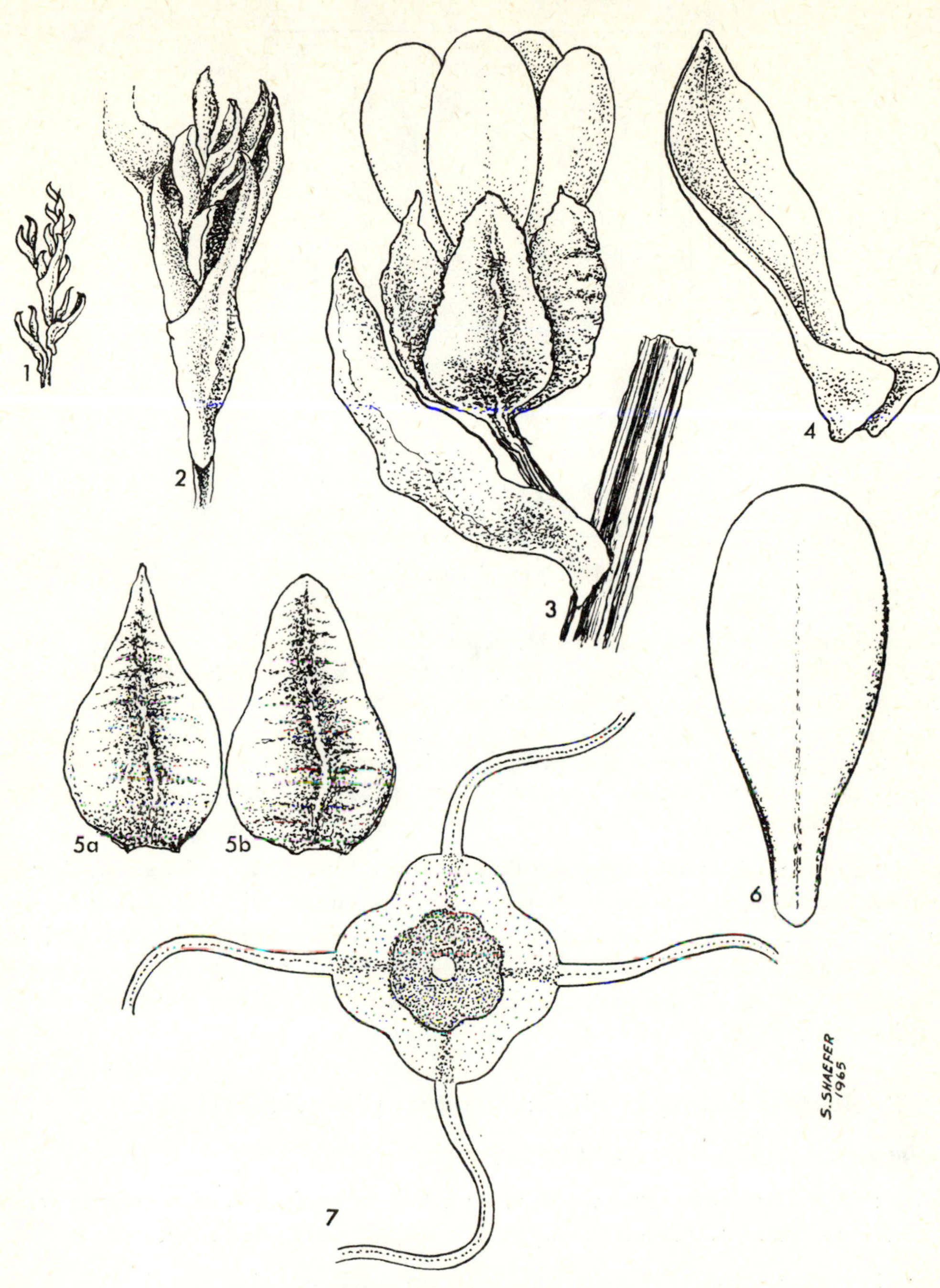

Plate XXXI *T. dalmatica*
1. Young twig (x 5); 2. id (x 10); 3. Flower (x 15); 4. Bract (x 15); 5a. Outer sepal (x 15); 5b. Inner sepal (x 15); 6. Petal (x 15); 7. Androecium (x 20).

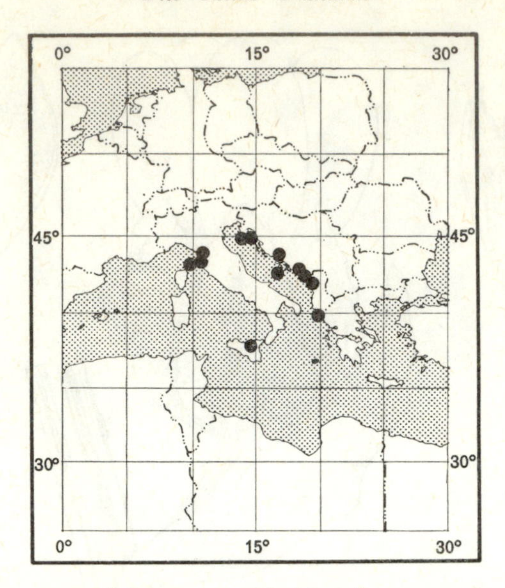

Map. 31: *T. dalmatica*

Selected specimens: YUGOSLAVIA: *Marchesetti*, Dalmazi Is. Lesina .3.1882 (FI); *Rohlena*, breh Moraiu u Donji Zete .5.1903 (PRC). ITALY: *Vaccari*, Toscana Monte Argentoro le Scorpacciate 14.4.1905 (FI); *Engelhardt*, an den Strandbaden von Gravo bei Triest 13.5.1934 (B); *Portenschlag* 572, in Dalmatia Ragusae in horto P.P. Franciscanorum (holotype of *T. praecox* Portenschlag ex Bge., W; isotypes P, PRC); '*T. pachystachya* Presl' in Sicilia campestribus humidis coliturque in hortis .4.1817 (PRC). FRANCE: *Müller* 210, Toulon 26.4.1851 (F, G, P).

Observations: (a) From the young flowers of the type of *T. africana* var. *ligustica* it cannot be decided whether this taxon is synonymous with *T. dalmatica*. The identity is based on external characters only. (b) *T. dalmatica* can be distinguished very well from *T. africana* by its typically unguiculate petals (not ovate to broadly trullate-ovate as in *T. africana*). *T. dalmatica* seems to be closely related also to *T. africana* var. *rungsii*.

32. **T. elongata** Ledeb., Fl. Alt., 1:421 (1829) [Plate XXXII]

T. salsa Stev. ex Bge., Tentamen, 24 (1852), pro syn. *T. meyeri* Boiss.

Type: CHINESE TURKESTAN: *Meyer* 445, legi in salsis deserti orientalis versus lacum Noor-Saissan 18.5.1826 (holotype LE; isotypes CGE, K, P, W).

Small tree or shrub, 1–5(6) m tall, with brown bark, glabrous. Leaves sessile with narrow base, 1.5–3 mm long, the first-produced ones broad, somewhat cordate, thick or fleshy. Vernal inflorescences simple, vernal-aestival and aestival ones rarely compound of racemes. Racemes usually 6–9 cm long, also up to 13 cm long, 7–8 mm broad, with, exceptionally, the lowermost flowers pentamerous or the uppermost pentandrous. Bracts exceeding flowers, narrowly triangular, subulate with a

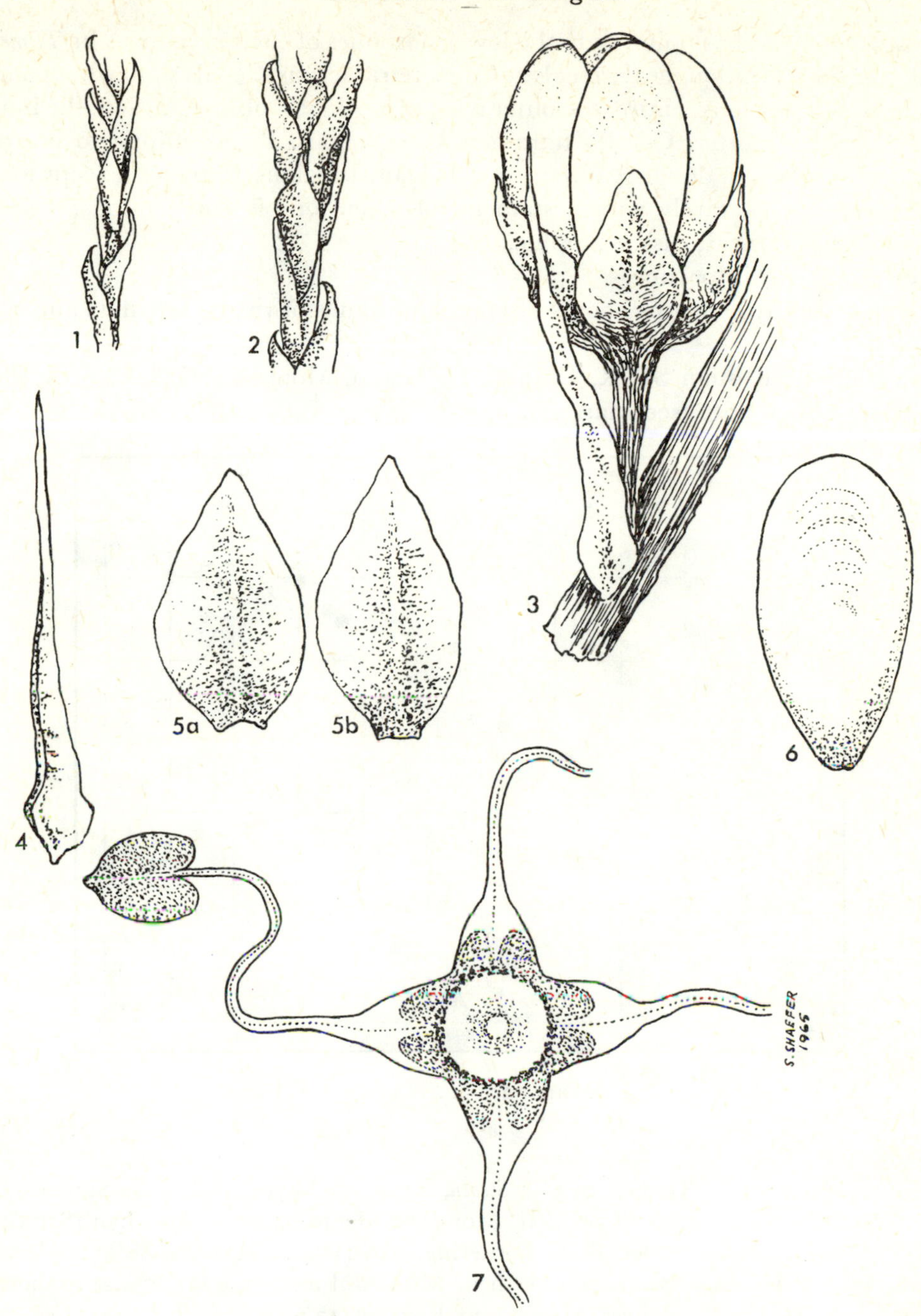

Plate XXXII *T. elongata*
1. Young twig (x 5); 2. id (x 10); 3. Flower (x 10); 4. Bract (x 10); 5a. Inner sepal (x 20); 5b. Outer sepal (x 20); 6. Petal (x 15); 7. Androecium (x 20).

long acute diaphanous point, only the lowermost ones of each raceme as in *T. meyeri*. Pedicel as long as or longer than calyx. Calyx tetramerous. Sepals somewhat connate at base, 1.5–2 mm long, entire, the outer 2 acute to slightly obtuse, keeled, the inner 2 obtuse, slightly shorter. Corolla tetramerous, caducous. Petals elliptic to narrowly obovate, 2.25–3 mm long. Androecium of 4 antesepalous stamens, except for the uppermost flowers, rarely with 5 stamens, of which one is antepetalous; insertion of filaments peridiscal; disk synlophic.

Flowering: April to June.

Habitat: Sandy or loamy shores of lakes and banks of rivers, salt flats and wadis in deserts.

Distribution: Russian SFSR, Chinese Turkestan, Mongolia, Kazakh SSR, Turkmen SSR, Uzbek SSR (see Map 32).

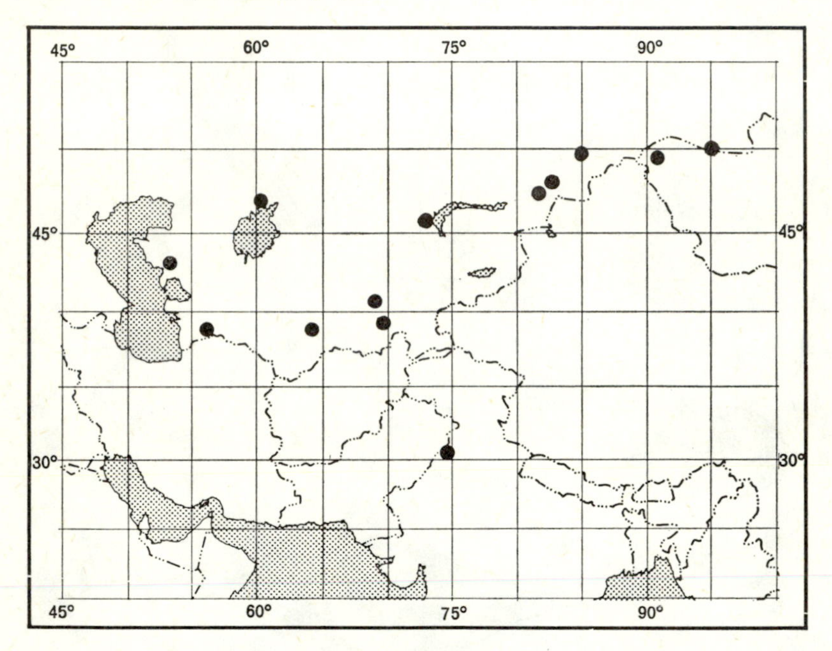

Map 32: *T. elongata*

Selected specimens: Russian SFSR: *Steven*, Kislar (holotype of *T. salsa* Stev. ex Bge., P; isotype K). Mongolia: *Przewalski*, Mong. occidentalis, mont. Alaschan 1872 (E, K, P, W); *Chaney* 620, Tsagan Nor, dune bordering lake (UC). Kazakh SSR: *Galaskakov*, Betnak-Dala W. shores of Balkhash Lake near Mun-Aral along the sandy loamy shores of the lake 7.6.1949 (LE); *Smirnov*, na Urzum Karagai 13.6.1938 (LE); *Gorschkova* 3042, Kazakhstania ripa septentr. maris Aralskoe, insula Zopouadnyi 29.5.1930 (BM, G, K, LE, US, W). Turkmen SSR: *Androssow* 1869, Turkestania Dominium Buchara ad fl. Amu-Darja pr. Farab 18.4.1901 (B, G, LE, PRC, S, WU); *Sedmuradov* 39, Transcaspia Kazandzhik station 7.6.1910 (LE). Uzbek SSR: *Popov* & *Vvedensky* 312, Jaxartica in deserto Mirza-Tachul (Goldanaja Step) in salsuginosos prope fontem Kamysty-Kuduk 17.5.1932 (B, E, G, K, S, W).

Observations: (a) The isotypes which the author has cited for *T. elongata* Ledeb. are uncertain. There is no indication that they are duplicates of the holotype, though they are labelled '*T. elongata* Nob.' in Ledebour's own handwriting. (b) *T. elongata* is sometimes confused with *T. meyeri*. Very good external characters that are easily seen are the typical broad and fleshy first-produced leaves and the very typical long acuminate bracts with acute diaphanous points. (c) Hiekisck (1884) reports that *T. elongata* is characteristic of the valleys of the Nan-Schan.

33. **T. hampeana** Boiss. & Heldr. emend. Boiss., Fl. Or., 1:767 (1867) [Plate XXXIII]

T. hexandra Hampe, Flora, 25:62 (1842), nom. nud.
T. hampeana Boiss. & Heldr., in: Boiss., Diagn. Pl. Or. Nov., I, 10:8 (1849).
T. hampeana Boiss. & Heldr. var. *phalerea* Bge., Tentamen, 20 (1852).
T. hampeana Boiss. & Heldr. var. *marmorissa* Bge., *op. cit.*, 21.
T. longipes Presl ex Bge., *loc. cit.*, pro syn.
T. hampeana Boiss. & Heldr. var. *composita* Boiss., *loc. cit.*
T. variabilis Bge. ex Koch, Dendrol., 1:457 (1869).
T. haussknechtiana Ndz., De Genere Tamarice, 5 (1895).
T. phalerea (Bge.) Ndz., *loc. cit.*
T. hampeana Boiss. & Heldr. var. *aegaea* Turrill, Hooker's Ic. Pl., 32: t. 3153 (1932).
T. africana Poir. var. *philistaea* Zoh , Trop. Woods, 104:39 (1956).

Lectotype: Greece: *Heldreich* 64, Phalère au bord des fossés 14.4.1844 (G; isolectotypes FI, G, K, OXF, P, W).

Tree with brown to reddish-brown bark, entirely glabrous. Leaves sessile with narrow base, 1.75–4 mm long. Inflorescences simple or loosely composed of racemes. Racemes 2–13 cm long, 10–12 mm broad. Bracts diaphanous, oblong, acute, slightly boat-shaped, usually shorter than pedicels, but those in upper part of vernal racemes or those of aestival racemes sometimes equal to or longer than pedicels. Pedicel longer than calyx. Calyx tetra-pentamerous.[26] Sepals trullate-ovate, acuminate, with obtuse apex, 2–2.5 mm long, the outer 2 keeled, subentire, more acute than the more or less regularly and slightly denticulate inner ones. Corolla tetra-pentamerous, caducous. Petals ovate-elliptic, 2.75–4 mm long. Androecium haplostemonous to partially diplostemonous, of 4–5 antesepalous and 0–2 antepetalous stamens; insertion of antesepalous filaments peridiscal; disk fleshy, paralophic.

Flowering: April to May.

Habitat: Maritime swamps, banks of pools, fields.

Distribution: Greece, Turkey (rare), Israel (introduced ?) (see Map 33).

Selected specimens: Greece: *Spruner*, Phalerum 1841 (G); *Spruner*, Attica 1843 (paralectotypes of *T. hampeana* Boiss. & Heldr., G); *Spruner* 46, Attica (holotype of *T. hexandra*

26 When hexamerous, it is due only to partial pleiomery of one sepal.

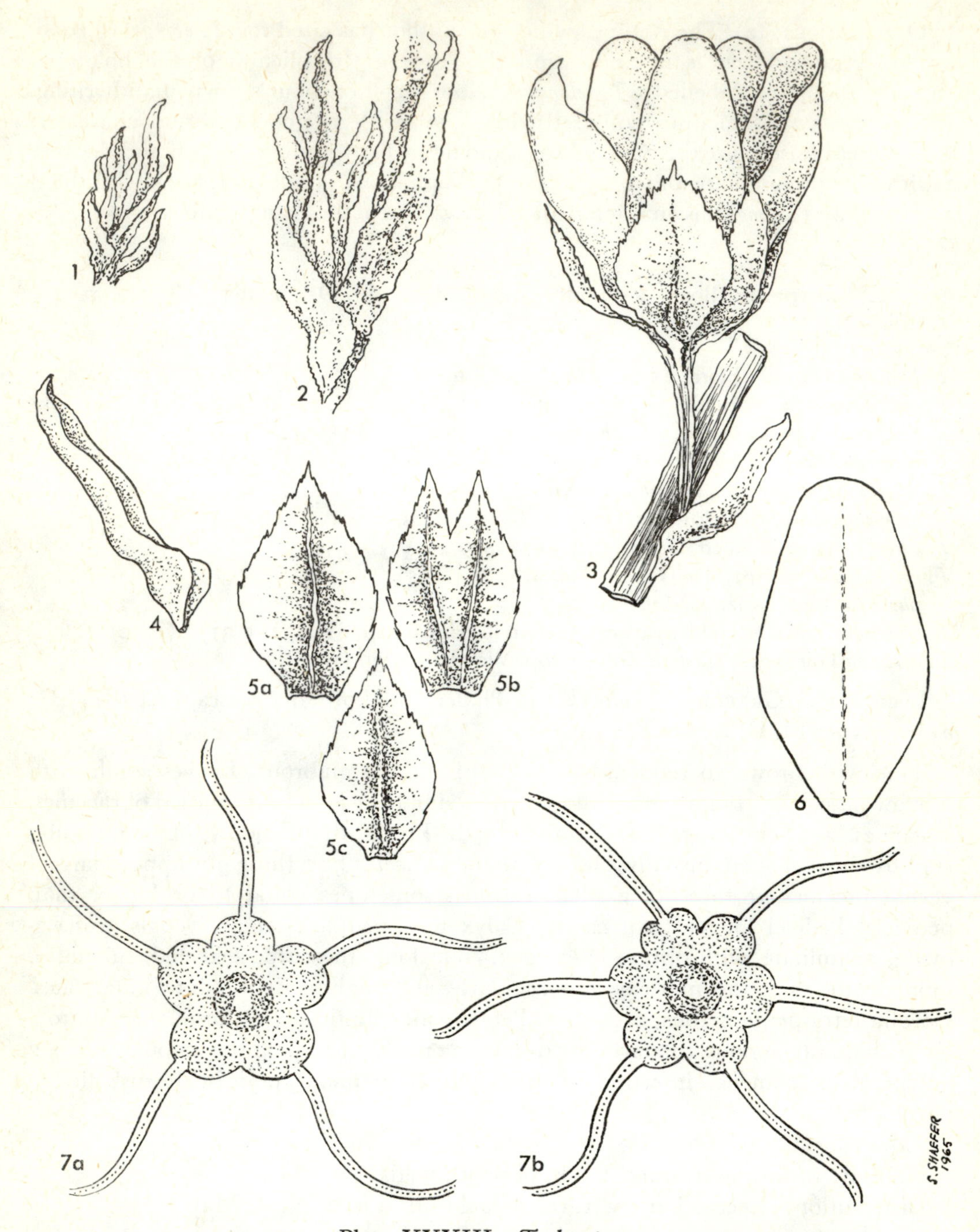

Plate XXXIII *T. hampeana*

1. Young twig (x 5); 2. id (x 10); 3. Flower (x 10); 4. Bract (x 15); 5a. Inner sepal (x 15); 5b. id, in meiomeric state (x 15); 5c. Outer sepal (x 15); 6. Petal (x 15); 7a. Androecium of 5 antesepalous stamens (x 20); 7b. id, with one antepetalous stamen (x 20).

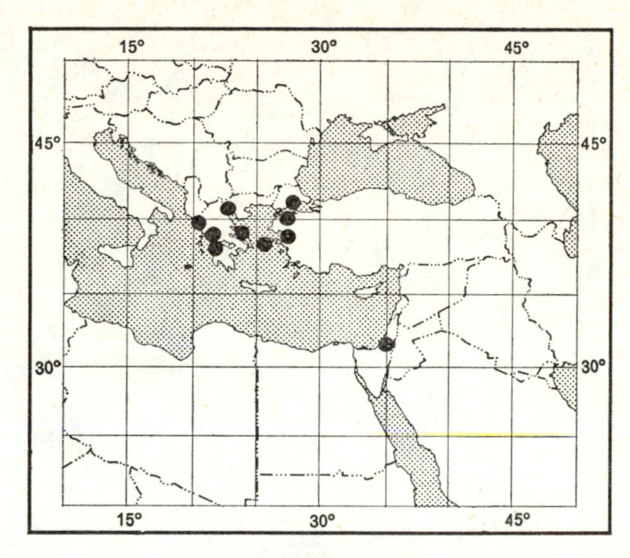

Map 33: *T. hampeana*

Hampe, W; isotypes G, UPS); *Haussknecht*, Laurion in lit. mar. 11.5.1885 (isotypes of *T. haussknechtiana* Ndz., K, W); *De Heldreich*, Attica ad Phalerum 3.5.1854 (syntypes of *T. phalerea* Ndz., K, L, P, W, WU); *Pichler*, Attica in foss. et palud. pr. Valeer. .5.1876 (syntypes of *T. phalerea* Ndz., FI, G, K, P, PRC); *Rechinger fil.* 4002, insula Samos in limosis ad ruinam templi Herae prope Colonna 12.5.1934 (BM, W); *Dimonie*, Macedonia in arenosis maritimis op. Thessalonica .4.1909 (FI, G, K, WU); *Spreizenhoffer* 767, Corcyra (Corfu) ufer der Potamo 11.5.1878 (FI, K, W, WU); *Baenitz*, Corcyra Kastrades (Corn) an der Strasse nach Gasturi in der Nähe des Hyllaeischen Hafens 22.4.1896 (B, L, OXF, P, PRC, US, WU); *Bornmüller* 279, insula Zakynthos (Zante) ad vias et in locis arenosis Kmonero 28.4.1926 (B, G, K, S); *Tedd* 230A, Bouloustra 1.6.1930 (holotype of *T. hampeana* var. *aegaea* Turrill, K, paratypes: *Tedd* 230C, eastern shores of Lake Boru, sands at water's edge 6.6.1931 (K), *Tedd* 250, Puyuk Osmanli 1.6.1930 (K). TURKEY: *Balansa* 133, Echelle de Papa, Golfe de Smyrne 24.4.1854 (holotype of *T. hampeana* var. *composita* Boiss., G; isotypes BM, K, OXF, W); *Groves*, in paludosis prope Gallipoli (FI, UC); *Lefèvre*, in Asia Minore ad Marmarissam .4.1826 (holotype of *T. hampeana* var. *marmorissa* Bge., FI; isotype P); *D. Zohary* 4109, N. W. Turkey 2 km W of Ipsala, banks of ditches 11.8.1962 (HUJ). ISRAEL: *Joffe* 1064 & 1065, S. Shefela, Ashkelon ruins 15.3.1954 (holotype of *T. africana* var. *philistaea* Zoh., HUJ).

Observation: One of the isolectotypes of *T. hampeana* which is in W bears a label in Bunge's handwriting: '*T. variabilis* Bunge'. This is the holotype of *T. variabilis* Bge. ex Koch.

34. **T. meyeri** Boiss., Diagn. Pl. Or. Nov., I, 10:9 (1849) [Plate XXXIV]

T. meyeri Bge., Mém. Acad. St. Pétersb., 7:291 (1851), nom. illegit.
T. tetragyna Ehrenb. var. *meyeri* (Boiss.) Boiss., Fl. Or., 1:768 (1867).

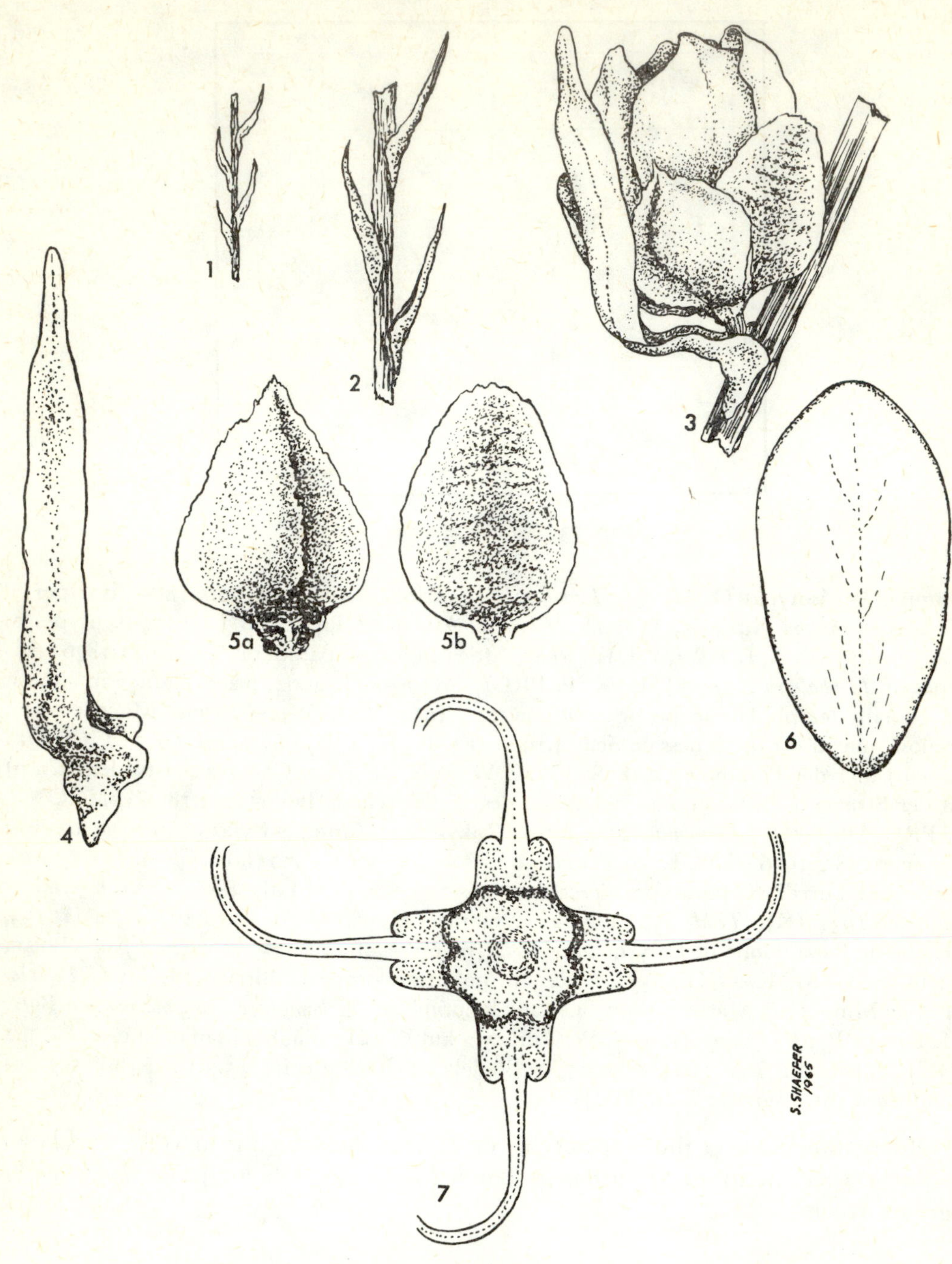

Plate XXXIV *T. meyeri*
1. Young twig (x 5); 2. id (x 10); 3. Flower (x 10); 4. Bract (x 18); 5a. Outer sepal (x 18); 5b. Inner sepal (x 18); 6. Petal (x 18); 7. Androecium (x 25).

Type: Russian SFSR: *Meyer* 1460, ad mare Caspium 1829 (holotype G, fragment of type K, P).

Small tree or shrub, up to 3–5(8) m tall, with reddish-brown bark, usually glabrous except for restricted places. Leaves sessile with narrow base, 1.5–4 mm long. Inflorescences usually simple, rarely loosely composed of racemes. Racemes 4–10 cm long, 6–7 mm thick, with glabrous to papillose rachis. Bracts narrowly oblong, the lower ones blunt with a short truncate obtuse point, the upper slightly acuminate, equalling to exceeding flowers, papillose at least on inner side with smooth margins. Pedicel shorter than calyx. Calyx urceolate, tetramerous. Sepals usually entire, sometimes faintly or inconspicuously denticulate, 2–2.25 mm long, ovate in outline, the outer 2 acute and keeled, the inner obtuse and narrower. Corolla tetramerous, caducous. Petals 3–3.5 mm long, obovate to elliptic-obovate. Androecium haplostemonous, of 4 antesepalous stamens, occasionally with one additional antepetalous stamen in uppermost flowers of racemes; insertion of filaments peridiscal; disk paralophic.

Flowering: April to May.

Habitat: Along streams.

Distribution: Russian SFSR, Turkmen SSR, Tadzhik SSR, Iran, Afghanistan (see Map 34).

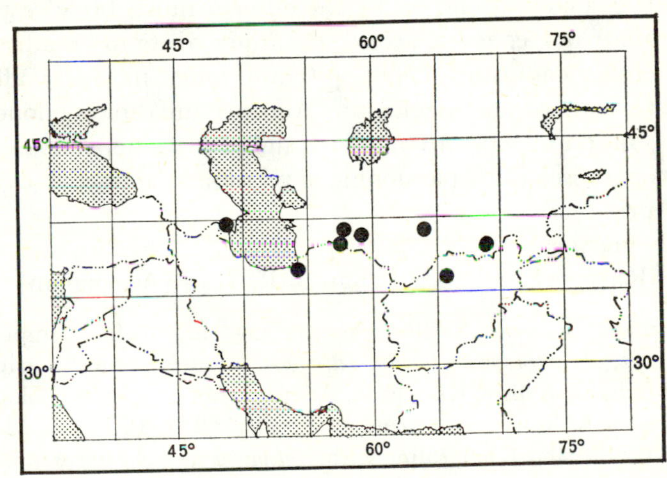

Map 34: *T. meyeri*

Selected specimens: Turkmen SSR: *Michelson* 260, reg. Transcaspica prope urb. Czardshuj 16.4.1910 (B, K, LE, P, S, UC); *Gorschkova* 3049, Turcomania in locis ad ripam fl. Murgab pr. st. viae ferreae Imam-Baba 25.5.1930 (BM, G, K, LE, S, US, W.). Tadzhik SSR: *Russanov* 72, Amu-Darja River, Khabu Rabat near Bauta 21.4.1939 (LE). Iran: *Sintenis* 1494, Persia borealis prov. Asterabad Bender-Ges in maritimis 30.3.1901 (E, G); *Sabeti* 436, Sistan 1.4.1950 (W, Esfand. Herb.). Afghanistan: *Pabot* 810, 40 km S. Ankhoi 2.5.1958 (Herb. Pab.); *Aitchison* 391, Hari-Rud Valley 7.5.1885 (K).

Observation: The species is sometimes confused with *T. brachystachys* but differs markedly from it in its more delicate leaves, its outer acute and keeled sepals and its typical paralophic disk.

35. **T. octandra** (M.B.) Bge., Tentamen, 17 (1852) [Plate XXXV]

T. taurica Pall. var. *octandra* Pall., Nova Acta Acad. Petrop., 10:36 (1797), nom. nud.
T. tetrandra Pall. ex M.B. var. *octandra* M.B., Fl. Taura.-Cauc., 3:252 (1819).
T. tetragyna Ehrenb. var. *pallida* Trautv., Acta Horti Petrop., 2:533 (1874).
T. tetragyna Ehrenb. var. *heterantha* Kuntze f. *longibracteata* Kuntze, *ibid.*, 10:175 (1887).
T. octandra (M.B.) Bge. var. *duplex* Regel & Mlokoss., in: Kusn., Busch & Fomin, Fl. Cauc.-Crit., 3 (9):110 (1909).

Type: Russian SFSR: *Marschall von Bieberstein*, '*Tamarix tetrandra* ex planitiebus Caucaso subadjacentibus. N.B.: an distincta a tauria floribus majoribus interdum 8-andris pedicellis et bracteis longioribus? 1813' (holotype LE; isotype P).

Shrub or very small bushy tree with brown to yellowish-brown bark, younger parts glabrous except for some organs. Leaves sessile with narrow base, 2–5 mm long. Vernal inflorescences simple and dense, aestival inflorescences simple to loosely compound. Racemes 4–9 cm long, 8–12 mm broad, rachis glabrous. Bracts narrowly oblong, blunt with a short obtuse point, equalling to exceeding flowers, more or less papillulose on both sides. Pedicel shorter than calyx. Calyx tetramerous, urceolate. Sepals 2–3.5 mm long, entire, the outer 2 much broader than the inner, very broadly trullate-ovate, acute, keeled, the inner ovate to broadly ovate, obtuse. Corolla tetramerous, caducous. Petals 4–6 mm long, narrowly elliptic-ovate to narrowly obovate, more or less unguiculate. Androecium diplostemonous or partially diplostemonous, of 4 antesepalous stamens and of 1–4 antepetalous stamens; insertion of filaments peridiscal, disk paralophic in relation to antesepalous stamens.

Flowering: April.

Habitat: Salt marshes.

Distribution: Russian SFSR, Azerbaijan SSR, Iran (Azarbaijan) (see Map 35).

Selected specimens: Russian SFSR: *Parreytz* 444, in tauriae campis 1829 (PRC). Iran: *Szowitz* 132, *T. tetragyna octandra, tetragyna* bracteis pedicellis triplo longioribus, in camp salso ad Schabanli distr. Khoi prov. Aderbeidzan 27.4.1828 (holotype of *T. octandra* Bge., P, fragment K; isotypes ? 'Armenia *Szowitz*', BM, E, FI, K, L, P, S, US, W); *Baum*, Between Marand and Khoi, saline with *Halimione* and *Aeluropus* 3.7.1965 (HUJ).

Observation: The author could not find any of Pallas' authentic material of *T. taurica* var. *octandra* Pall., nor did he see the type of *T. tetragyna* var. *pallida* Trautv., but its identity is beyond question from the description. The same is true of *T. tetragyna* var. *heterantha* f. *longibracteata* Kuntze and *T. octandra* var. *duplex* Regel & Mlokoss.

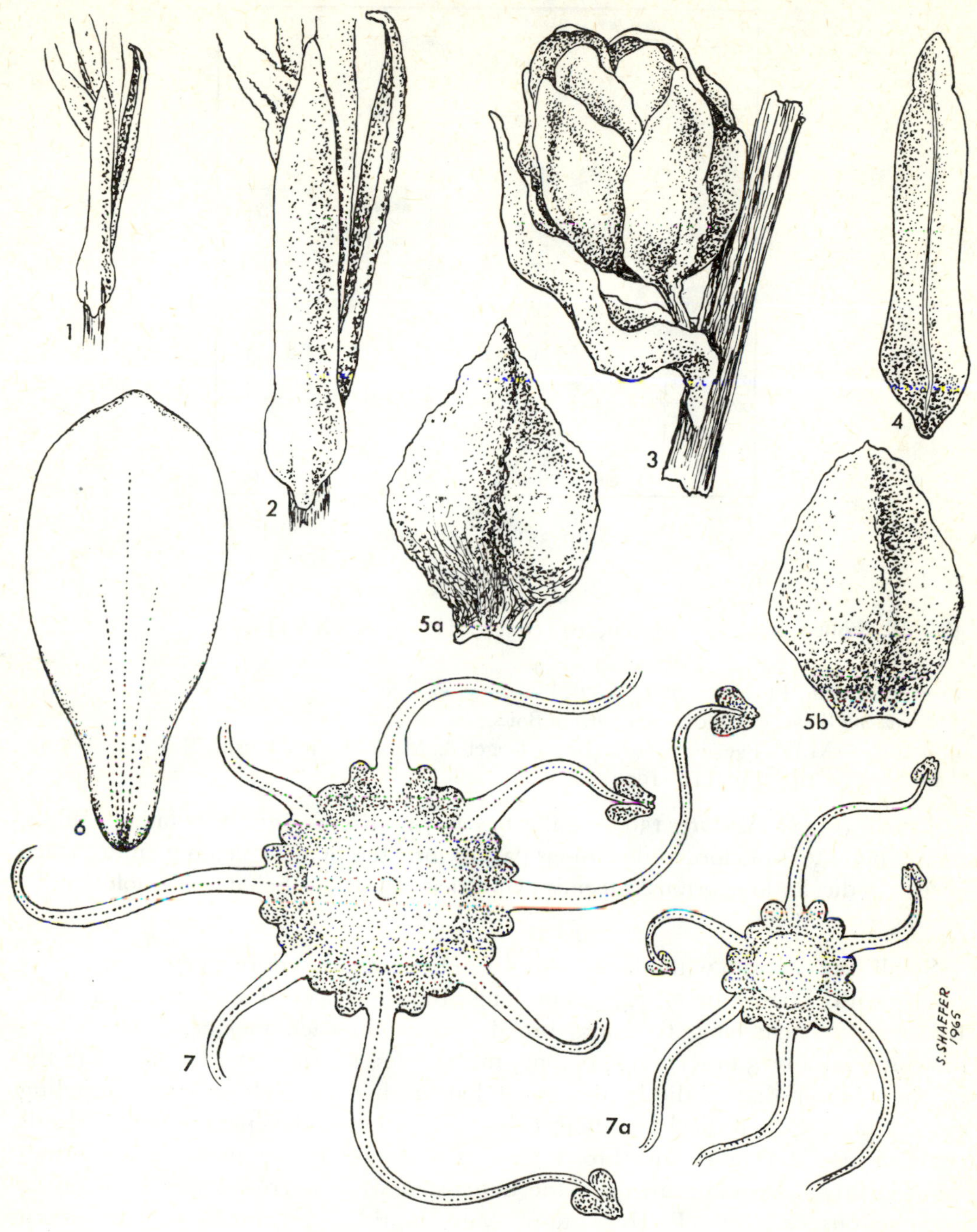

Plate XXXV *T. octandra*
1. Young twig (x 5); 2. id (x 10); 3. Flower (x 10); 4. Bract (x 15); 5a. Outer sepal (x 15); 5b. Inner sepal (x 15); 6. Petal (x 15); 7. Androecium (x 20); 7a. id, with less than 4 antepetalous stamens (x 15).

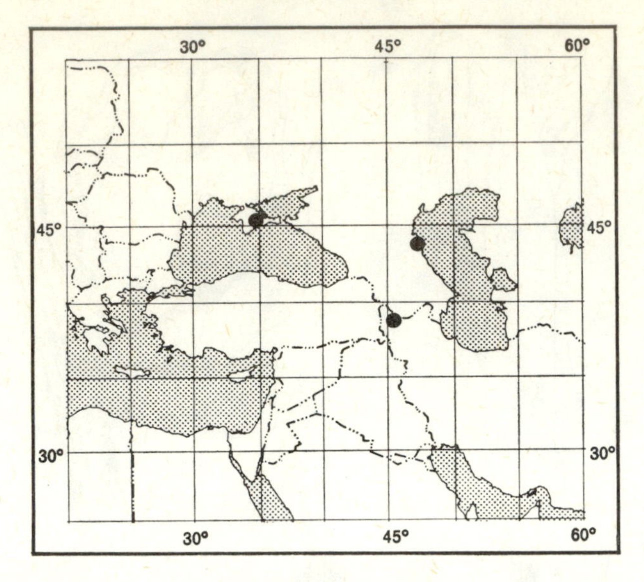

Map 35: *T. octandra*

36. **T. rosea** Bge., Tentamen, 19 (1852) [Plate XXXVI]

T. syriaca Boiss., Fl. Or., 1:767 (1867), non Stev. ex Bge.
T. hampeana Boiss. & Heldr. var. *smyrnea* Boiss., *loc. cit.*
T. octandra (M.B.) Bge. var. *rosea* (Bge.) Regel & Mlokoss., in: Kusn., Busch & Fomin, Fl. Cauc.-Crit., 3(9):110 (1909).

Type: IRAN: *Szowitz* 136, 'T. floribus 6–7-andris 4–5-gynis, petalis ovalibus, calyce patente 3-plo longioribus roseis patulis, bracteis ovatis apiculo membranaceo obtuso pedicello longioribus, racemis cylindricis densis, ad Schabanli' (holotype P; isotype G).

Small tree or shrub with blackish-brown to brown bark, entirely glabrous. Leaves sessile with narrow base, 1.5–3 mm long. Inflorescences simple or compound. Racemes 5–12 cm long, 6–12 mm broad. Bracts: in each raceme, some flowers have at least 2 or 3 bracts each, oblong, more or less short, acute, the lowest in the raceme shorter than pedicels, the upper longer than pedicels or even exceeding calyces in length. Pedicel equalling calyx or shorter. Calyx pentamerous. Sepals 1–2 mm long, the inner with broad scarious margins, trullate-ovate, acute, sometimes faintly denticulate at apex, the outer somewhat narrower, keeled. Corolla pentamerous, caducous. Petals ovate to ovate-elliptic, 2–3.5 mm long. Androecium haplostemonous to rarely partly diplostemonous, of 5 antesepalous stamens and 0–4 antepetalous stamens; insertion of filaments hypodiscal;[27] disk hololophic (the antepetalous stamens arise from the nectariferous lobes).

27 Sometimes only slightly hypodiscal.

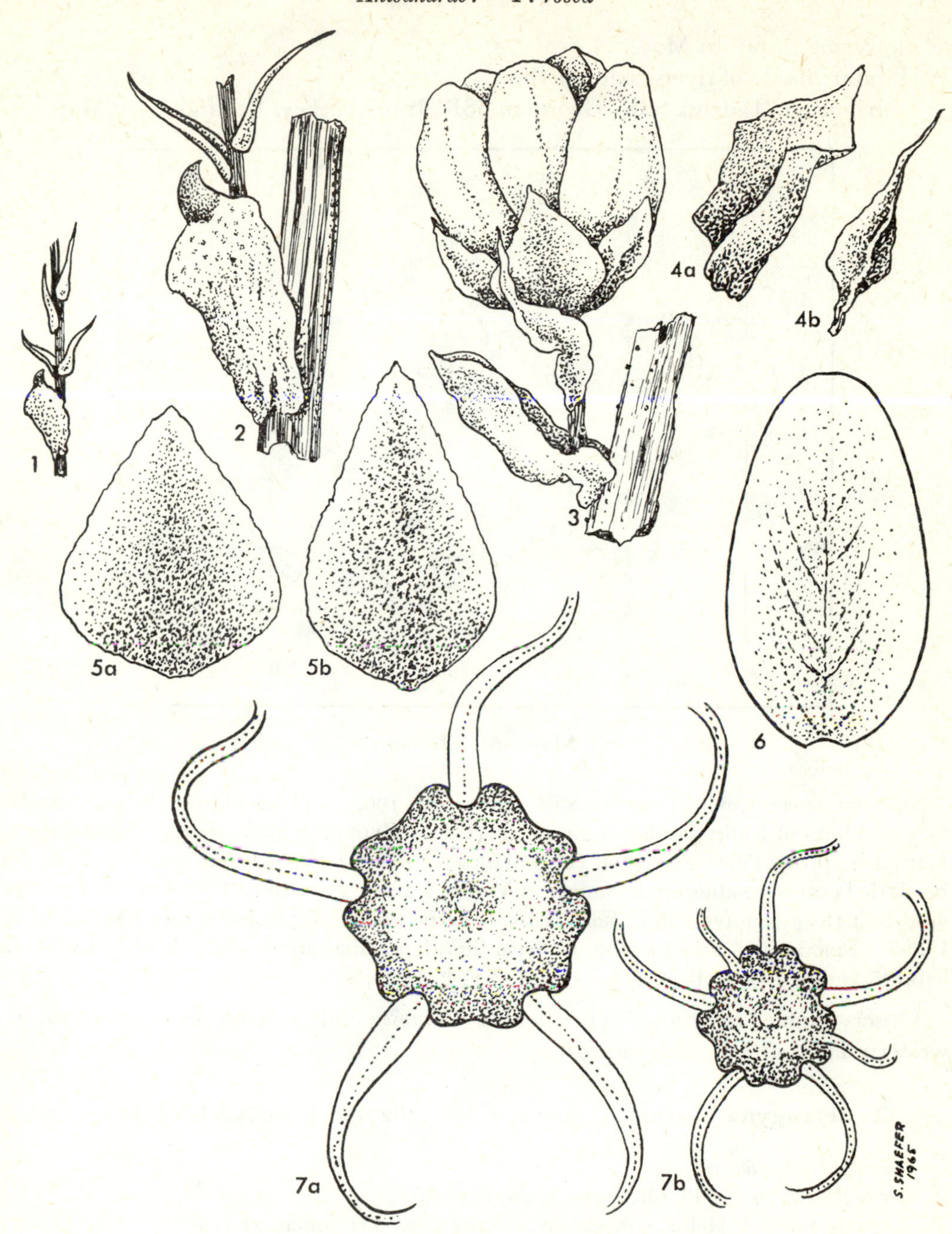

Plate XXXVI *T. rosea*
1. Young twig arising from the axil of a bract-like leaf (x 5); 2. id (x 10);
3. Flower with its 2 bracts (x 10); 4a. Lower bract (x 25);
4b. Upper bract (x 25); 5a. Inner sepal (x 25); 5b. Outer sepal (x 25;)
6. Petal (x 20); 7a. Haplostemonous androecium (x 25);
7b. Partial diplostemonous androecium (x 12).

Flowering: April to May.
Habitat: Banks of rivers, salt marshes.
Distribution: Tadzhik SSR, Georgian SSR, Iran, Turkey, Lebanon (see Map 36)

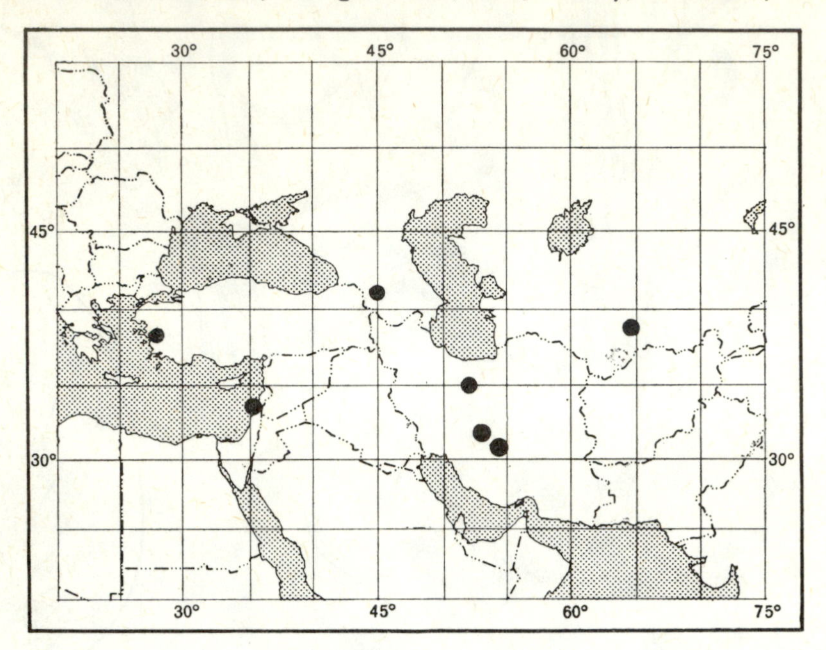

Map 36: *T. rosea*

Selected specimens: TADZHIK SSR: *Varizeva* 196, S. Tadshikistan near Issan-Ba village, banks of Kafirnijan River 24.4.1948 (LE). GEORGIAN SSR: *Konig*, Transcaucasia Karajazg (prov. Tiflis) 21.5.1908 (L). TURKEY: *Balansa* 130, Marais salés situés sur les bords de la route conduisant de Smyrne à Vourla à 2 lieues environ à l'est de cette première ville 29.4.1854 (lectotype of *T. hampeana* var. *smyrnea* Boiss., G; isolectotypes BM, E, K, P, UPS). LEBANON: *Gaillardot* 1102, in Syria litorali ad ripas amnis Nahr Aoule ad boream urbis Sidonis 5.3.1858 (B, BM, E, G, K, OXF, P, PRC, S).

Observation: Transitional types between *T. rosea* and *T. hampeana* occur in south-western Turkey (e.g., *Balansa* 130).

37. **T. tetragyna** Ehrenb., Linnaea, 2:258 (1827) [Plate XXXVII]

T. effusa Ehrenb., *loc. cit.*
T. deserti Boiss., Diagn. Pl. Or. Nov., I, 10:9 (1849).
T. hampeana Boiss. & Heldr. var. *syriaca* Stev. ex Bge., Tentamen, 21 (1852).
T. syriaca Stev. ex Bge., *loc. cit.*, pro syn.
T. gennessarensis Zoh., Trop. Woods, 104:36 (1956).

Type: EGYPT: *Ehrenberg*, 'ad lacum Menzaleh, 4.1820–1826' (isotype K).

Small tree or shrub with purple to blackish-brown bark, younger parts papillose to occasionally glabrous, especially papillose on bracts and rachis of racemes.

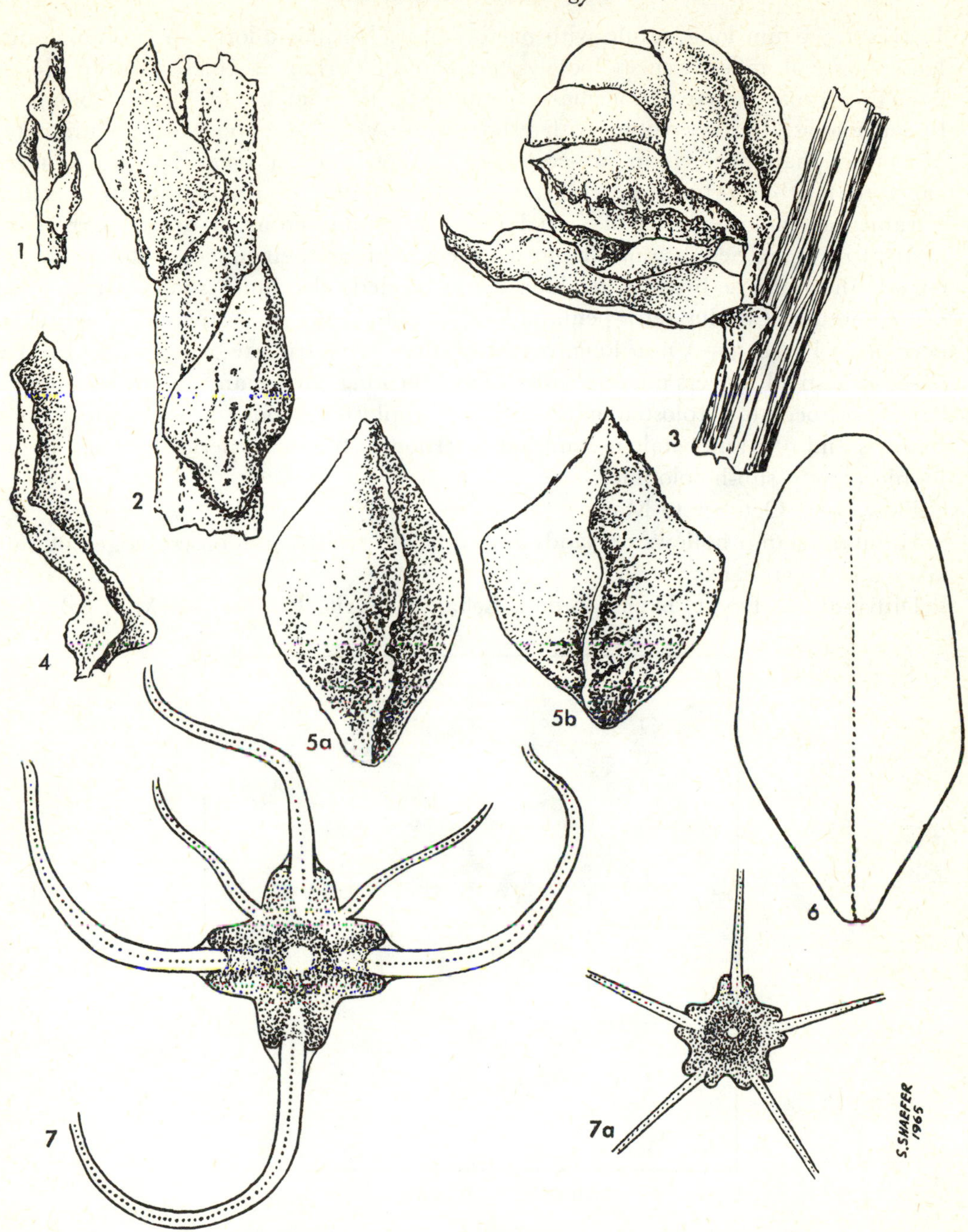

Plate XXXVII *T. tetragyna*
1. Young twig (x 5); 2. id (x 20); 3. Flower (x 15); 4. Bract (x 20);
5a. Inner sepal (x 20); 5b. Outer sepal (x 20); 6. Petal (x 20);
7. Partial diplostemonous androecium with 4 antesepalous stamens (x 30);
7a. Androecium with 5 antesepalous stamens (x 15).

Leaves 1.5–6 mm long, sessile with narrow base. Vernal inflorescences simple and loose, aestival inflorescences loosely compound. Vernal racemes 5–15 cm long, 8–10 mm broad, aestival racemes 2–5 cm long, 5–7 mm broad. Bracts oblong to linear-oblong, the lowest blunt with a short obtuse point, the upper more acuminate, all longer than pedicels to sub-equalling calyces. Pedicel somewhat longer than calyx to much shorter than calyx. Calyx urceolate, tetra-pentamerous, usually tetramerous in lower part of vernal racemes, pentamerous in upper part or in aestival racemes. Sepals 2 mm long, with few teeth at their apices or finely denticulate, the outer 2 broadly trullate, acute, keeled, the inner somewhat shorter, ovate, obtuse. Corolla tetra-pentamerous, pentamerous in aestival racemes, sub-persistent. Petals 3.5–5 mm long, narrowly obovate to trullate-ovate with cuneate base; in aestival flowers much smaller, 2.25 mm long, ovate and also with cuneate base. Androecium haplostemonous to partly diplostemonous, of 4–5 antesepalous stamens and 0–4 antepetalous stamens; insertion of filaments peridiscal; disk paralophic (rarely sub-hololophic).

Flowering: October to May.

Habitat: Damp hollows on sandy soil, ditches, wadis and oases; edges of salt lakes and river banks.

Distribution: Egypt, Israel, Jordan, Lebanon, Syria, Cyprus (see Map 37).

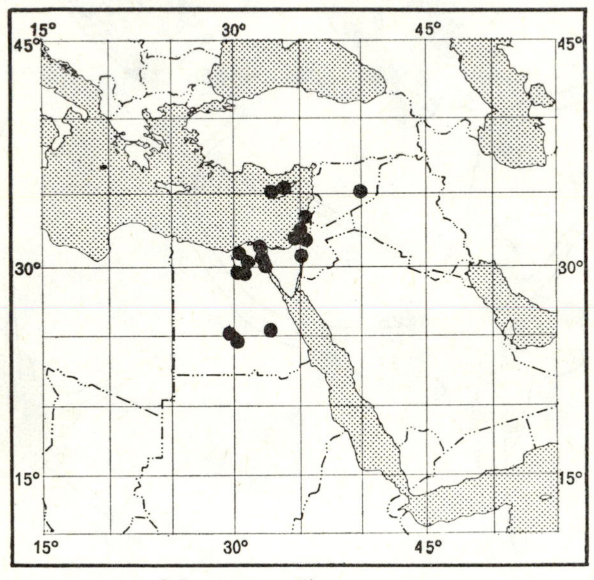

Map 37: *T. tetragyna*

Selected specimens: EGYPT: *Ehrenberg*, '*T. tetragyna* ad San. .6.1821' (isotype of *T. effusa* Ehrenb., S); *Husson* 387, desert Arabique terrain salsugineux .3.1844 (lectotype of *T. deserti* Boiss.), paralectotypes: *Husson*, rocher granitique (G), *Boissier*, Chaîne du Tih Arabia Petraea .3.1846 (G; isoparalectotype G); *Letourneux* 232, ad ripas lacus prope Ramle, .3.1880, G, K, P, PRC, S, W); *Bornmüller* 10443 & 10445, Alexandria Sidi Gaber 8.4.1908 (B, BM, E, G, P, W, WU); *Hassib*, Dakhla Oasis Ezbet Sheikh Wadi 10.2.1931

(CAI). ISRAEL: *Rimon* 606, Kinroth distr. bank of the Jordan River near Beith Zera 15.3.1954 (holotype of *T. gennessarensis* Zoh., HUJ); *Bornmüller* 236, Jericho ad fluvium Jordanum 31.3.1897 (B, E, G, K, OXF, P, PRC, W, WU); *D'Angelis* 552, Negev Revivim ditches 22.3.1952 (BM, DEL, E, G, HUJ, K, L, OXF, S, SMU, UC, US, W). JORDAN: *Kassapligil* 2832, Deir Alla Zarqa River 4.2.1956 (UC). LEBANON: *Mouterde* 10105, sables au sud de Beyrouth vers à prison 27.3.1951 (Herb. Mouterde). SYRIA: *Labillardière*, Syrie (holotype of *T. syriaca* Stev., P; isotype K); *Davis* 5903, Palmyra small dunes at N. of Salt Lake 18.4.1943 (E). CYPRUS: *Kotschy* 243, in viciniis urbis Larnaca lucas format ad aquas stagnantes 25.4.1862 (BM, G, K, L, P, PRC, S, W).

Observations: (a) See observation (c) on *T. africana*. (b) The specimens of the type collection of *T. deserti* Boiss. also have pentandrous flowers. This does not agree with the original description of Boissier (1849).

38. **T. tetrandra** Pall. ex M.B. emend. Willd., Abh. Akad. Berlin Physik, 1812–1813:81(1816) [Plate XXXVIII]

T. tetrandra Pall., Neue N. Beitr., 7:430 (1796), nom. nud.
T. taurica Pall., Nova Acta Acad. Petrop., 10:376 (1797), nom. nud.
T. tetrandra Pall. ex M.B., Fl. Taur.-Cauc., 1:247 (1808), p.p. (pars altera = *T. laxa* Willd.).
T. hohenackeri Bge. var. *taurica* Regel & Mlokoss., in: Kusn., Busch & Fomin, Fl. Cauc.-Crit., 3(9):102 (1909).

Type: RUSSIAN SFSR: *Pallas*, '*T. tetrandra* e tauria' (holotype BM; isotypes B, BM, P, PRC).

Shrub or small tree, 2–3 m high, with black to greyish-black bark, younger parts glabrous to minutely papillose. Leaves sessile with narrow base, scarious-margined, 4–5 mm long. Vernal inflorescences simple or loosely compound, aestival inflorescences rare. Racemes[28] 3–6 cm long, 6–7 mm broad. Bracts oblong, blunt or with a short obtuse point, longer than pedicels, herbaceous at least up to half their length, diaphanous at apex. Pedicel shorter than calyx. Calyx tetramerous, urceolate. Sepals 2–2.5 mm long, entire, sometimes faintly and minutely papillose on back, the outer 2 larger, keeled, acute, sometimes with few teeth, trullate-ovate, the inner smaller, obtuse, ovate. Corolla tetramerous, subpersistent. Petals 2.25–2.5 mm long, ovate, sometimes broadly ovate or occasionally ovate-elliptic (parabolical) and then 3 mm long (only in the few and occasional pentamerous flowers). Androecium haplostemonous, of 4 antesepalous stamens, antepetalous stamens usually or occasionally 1–3; insertion of antesepalous filaments peridiscal; fleshy disk paralophic to para-synlophic.

Flowering: March to June.
Habitat: Temporary river beds, river banks.
Distribution: Russian SFSR, Turkey, Greece, Cyprus, Syria (see Map 38).

28 Less densely flowered than in *T. parviflora*.

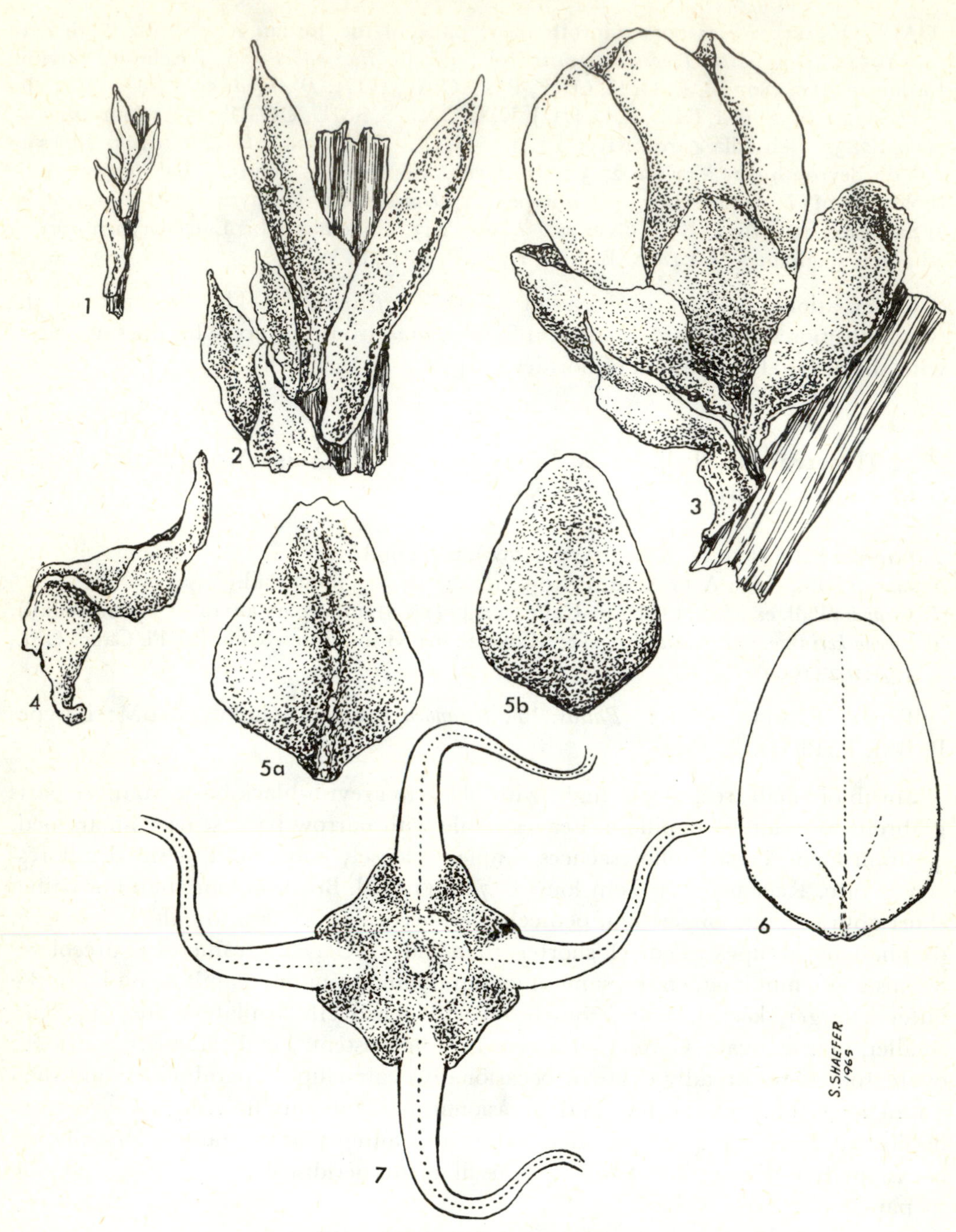

Plate XXXVIII *T. tetrandra*

1. Young twig (x 5); 2. id (x 10); 3. Flower (x 15); 4. Bract (x 15); 5a. Outer sepal (x 15); 5b. Inner sepal (x 15); 6. Petal (x 15); 7. Androecium (x 20).

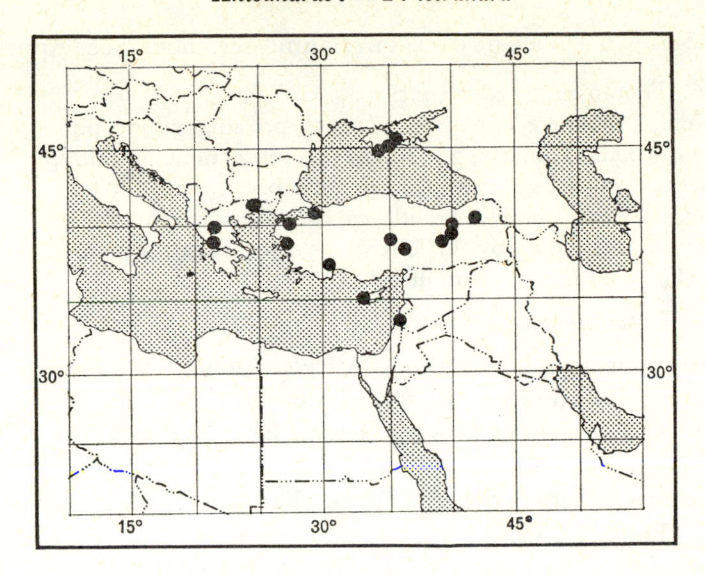

Map 38: *T. tetrandra*

Selected specimens: RUSSIAN SFSR: *Marschall de Bieberstein*, '*T. tetrandra* ex taurica 1800' (lectotype of *T. tetrandra* M.B., LE; isolectotype L); *Callier* 92, Tauria Bachärnder bei Sudak 18.6.1896 (B, BG, FI, G, K, P, PRC, S, UPS, W, WU); *Golde* 1959, in maritimis pr. Simers 7.6.1904 (isotype of *T. hohenackeri* var. *taurica* Regel & Mlokoss., PRC); *Fiek*, Krim zwischen Felsen an der Meeresküste bei Alupka 5.6.1886 (PRC). TURKEY: *Balansa* 833, bords du Guzel-Dere a l'ouest de Mersina (Cilicie) 2.4.1855 (BM, FI, G, K, OXF, US, W); *Davis* & *Hedge* 29247, Prov. Tunceli, Tunceli Pulumur c. 30 miles from Tunceli river bed bank 7.6.1957 (E, K, W). GREECE: *Sintenis* 1374, Thessalia Kampaka ad fluvium prope Murgani Chan 1.6.1896 (B, BG, G, K, PRC, S, W, WU). CYPRUS: *Kotschy* 258, inter Limassol et Larnaca 9.4.1859 (G, L, P, S, W). SYRIA: *Mouterde* 11202, Nahr el Kelb 27.3.1955 (Herb. Mouterde).

Observations: (a) In *T. tetrandra* the bark is black or deep purplish-black, while it is brown or reddish-brown in *T. parviflora*. Racemes and flowers are larger in *T. tetrandra*. Sepals are not adnate as in *T. parviflora*. (b) The author has not seen any type of *T. taurica* Pall. (c) Georgi (1800) writes of *T. tetrandra* Pall.: 'Mit vier Staubfäden, übrigens noch unbeschrieben. In Taurien'. If this counts as a description, it has priority over that of Bieberstein.

Series 6. ARBUSCULAE Baum (ser. nov., see Appendix)

Leptobotryae Bge., Tentamen, 6 (1852), p.p.
Haplostemones Ndz., De Genere Tamarice, 5 (1895), pro subsectione, p.p.
Tetrascopae Arendt, Beitr. Tamarix, 35 (1926), pro subsectione, p. min. p.
Tetrastemones Arendt, *op. cit.*, 46, pro subsectione, p.p.
Tetrastemones Grex *Semiamplexicaules* Arendt, *op. cit.*, 48, p. min. p.
Laxiusculae Gorschk., in: Komarov, Fl. URSS, 15:299 (1949), p.p.
Parviflorae Gorschk., *op. cit.*, 301, nom. illegit.

Type species: *T. parviflora* DC.

Bracts slightly shorter to longer than pedicels. Flowers tetrameruso, tetrandrous Petals 2–2.25 mm. Racemes simple, 3–5 mm broad.

Included species: *T. androssowii* Litw., *T. kotschyi* Bge., *T. parviflora* DC.

39. **T. androssowii** Litw., Sched. Herb. Fl. Ross., 5:41(1905), 'androssowi' orth. mut. [Plate XXXIX]

Type: TURKMEN SSR: *Androssow* 1317, Turkestania, dominium Buchara, in locis arenoso-salsis pr. p. Farab, ad fl. Amu-Darya 1.3.1903 (holotype LE; isotypes G, PRC, S, WU).

Shrubby tree, 2–5 m tall, with brown to reddish-brown bark, younger parts glabrous. Leaves sessile with narrow base, 1–2 mm long. Inflorescences simple, usually vernal. Racemes 1.5–4 cm long, 3–5 mm broad, loosely flowered. Bracts slightly shorter than pedicels or equalling them, rarely the upper ones slightly longer, boat-shaped, diaphanous, acute, entire or more or less denticulate at apex. Pedicel as long as calyx. Calyx tetramerous. Sepals 1 mm long, more or less finely and irregularly denticulate to subentire, with scarious margins, the 2 outer keeled and somewhat connate at base, trullate-ovate, more or less acute, the inner 2 ovate, shorter and broader than the outer, not keeled, obtuse or occasionally acute. Corolla tetramerous, subpersistent. Petals 2–2.25 mm long, elliptic-ovate to obovate, entire or somewhat denticulate at apex. Androecium haplostemonous, of 4 antesepalous stamens; insertion of filaments peridiscal; disk synlophic to para-synlophic.

Flowering: April to May.

Habitat: River banks, wadis.

Distribution: Azerbaijan SSR, Turkmen SSR, Uzbek SSR, Iran, Afghanistan, Pakistan (see Map 39).

Selected specimens: AZERBAIJAN SSR: *Kanjagin*, Transcaussia in fauce ad ripam fl. Araxis 3.5.1934 (B, LE). TURKMEN SSR: *Sintenis* 93, regio Transcaspica, Aschabad ad ripas prope Babuschkinassad 21.4.1900 (B, E, K, L); *Litwinow* 994, pr. st. Tezhen 8.4.1898 (B, E, G, LE, WU); *Michelson* 120, near Farab 10.4.1910 (B, HUJ, LE, US); *Sokolov* 68 & 102, Kara-Kum Askhabad Railway Repetek Station 23.4.1952 (LE). UZBEK SSR: *Popov* 313, montes meridionales, Tian-Schan occidentalis in valle fl. Keles in loco Kaplanbek 27.4.1926 (E, K, LE, S). IRAN: *Rechinger, Aellen* & *Esfandiari* 2648, Ghom 19.4.1948 (Herb. Esfand., W). PAKISTAN: *Rechinger*, Quetta inter Sarawan et Quila Abdullah, 1400 m, 8.5.1965 (W).

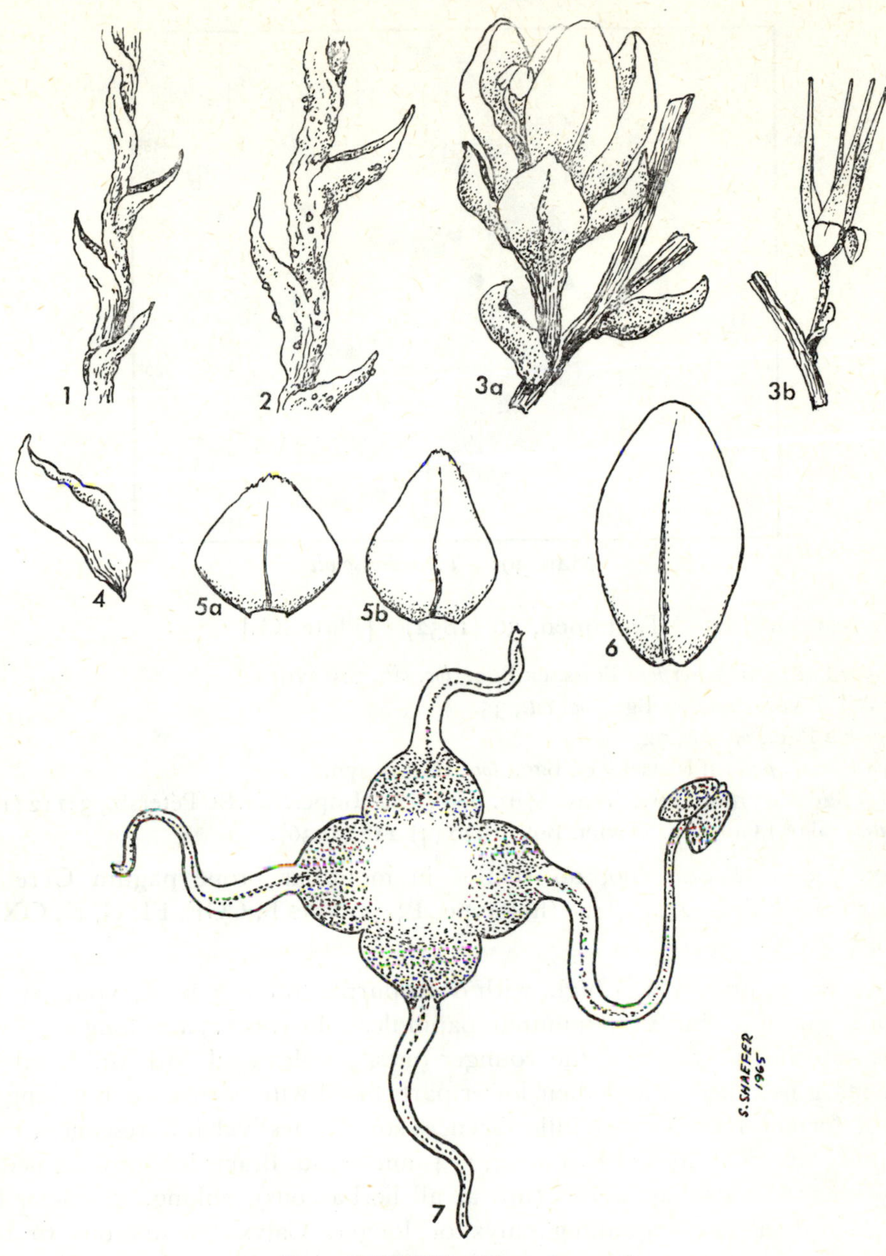

Plate XXXIX *T. androssowii*
1. Young twig (x 10); 2. id (x 10); 3a. Flower (x 10); 3b. Fruit (x 5); 4. Bract (x 20); 5a. Outer sepal (x 20); 5b. Inner sepal (x 20); 6. Petal (x 19); 7. Androecium (x 30).

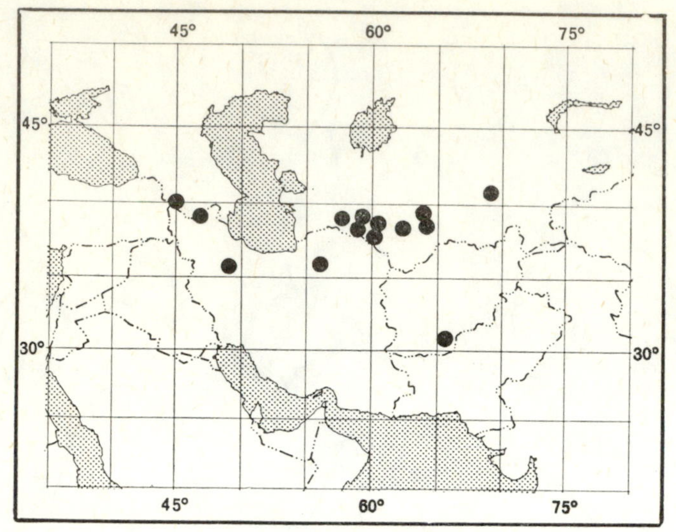

Map 39: *T. androssowii*

40. **T. kotschyi** Bge., Tentamen, 30 (1852) [Plate XL]

T. tetrandra Pall. var. *parviflora* Boiss. ex Bge., *loc. cit.*, pro syn.
T. laxa Willd. var. *araratica* Bge., *op. cit.*, 35.
T. leptopetala Bge., *op. cit.*, 72.
T. gallica L. var. *pallasii* Kotschy ex Bge., *loc. cit.*, pro syn.
T. kotschyi Bge. var. *rosea* Litw., Trav. Mus. Bot. Acad. Imper. Sci. St. Pétersb., 3:112 (1907).
T. araratica (Bge.) Gorschk., Soviet. Bot., 1936 (4):118 (1936).

Type: IRAN: *Kotschy* 100, ad rivulos in montibus prope pagum Gere inter Abuschir et Schiras 24.3.1842 (holotype P; isotypes B, CGE, FI, G, K, OXF, P, PRC, S, UPS, W).

Small tree or shrub c. 2 m high, with deep purple to black bark, younger parts practically glabrous, but very minutely papillulose. Leaves 1.5 mm long with white scarious margins, particularly the younger ones, amplexicaul with the basal parts of their margins coherent and their lower parts fused with cortex, each arising near centre of former leaf. Vernal inflorescences simple, aestival inflorescences loosely compound, rare. Racemes 1–3 cm long, 3–4 mm broad. Bracts longer than pedicels, entirely membranous-diaphanous (not at all herbaceous), oblong, spoon- or boat-shaped, blunt. Pedicel equalling calyx or longer. Calyx tetramerous to rarely pentamerous. Sepals 0.75–1 mm long (the outer 2), broadly trullate, entire, acute, keeled, the inner trullate-ovate, obtuse, eroded-denticulate, 0.5–0.75 mm long. Corolla tetramerous to rarely pentamerous, caducous. Petals narrowly elliptic to obovate-elliptic, 2–2.25 mm long. Androecium haplostemonous, of 4, rarely 5, antesepalous stamens; insertion of filaments peridiscal; disk synlophic to para-synlophic.

Flowering: March to May.

Habitat: Sandy soil, river banks, wadis.

Distribution: Turkmen SSR, Azerbaijan SSR, Iran, Afghanistan (see Map 40).

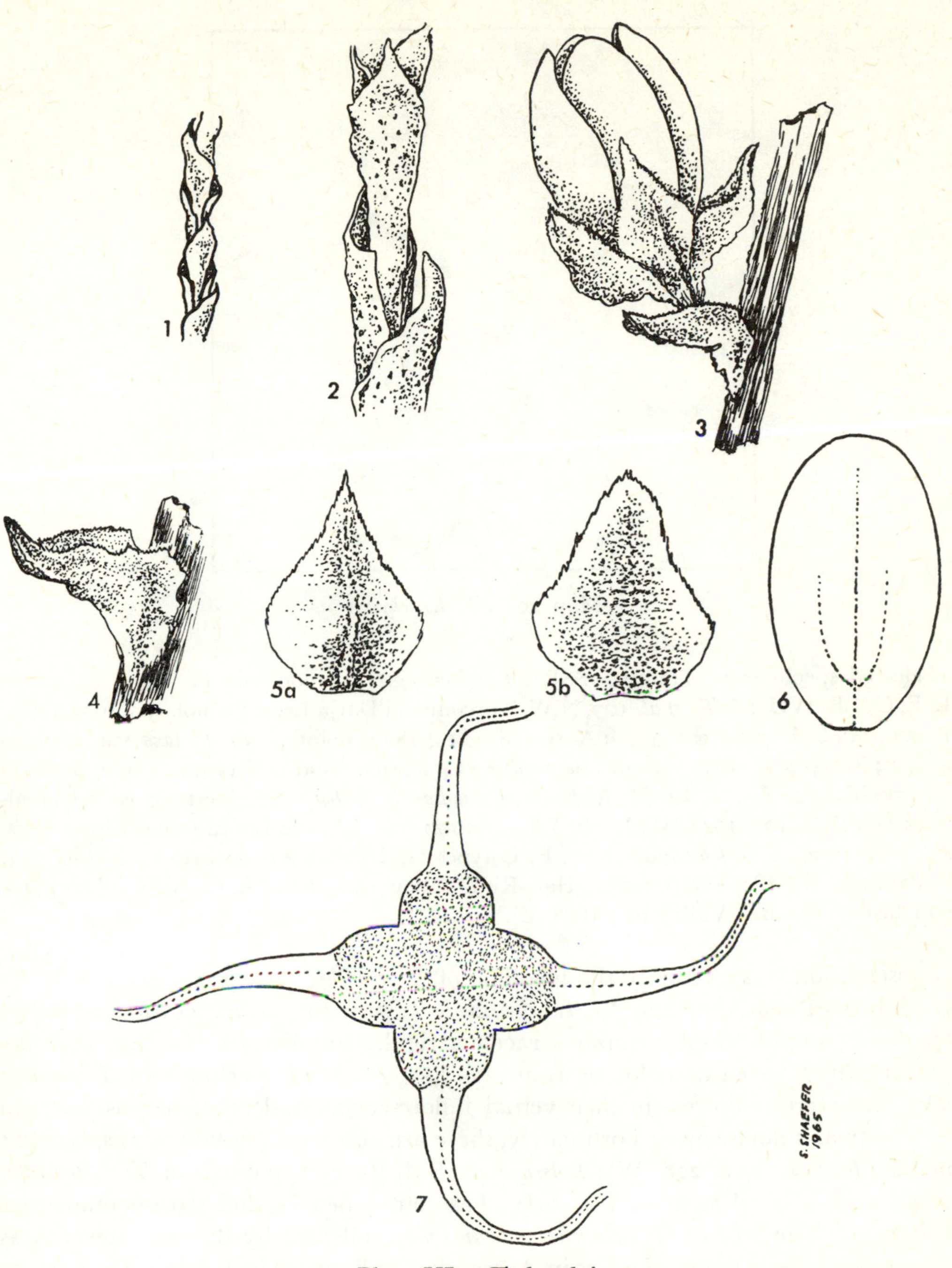

Plate XL *T. kotschyi*
1. Young twig (x 5); 2. id (x 10); 3. Flower (x 10); 4. Bract (x 20); 5a. Outer sepal (x 20); 5b. Inner sepal (x 20); 6. Petal (x 20); 7. Androecium (x 30).

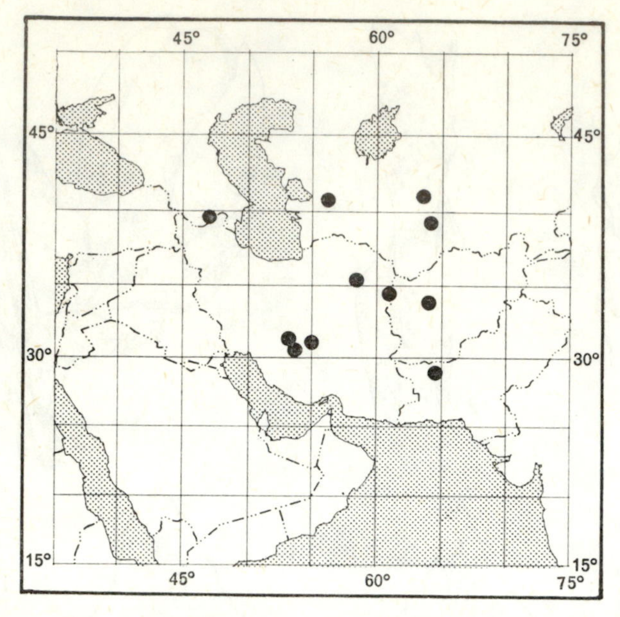

Map 40: *T. kotschyi*

Selected specimens: TURKMEN SSR: *Litwinow* 980, Turcomania pr. Czuli 23.4.1898 (B, E, G, LE, WU); *Rodin et al.* 107, N.W. Turcomania Darja Leka Canion 5.5.1953 (LE). AZERBAIJAN SSR: *Buhse* 114, am Araxes ufer 21.4.1847 (holotype of *T. laxa* var. *araratica* Bge., P; isotypes K, LE). IRAN: *Bornmüller* 3343, prov. Yesd et Kerman (inter Beyas et Kuschkuh) 15.4.1892 (B, PRC); *Rechinger, Esfandiari & Aellen* 3828, Kerman Kuh-Djabal-barez Djiroft 8–10.5.1948 (W, Herb. Esf.); *Kotschy* 728, in valle Loura mont. Elbrus Pers. bor. (holotype of *T. leptopetala* Bge., P; isotypes G, UPS). AFGHANISTAN: *Griffith* 954, Afghanistan (L, P); *Aitchison* 286, Hari-Rud Valley 19.4.1885 (E, G, K, S); *Lace* 3711, Baluchistan, Sarkhab Valley 19.4.1888 (E).

Observations: (a) Possibly the holotype of *T. kotschyi* Bge. is a specimen in W which bears Bunge's remark: '*T. kotschyi* m. a *T. parviflora, cretica*, et *tetrandra* differt jam foliis omnino amplexicaulibus racemis rigidis simplicibus.' In that case the specimen in P given as holotype is an isotype. (b) Most specimens of *T. kotschyi* have tetramerous flowers in their vernal inflorescences and pentamerous ones in their aestival inflorescences. Fortunately, there are also tetra-pentamerous specimens such as: *Ruttner* 227 & 228 (W); *Kotschy* 646 (BM, P); even the type of *T. leptopetala*, which is an aestival form of *T. kotschyi*, has tetra-, penta-, and tetra-pentamerous flowers. (c) The type of *T. laxa* var. *araratica* was collected by Buhse in lower Aras (Araxes), which is today entirely in Azerbaijan SSR or on its border with Iran, as we see from Buhse (*Itinerary*, in *Die Flora des Alburs*, etc. p. X, 1899).

41. **T. parviflora** DC., Prodr., 3:97 (1828) [Plate XLI]

T. laxa Willd. var. *subspicata* Ehrenb., Linnaea, 2:254 (1827).
T. petteri Presl ex Bge., Tentamen, 32 (1852), pro syn.
T. cretica Bge., *op. cit.*, 33.
T. parviflora DC. var. *cretica* (Bge.) Boiss., Fl. Or., 1:770 (1867).
T. rubella Batt., Bull. Soc. Bot. France, 44:256 (1907).
T. lucronensis Sennen & Elias, Bol. Soc. Iberica Ci. Nat., 27:133 (1928).

Type: TURKEY: *Castagne* 24, Constantinople 1822 (holotype G-DC.; isotype G-Boiss.).

Low tree or shrub, 2–3 m high, with brown to deep purple bark, entirely glabrous. Leaves sessile with narrow base, 2–2.5 mm long. Vernal inflorescences simple, aestival inflorescences rare. Racemes 1.5–4 cm long, 3–5 mm broad, densely flowered. Bracts triangular-acuminate, blunt, boat-shaped, almost completely diaphanous (not herbaceous), longer than pedicels. Pedicel much shorter than calyx. Calyx tetramerous. Sepals adnate at base, erose-denticulate, 1.25–1.5 mm long, the outer 2 trullate-ovate, acute and keeled, the inner ovate, obtuse. Corolla tetramerous, subpersistent. Petals parabolic, ovate, 2 mm long, subentire or faintly erose. Androecium haplostemonous, of 4 antesepalous stamens; insertion of filaments peridiscal; disk synlophic.

Flowering: March to June.

Habitat: Pebbly river banks, damp sites near river banks, well-watered forests on lime.

Distribution: Turkey, Greece, Crete, Albania, Yugoslavia, Italy, Corsica, Spain (introduced ?), Algeria (introduced ?) (see Map 41).

Selected specimens: TURKEY: *Davis* & *Polunin* 25677, prov. Antalya Aksu Cay between Antalya and Serik River bank 6.4.1956 (E, K); *Bornmüller* 13445, Bithynia ad Bilecik in declivitatibus vallis fluvii Kara-Su ad ripas 17–24.5.1929 (B, BM, E, G, S, W); *Aznavour* 414, entre la vallée de Domuzdere 26.9.1892 (E). GREECE: *Heldreich* 928, Attica ad sepes et margines vinetorum in valle Cephisi versus Phalerum, l. d. Moschato 23.4.1886 (B, G, K, PRC, W, WU); *Bornmüller* 3653, Macedonia Demir-Kapu ad fossas prope Makensentunnel 24.4.1918 (B, PRC). CRETE: *Dörfler* 185, Distr. Apokorono am Kiliaris Flusse bei Kalyves 29.3.1904 (B, WU); *Sieber*, *T. gallica* Armiro (holotype of *T. cretica* Bge. and of *T. laxa* var. *subspicata* Ehrenb., W; isotypes K, OXF, P, PRC). ALBANIA: *Junkmann*, Umgebung von Skodra, am linken Ufer der Buna (Bojna) 6.6.1916 (WU). YUGOSLAVIA: *Petter* 921, Dalmatia in Menge an den Ufern der Narenta bei Fort 'Opus (type of *T. petteri* Presl ex Bge., W); *Baenitz*, Istria Pola in Kaiserwalde nähe dem Jägerhäuse 17.5.1903 (B, US, W); *Baenitz* 9749, Dalmatia Gravosa Ufer der Ombla an Papi's Besitzung 10.4.1898 (B, PRC, US, WU). ITALY: *Minio*, estuario Veneto (Cavallino, Treforti, Lido, Torcello) 9.5.1940 (FI). CORSICA: *Reverchon* 6538, Evisa 24.4.1885 (B, BG, G, P, PRC, S, US, W, WU). SPAIN: *Elias* 4989, Logrono, Herrera, bosquet d'un ancien monastère 29.5.1923 (holotype of *T. lucronensis* Sennen & Elias, MA; isotypes G, W). ALGERIA: *Battandier* (13547), C. Batna .4.1892 (holotype of *T. rubella* Batt., RAB; isotypes FI, G); *Battandier*, cultivé des bouttures apportées de Batna 9.1.1906 (G, P, S).

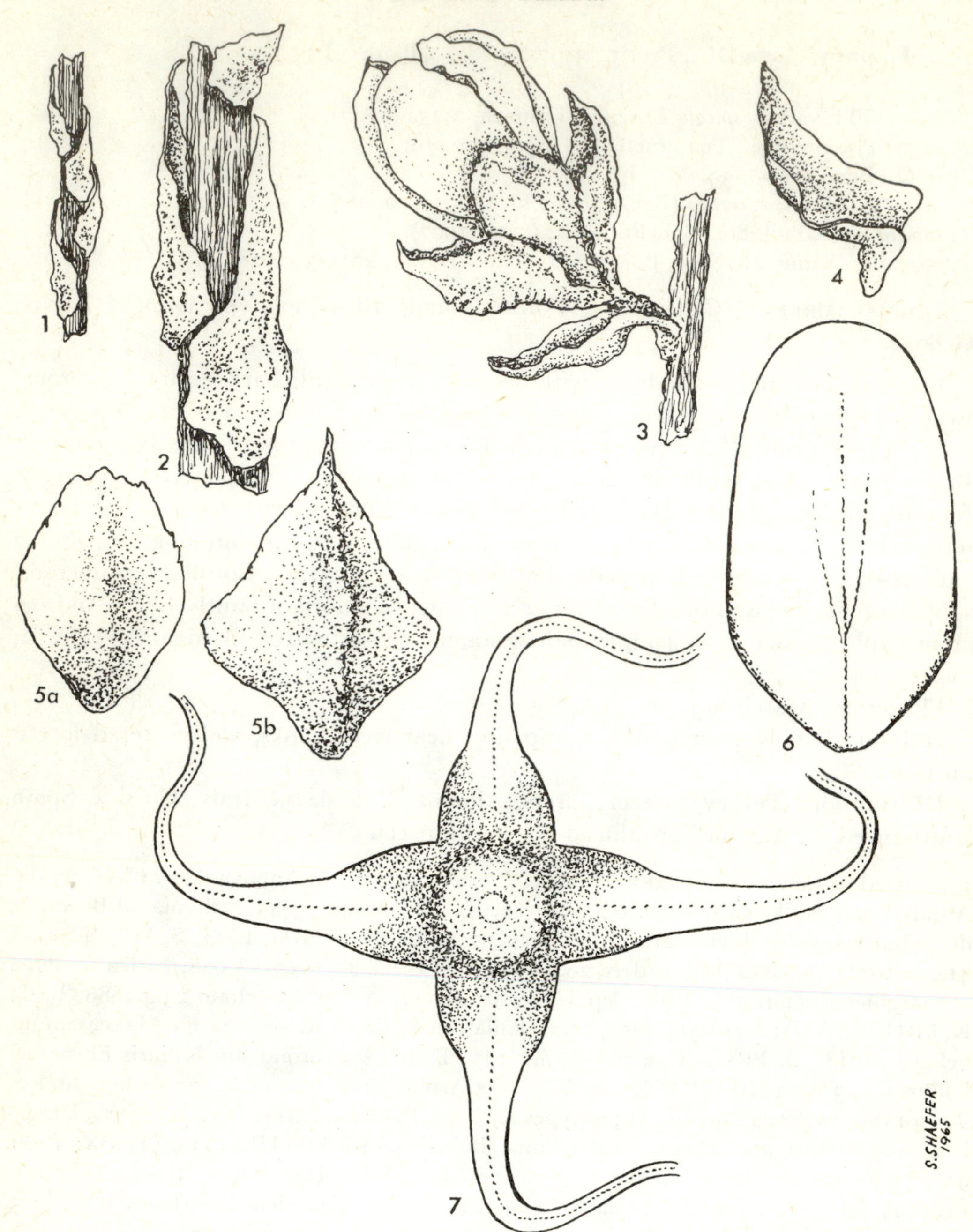

Plate XLI *T. parviflora*
1. Young twig (x 5); 2. id (x 10); 3. Flower (x 10); 4. Bract (x 10); 5a. Inner sepal (x 25); 5b. Outer sepal (x 25); 6. Petal (x 20); 7. Androecium (x 30).

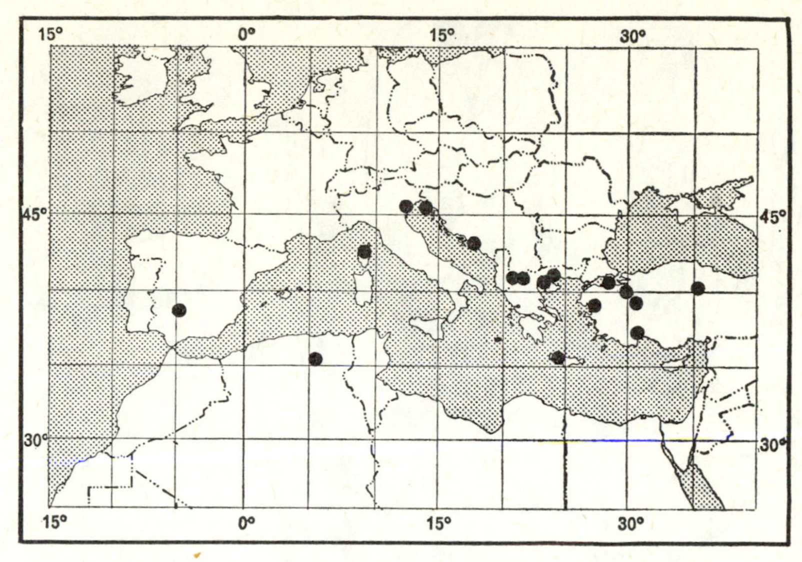

Map 41: *T. parviflora*

Observations: (a) Extensively cultivated as an ornamental plant and introduced into the New World. (b) See also observation (a) on *T. tetrandra.*

Series 7. FASCICULATAE Baum (ser. nov., see Appendix)

Leptobotryae Bge., Tentamen, 6 (1852), p.p.
Laxiusculae Gorschk., in: Komarov, Fl. URSS, 15:299 (1949), p.p.

Type species: *T. polystachya* Ledeb.

Bracts much shorter than to equalling pedicels. Flowers tetramerous, tetrandrous. Petals 2–2.25 mm long. Racemes 3–6 mm wide with the uppermost flowers contracted into an umbel-like, fasciculate apical cluster.

Included species: *T. litwinowii* Gorschk., *T. polystachya* Ledeb.

42. **T. litwinowii** Gorschk., Sched. Pl. Herb. Fl. SSSR, 10(61–64):24 (1936) [Plate XLII]

T. laxa Willd. var. *transcaucasica* Bge., Tentamen, 35 (1852).
T. laxa Willd. var. *parviflora* Litw., Sched. Herb. Fl. Ross., 5:79 (1905).

Type: TURKMEN SSR: *Androssow* 3048, in arenosis pr. pag. Farab 23.4.1903 (holotype LE; isotypes BM, G, K, S, US, W).

Small, low tree, 2–3 m high, with reddish-brown to brown bark, entirely glabrous. Leaves sessile with narrow base, 1.5–2.5 mm long. Vernal inflorescences simple and

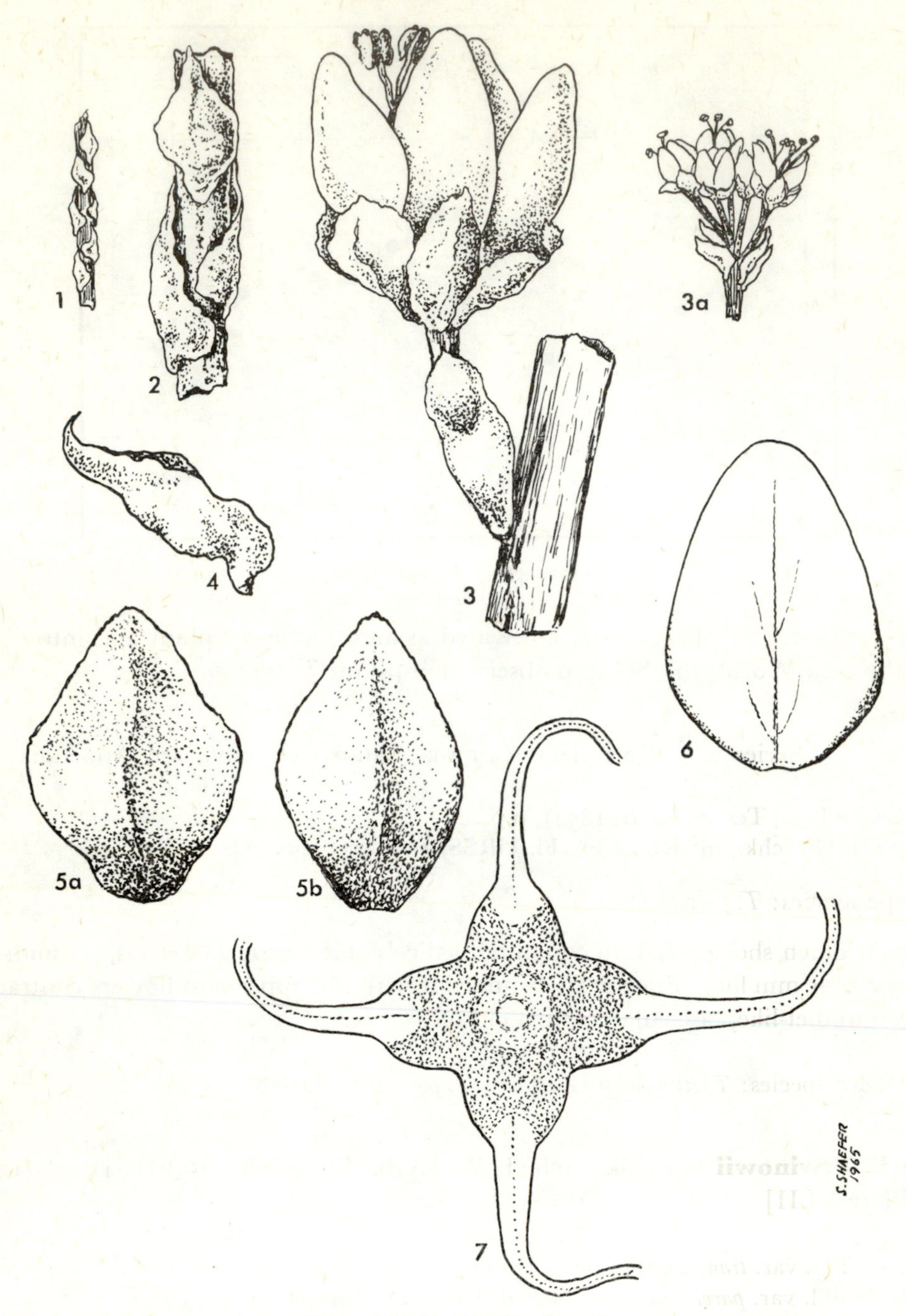

Plate XLII *T. litwinowii*

1. Young twig (x 5); 2. id (x 10); 3. Flower (x 10); 3a. Apex of raceme (x 5); 4. Bract (x 20); 5a. Inner sepal (x 20); 5b. Outer sepal (x 20); 6. Petal (x 20); 7. Androecium (x 30).

dense. Racemes 1–3 cm long, 3–5 mm broad, fasciculate at apex, i.e., with 2–4 flowers arising from almost the same point. Lower bracts often almost equalling pedicels, others or sometimes all scarcely longer than pedicels, oblong and blunt with upper part scarious, their lower half herbaceous. Pedicel as long as calyx. Calyx tetramerous. Sepals 1 mm long, connate at base, entire, the outer 2 ovate, acute, keeled, the inner ovate, obtuse and scarcely longer. Corolla tetramerous, caducous. Petals about 2 mm long, ovate, ovate-elliptic or elliptic. Androecium haplostemonous, of 4 antesepalous stamens; insertion of filaments peridiscal; disk para-synlophic to synlophic.

Flowering: April.

Habitat: River banks, sand.

Distribution: Turkmen SSR, Iran (see Map 42).

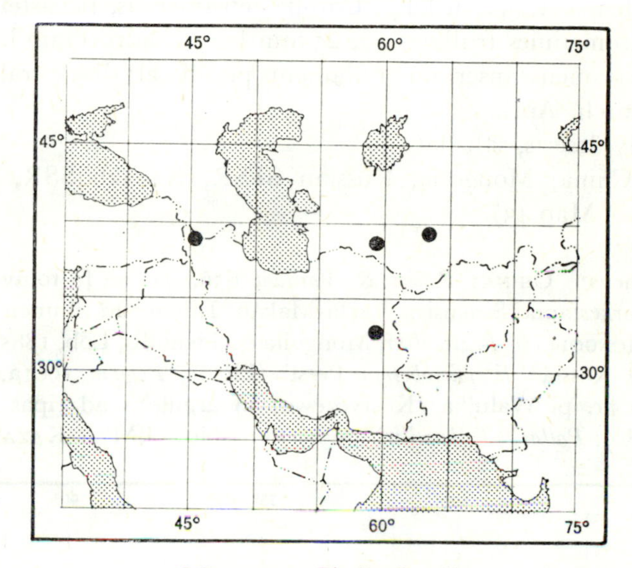

Map 42: *T. litwinowii*

Selected specimens: TURKMEN SSR: *Androssow* 1418, Turkestania Dominatio Buchara ad fluv. Amu-Darja pr. Farab 6.4.1900 (fl.), 1.5.1900 (fr.) (holotype of *T. laxa* var. *parviflora* Litw., LE; isotypes B, G, PRC, S, WU); *Litwinow* 993, P. Askhabad in salsis 2.4.1897 (B, E, G, LE, WU). IRAN: *Szowits* 137, Persia distr. Khoi provinc. Aderbeidshan (holotype of *T. laxa* var. *transcaucasica* Bge.; isotypes BM, G, K, L, S, US); *Bunge*, Persia pr. Ssertschah .3.1859 (FI, G, L, P, US).

Observation: Often confused with *T. androssowii* and *T. parviflora*. Immediately recognizable by the fasciculate tips of its racemes. It differs, of course, from the latter in many characteristics, as may be seen from the descriptions.

43. **T. polystachya** Ledeb., Fl. Ross., 2:133(1843) [Plate XLIII]

T. laxa Willd. var. *polystachya* (Ledeb.) Bge., Tentamen, 35 (1852).

Type: Kazakh SSR: *Karelin*, ad Novo Alexandrowsk 1.5.1840 (holotype P).

Small bushy tree or shrub with reddish-brown bark, entirely glabrous. Leaves sessile with narrow base, 1–2 mm long. Vernal inflorescences simple and dense, aestival inflorescences loosely compound. Racemes 1–2 cm long, 6 mm broad, with 3–6 flowers in umbel-like cluster at the tip of each raceme. Bracts usually much shorter than pedicels, spoon-shaped, blunt, diaphanous or scarious especially at apex. Pedicel longer than calyx. Calyx tetramerous. Sepals trullate-ovate, acute, more or less keeled, 1–1.25 mm long. Corolla tetramerous, persistent. Petals ovate to elliptic-ovate, sometimes trullate, 2–2.25 mm long. Androecium haplostemonous, of 4 antesepalous stamens; insertion of filaments peridiscal; disk paralophic.

Flowering: March, April.

Habitat: Sandy deserts, salt flats.

Distribution: China, Mongolia, Russian SFSR, Kazakh SSR, Tadzhik SSR, Turkmen SSR (see Map 43).

Selected specimens: China: *Pelliot* & *Vaillant* 636, To'ien fo touiy 22.4.1908 (P); *Hummel* 5559, Turkestania Sinensis, Takla-Makan Dilpar ad flumen Kantze Darya 11.5.1934 (S). Mongolia: *Przewalski*, Mongolia occidentalis, Lob, 1885 (K, P). Russian SFSR: *Lessing*, in Rossia '*T. microbotrya* Presl' (PRC); *Bogdan* 1961a, prov. Samara, distr. Nowo-Usen pr. p. Walujka (Kostyczewo) in argillosis ad ripas fluvii 26.5.1903 (G, LE, PRC, WU); *Pallas*, '*T. humilis*' e deserto caspico (BM). Kazakh SSR: *Karelin*

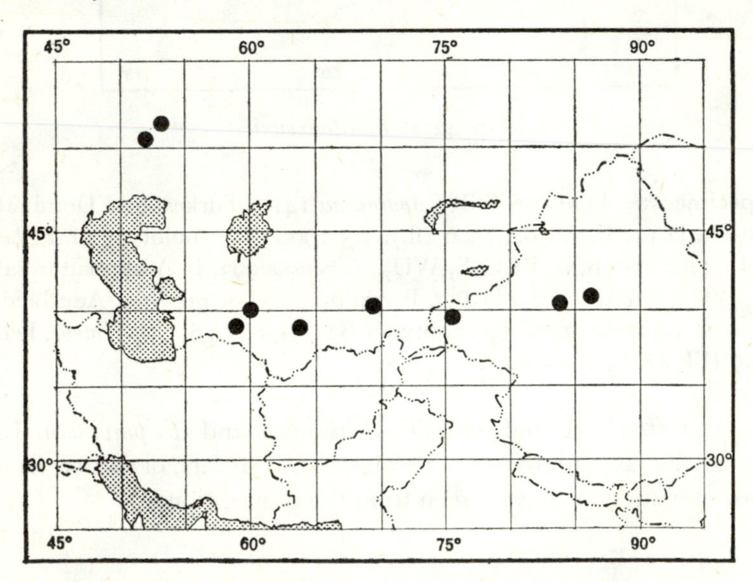

Map 43: *T. polystachya*

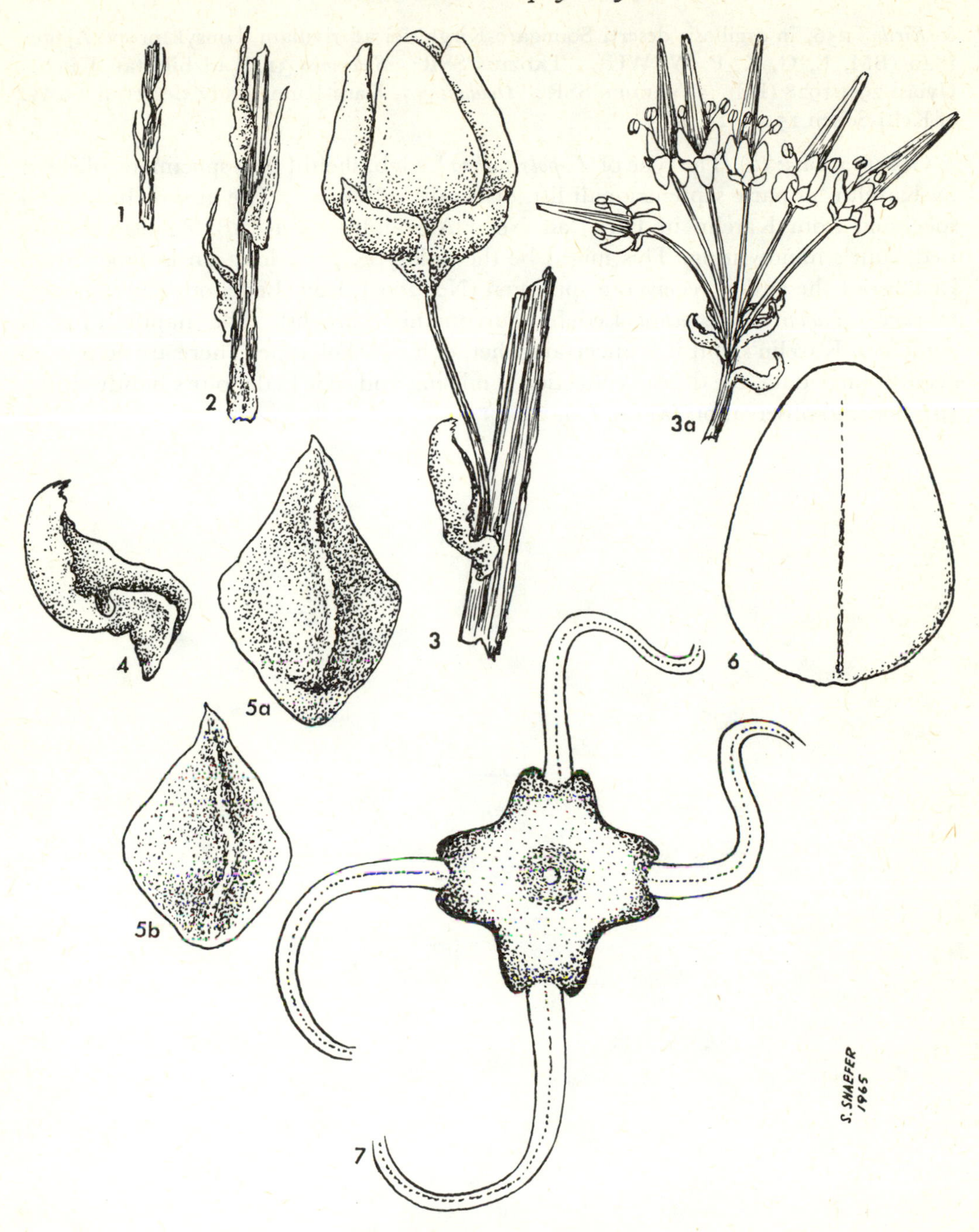

Plate XLIII *T. polystachya*
1. Young twig (x 5); 2. id (10); 3. Flower (x 10); 3a. Apex of raceme (x 5); 4. Bract (x 15); 5a. Outer sepal (x 35); 5b. Inner sepal (x 35). 6. Petal (x 20); 7. Androecium (x 30).

& *Kiriloff* 256, in argillosis deserti Soongaro-Kirghisici ad rivulam Tonsyk prope Ajagus 1840 (BM, E, G, K, P, W, WU). TADZHIK SSR: *Sidorenko* 31, Tadzhikistan Kischel Uyiass 29.4.1928 (LE). TURKMEN SSR: *Dubiansky* 7, Kara-Kum sandy deserts 8 km W. of Kelti-Schen 15.4.1925 (LE).

Observations: (a) The type of *T. polystachya* Ledeb. should be a specimen collected by Karelin 'in parte septentrionali litt. orientalis m. Caspii'. The author has seen a specimen from Karelin in Paris 'ad Novo Alexandrowsk 1.5.1840', also bearing Ledebour's handwriting. This might be the holotype, or at least an isotype. From Leningrad the author received a specimen (No. 299.4, from Ledebour's herbarium) marked '*Tamarix polystachya* Ledeb. Turcomania borealis'. The handwriting is similar to Karelin's, but it is uncertain whether it is *the* holotype. There are no precise geographical data, the date of collection is missing and so is Ledebour's handwriting. (b) See also observation (a) on *T. laxa*.

Section Three. POLYADENIA (Ehrenb.) Baum (Comb. Nov.)

Polyadenia Ehrenb., Linnaea, 2:253 (1827), pro subgenere.
Vernales Bge., Tentamen, 5 (1852), p. min. p.
Aestivales Bge., *op. cit.*, 6, p.p.
Sessiles Ndz., De Genere Tamarice, 4 (1895), pro subgenere, p. min. p.
Amplexicaules Ndz., *op. cit.*, 10, pro subgenere, p.p.
Obdiplandrae Ndz., *loc. cit.*
Parallelantherae Ndz., in: Engler, Nat. Pflanzenfam., ed. 2, 21:289 (1925), pro sectione Gen. *Myricariae*.
Primitivae Arendt, Beitr. Tamarix, 33 (1926), p. min. p.
Pentamerae Arendt, *op. cit.*, 36, p.p.
Trichaurus (Arn.) Gorschk., Not. Syst. Leningrad, 7:96 (1937), pro subgenere.

Type species (lectotype): *T. passerinoides* Del. ex Desv. designated by Endlicher, 1840.

Leaves usually amplexicaul or nearly so, except for a few species in which they are sessile with narrow bases or vaginate. Racemes 6–10 mm, rarely up to 15 mm broad. Bracts shorter to longer than pedicels. Flowers pentamerous. Petals 2.5–6(7) mm long, if 2–2.5 mm then leaves vaginate (*T. stricta*) or disk para-synlophic (*T. komarovii*). Androecium diplostemonous, partially diplostemonous, triplostemonous, partially triplostemonous or rarely (in *T. salina*) also haplostemonous; 5 antesepalous stamens with long filaments, antepetalous stamens (0–) 1–10 with shorter filaments. Except for *T. komarovii*, none of the species has discal lobes.

Series 8. ARABICAE Baum (ser. nov., see Appendix)

Pleiandrae Bge., Tentamen, 7 (1852), p. min. p.
Platybasis Ndz., in: Engler, Nat. Pflanzenfam., ed. 2, 21:287 (1925), pro subsectione, p. min. p.
Decastemones Arendt, Beitr. Tamarix, 44 (1926), pro subsectione, p. min. p.

Type species: *T. pycnocarpa* DC.

Androecium of some flowers in each raceme partially triplostemonous (i.e., of at least one pair of shorter and antepetalous stamens between 2 antesepalous longer ones) to triplostemonous.

Included species: *T. aucheriana* (Decne.) Baum, *T. pycnocarpa* DC.

44. **T. aucheriana** (Decne.) Baum (comb. nov.) [Plate XLIV]

Trichaurus aucherianus Decne. ex Walpers, Repert. Bot. Syst., 2:115 (1843).
Trichaurus aucherianus Decne., in: Jacquem., *Voy. Inde*, 4:59 (1844).
Trichaurus brachycarpus Decne. ex Bge., Tentamen, 77 (1852), pro syn. *T. passerinoides* Del. ex Desv.
T. passerinoides Del. ex Desv. var. *buhseana* Bge., *op. cit.*, 78.

Type: Iran and Afghanistan: *Aucher-Eloy* 4509, in deserto Sinus Persici (holotype P; isotypes FI, G, K, W).

Tall or bushy shrub, 1–2 m high, with deep purple to blackish-purple bark; younger parts more or less papillose all over. Leaves fleshy, strongly auriculate, amplexicaul, remote, 1–1.5 mm long. Inflorescences usually simple and aestival. Racemes c. 2–6 cm long, 8–10 mm broad. Bracts shorter than pedicels, slightly cordate-clasping. Pedicel as long as to longer than calyx. Calyx pentamerous. Sepals broadly ovate, subentire, 1.5–2 mm long, the outer 2 trullate-ovate, acute, keeled, connate at base, smaller than the inner, the inner broadly trullate-ovate, obtuse and nearly denticulate. Corolla pentamerous, subpersistent. Petals obovate to obovate-elliptic, equi- to inequilateral, 3.5–5 mm long. Androecium diplostemonous to partially triplostemonous, of 5 antesepalous stamens with filaments gradually enlarging towards base and 5–8 (usually 5 or 6) somewhat shorter antepetalous stamens; insertion of filaments peridiscal.

Flowering: October to April, sometimes also July and August.

Habitat: Salt pans, salty canals and deserts, littoral swamps.

Distribution: Turkmen SSR, Afghanistan, Iran, Iraq, Saudi Arabia, Kuwait (see Map 44).

Selected specimens: Turkmen SSR: *Sintenis* 1313a, Krasnowodsk in maritimis 31.10.1900 (B, BM, E, G, K, P, S, UC, WU); *Rodin et al.* 3332, S. W. Turkmenistan N. of colony Bugdoily 11.5.1952 (LE); *Russanov* 100, near railway station of Aidin 17.6.1930 (LE). Iran and Afghanistan: *Bornmüller* 3350, Persia austr. ditionis Buschir in salsis ad pagum Boradschum 16.12.1892 (B, G, K, OXF, PRC, WU); *Bunge*, Ufer der Hamun See 21.2.1859 (ad ripas lacus Hamun) (FI, G, K, P, US); *Kotschy* 158, Pers. austr. ad fontes petroleum fundentes prope Dalechi .3.1842 (BM, E, FI, G, OXF, P); *Aucher-Eloy*, orient (cum Decne. autographum et Bge. autograph.; holotype of *Trichaurus brachycarpus* Decne., W); *Buhse* 1210a, Husseinon 21.4.1849 (holotype of *T. passerinoides* var. *buhseana* Bge., G). Iraq: *Polunin* 5006, Kurdistan Kut rd., opposite Salman Pak, edge of shallow pool, irrigation seepage 31.3.1958 (E); *Evans*, fields between the Chahalah and Tigris, Amara, Mesopotamia 22.4.1918 (E); *Noë* 935, am Tigris von Bagdad nach Bassora .5.1851 (G, P). Saudi Arabia: *Munt*, Kwajem DII (E); *M. Zohary* 11522, Ras Al Khaima (Trucial Oman) estuary near sea 20.2.1943 (HUJ). Kuwait: *Wilison* 148, swamps of Aradjan also Ain-el-Abed 5.1.1935 (K).

Observations: (a) *T. aucheriana* is closely related to *T. pycnocarpa*, the former being common and the latter rare. In the latter the adnation of the 2 androecial whorls is incomplete, while in *T. aucheriana* it is complete. In *T. pycnocarpa* the bases of the antesepalous filaments are abruptly and much enlarged and are overlapped by the

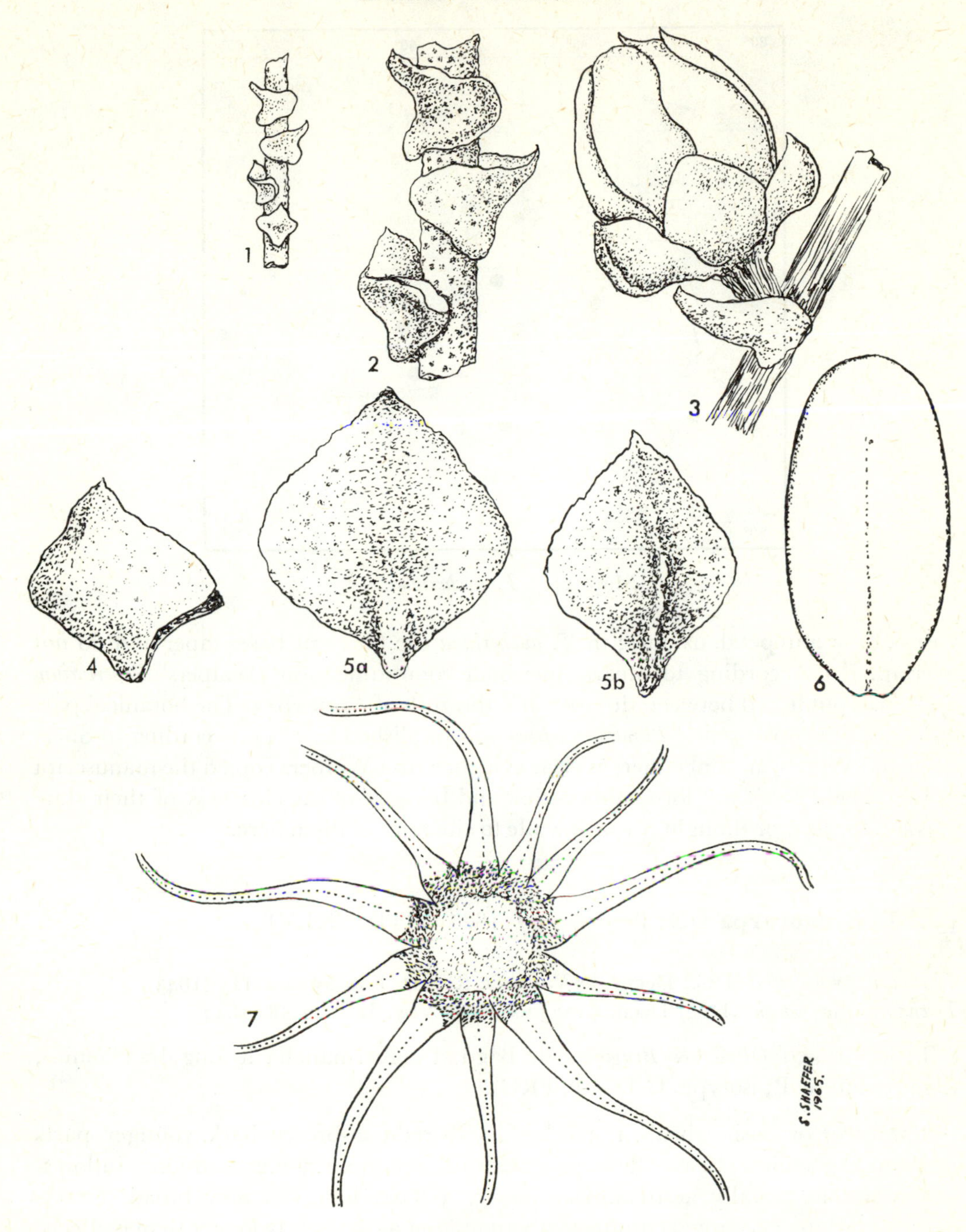

Plate XLIV *T. aucheriana*
1. Young twig (x 5); 2. id (x 10); 3. Flower (x 10); 4. Bract (x 20); 5a. Inner sepal (x 20); 5b. Outer sepal (x 20); 6. Petal (x 15); 7. Androecium (x 20).

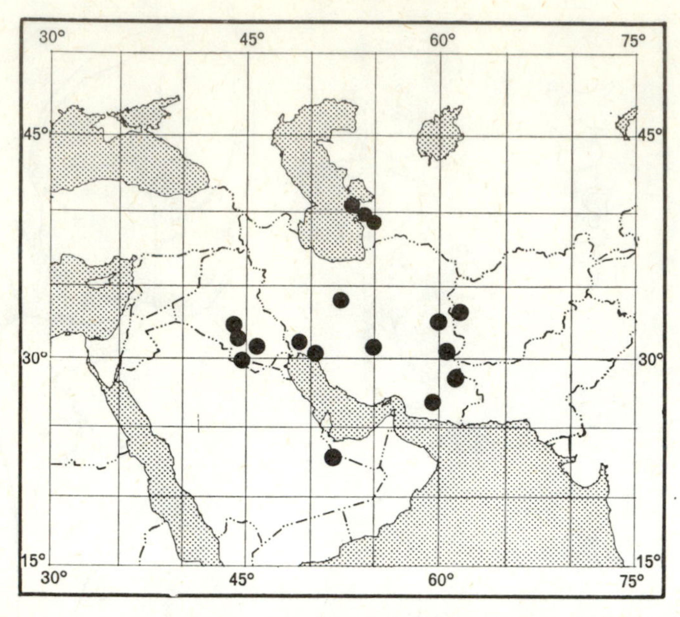

Map 44: *T. aucheriana*

bases of the antepetalous ones; in *T. aucheriana* the filament bases taper and do not overlap. (b) According to Stearn (personal communication) Walpers' *Repertorium* 2.115 was published between the 10th and the 30th of April 1843. The botanical part of Vol. 4 of Cambessedes' *Plantae Rariores* was published in 1844, according to Sherborn and Woodwar. Since there is clear evidence that Walpers copied the manuscript of Decaisne's *Trichaurus* for Cambessedes, and because of the closeness of their date of issue, the author thought it worth while to cite both of them here.

45. **T. pycnocarpa** DC., Prodr., 3:97 (1828) [Plate XLV]

Trichaurus pycnocarpus (DC.) Decne. ex Walpers, Repert. Bot. Syst., 2:115 (1843).
Trichaurus pycnocarpus (DC.) Decne., in: Jacquem., Voy. Inde, 4:58 (1844).

Type: IRAQ: *Olivier* & *Bruguière*, de Bagdad à Kermancha le long des chemins, 1822 (holotype P; isotypes G-DC, P, PRC).

Small tree or bushy shrub, 1–3 m high, with reddish-brown bark, younger parts glabrous to papillose. Leaves fleshy, amplexicaul, remote, 1.5–2.25 mm long. Inflorescences simple, usually aestival. Racemes c. 3–8 cm long, 10 mm broad. Bracts slightly cordate, clasping, acuminate, about as long as to slightly longer than pedicels. Pedicel equalling or shorter than calyx. Calyx pentamerous. Sepals 2–2.5 mm long, finely denticulate at apex, broadly ovate-elliptic, acute, almost always alike.[29] Corolla pentamerous, caducous. Petals usually inequilateral, broadly elliptic to

29 In *T. aucheriana* the outer differ from the inner in length and shape.

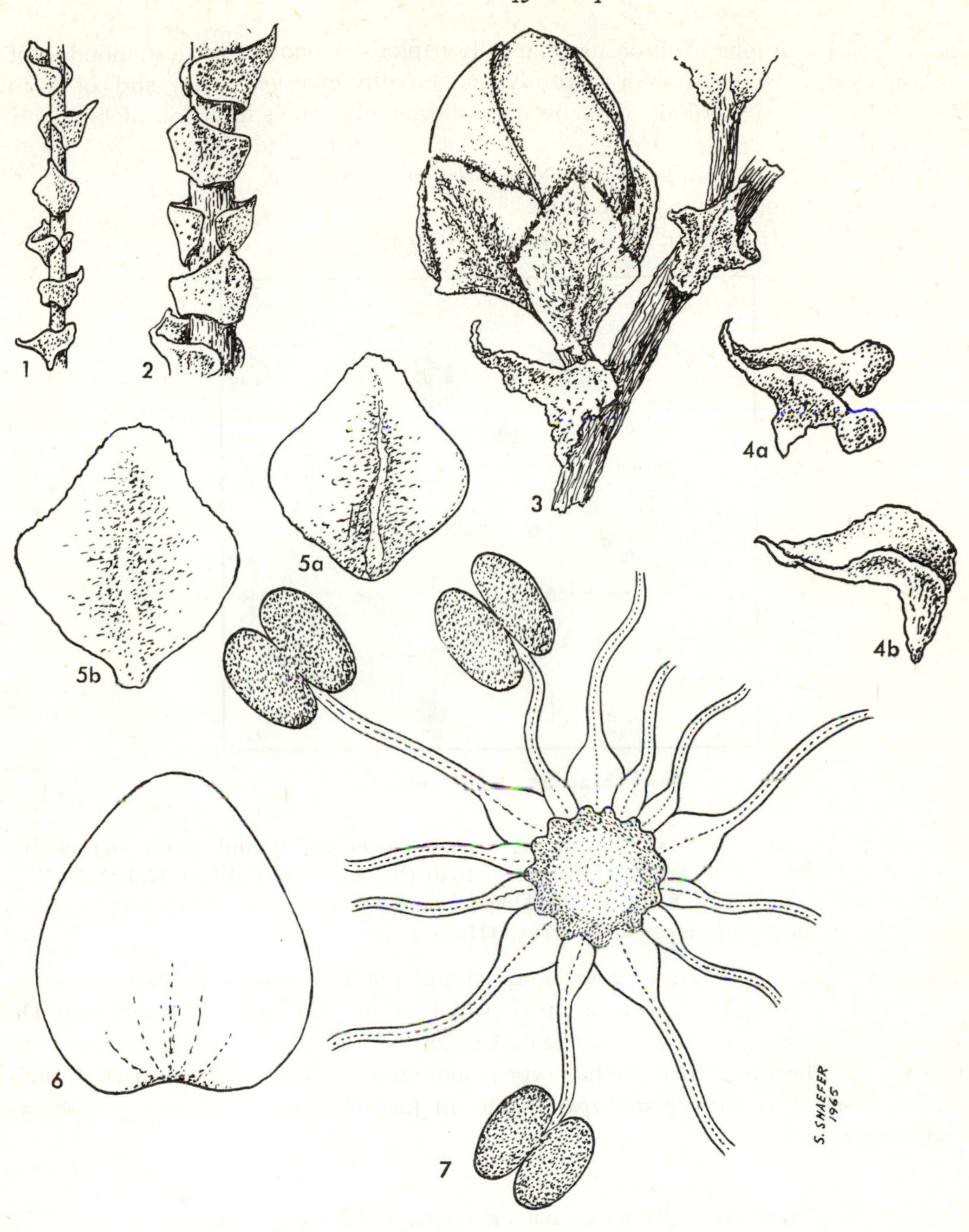

Plate XLV *T. pycnocarpa*
1. Young twig (x 5); 2. id (x 10); 3. Flower in bud (x 10); 4a. Lower bract (x 15); 4b. Upper bract (x 15); 5a. Outer sepal (x 15); 5b. Inner sepal (x 15); 6. Petal (x 10); 7. Androecium (x 20).

ovate, 4.5–5 mm long. Androecium partially triplostemonous to triplostemonous, of 5 antesepalous stamens with abruptly and broadly enlarged base, and of 5–10 (usually 8–9) antepetalous stamens with shorter filaments; insertion of filaments peridiscal.

Flowering: May, sometimes also November to March.

Habitat: Humid saline depressions.

Distribution: Iraq, Iran, Afghanistan (see Map 45).

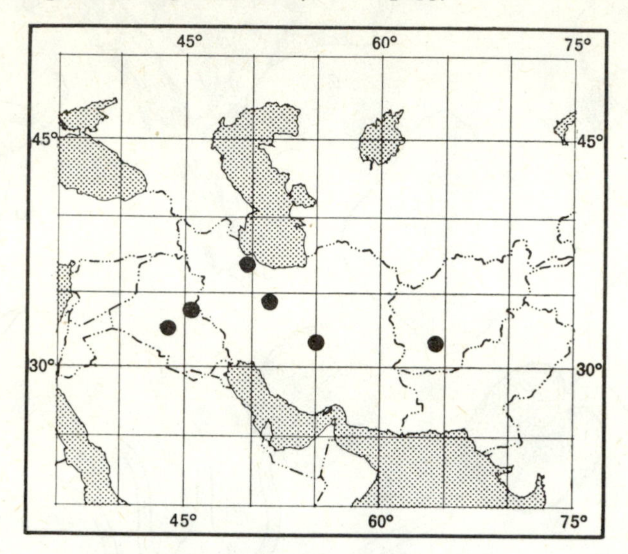

Map 45: *T. pycnocarpa*

Selected specimens: IRAQ: *Rechinger, Guest* & *Rawi* 75, humid saline depressions north of Dhithatha in Karbala Liwa 21.11.1956 (B, BM, DEL, E, FI, G, K, L, S, SMU). IRAN: *Gauba* 1439, Keredj Steppen Hügel bei Rawandeh 1.9.1938 (B). AFGHANISTAN: *Pabot* A 385, entre Marga et Lachkagar 13.4.1958 (Herb. Pabot).

Obervations: (a) There are four mounted sheets of *T. pycnocarpa* in Paris collected by Oliver & Bruguière. One of them, which is the holotype, is labelled in De Candolle's handwriting: '*T. pycnocarpa*'. A fragment from one of the four sheets is in G-DC. The sheet on which the holotype is mounted also bears a label in Decaisne's handwriting: '*Trichaurus pycnocarpus* Decne. in Jacquemont'. (b) See also observations on *T. aucheriana*.

Series 9. PLEIANDRAE Bge., Tentamen, 7 (1852)

Platybasis Ndz., in: Engler, Nat. Pflanzenfam., ed. 2, 21:287 (1925), pro subsectione, p. maj. p.
Stenobasis Ndz., *op. cit.*, 288, pro subsectione.
Indefinitae Arendt, Beitr. Tamarix, 33 (1926), pro subsectione, p.p.
Decastemones Arendt, *op. cit.*, 44, pro subsectione, p. maj. p.

Type species: *T. ericoides* Rottl.

Androecium partially diplostemonous to diplostemonous, rarely haplostemonous (i.e., of 5 antesepalous stamens), and then the uppermost flowers in the raceme have one or more additional antepetalous stamens.

Included species: *T. amplexicaulis* Ehrenb., *T. dubia* Bge., *T. ericoides* Rottl., *T. komarovii* Gorschk., *T. ladachenis* Baum, *T. macrocarpa* (Ehrenb.) Bge., *T. passerinoides* Del. ex Desv., *T. salina* Dyer, *T. stricta* Boiss.

46. **T. amplexicaulis** Ehrenb., Linnaea, 2:275 (1827) [Plate XLVI]

T. pauciovulata J. Gay ex Coss., Ann. Sci. Nat. Bot., IV, 4:283 (1855), nom. nud.
T. balansae J. Gay ex Coss., *loc. cit.*, nom. nud.
T. pauciovulata J. Gay ex Batt. & Trab., Fl. Alg., I(2):322 (1889).
T. balansae J. Gay ex Batt. & Trab., *loc. cit.*
T. balansae J. Gay ex Batt. & Trab. var. *longipes* Maire & Trab. ex Maire, Bull. Soc. Hist. Nat. Afr. N., 22:30 (1931).
T. trabutii Maire, *op. cit.*, 31 (1931).
T. balansae J. Gay ex Batt. & Trab. var. *microstyla* Maire, *op. cit.*, 29:403 (1938).

Bushy tree with reddish-brown bark; younger parts entirely glabrous. Leaves strongly amplexicaul, more or less imbricate,[30] dense, with a short point, fleshy, 1–1.5 mm long. Inflorescences usually densely compound[31] of racemes, pyramidal in outline, usually aestival. Racemes 3–7 cm long, 8–9 mm broad. Bracts not amplexicaul, those on lower part of the raceme shorter than pedicels, on the upper longer than pedicels. Pedicel slightly shorter than to equalling calyx. Calyx pentamerous. Sepals trullate-ovate, more or less acute, entire, 1.75–2 mm long, the outer 2 acute, keeled, connate at base, the inner obtuse, longer and larger. Corolla pentamerous. Petals broadly ovate-elliptic to broadly ovate, usually keeled at base, 3.25–4 mm long, subpersistent. Androecium[32] diplostemonous, of 5 antesepalous stamens with filaments much enlarged at their base into disk lobes and of 5 antepetalous stemens (or rarely fewer by abortion) with shorter filaments, with less to not at all enlarged bases. Gynaecium pauciovulate and smaller than in the allied species.[33]

Flowering: August to May, but usually January to May.

Habitat: Salt marshes, salt flats in deserts, salty banks of rivers, sandy depressions.

Distribution: Morocco, Algeria, Tunisia, Egypt (see Map 46).

Selected specimens: MOROCCO: *Maire* 253, in montibus Emmidir (Mouydir): Haci-el-Kheneg, in alveo amnis 310 m 28.2.1928 (holotype of *T. trabutii* Maire, P; isotypes FI, RAB, US); *Sauvage*, Dahiet-el-Moshrane 5.11.1948 (RAB); *Joly* 25, au sud du Kemkan

30 In *T. aucheriana* and *T. pycnocarpa* the leaves are remote.
31 In *T. passerinoides* not compound.
32 Androecium of 2 completely adnate whorls.
33 Allied species *T. aucheriana, macrocarpa, passerinoides, pycnocarpa.*

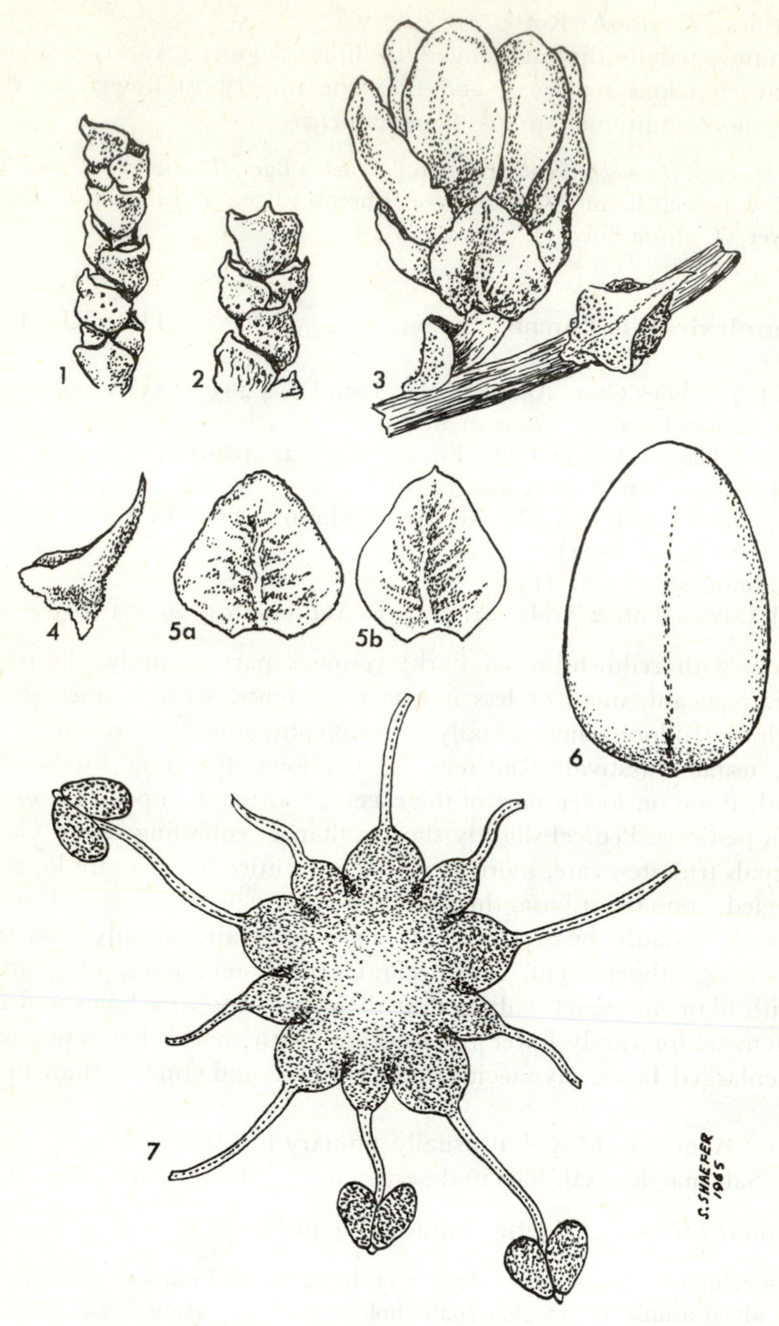

Plate XLVI *T. amplexicaulis*
1–2. Young twigs (x 7); 3. Flower (x 7); 4. Bract (x 10);
5a. Inner sepal (x 10); 5b. Outer sepal (x 10); 6. Petal (x 10);
7. Androecium (x 18).

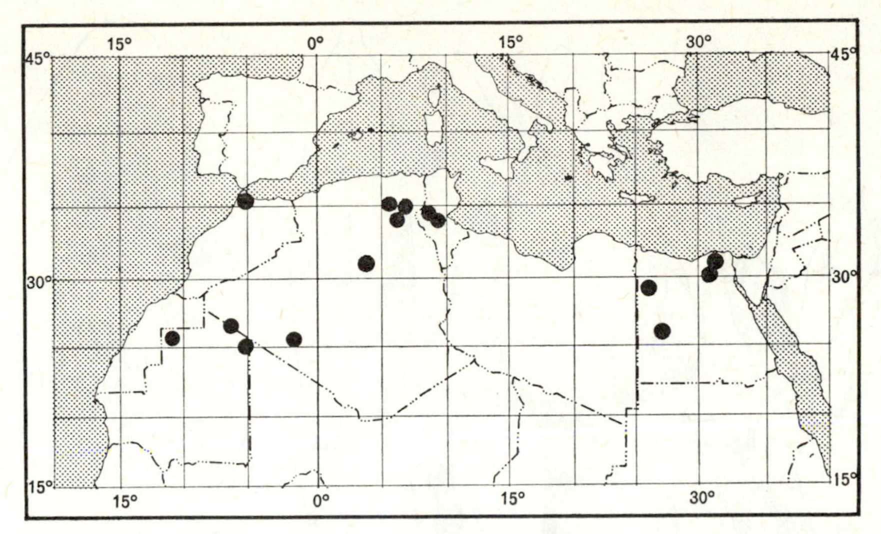

Map 46: *T. amplexicaulis*

Zzougouerm Erg El Hamra .2.1949 (RAB); *Rung* & *Sauvage* 42, Lennour (Sahara) O. Bou Dhir au S.W. de Metlani 9.11.1942 (RAB); *Adam* 13018, Mauritanie Houakchat 13.2.1957 (P); *Maire* 248, Mougdir, Tigelgemin, près de l'Agelman inférieur 440 m (holotype of *T. balansae* var. *longipes* Maire & Trab., P). ALGERIA: *Balansa*, forêt de Saada à 25 km au SSE de Biskra 11.4.1853 (holotype of *T. balansae* J. Gay, K); *Balansa* 987, bords des ruisseaux salés aux environs de la Fontaine Chaude près Biskra 20.3.1853 (BM, E, G, K, P, W); *Jamin* 240, ibid. 20.4.1852 (holotype of *T. pauciovulata* J. Gay, K; isotypes FI, P); *Maire*, in arenosis salsis ditionis Oued Rhir prope El Arfiane 19.3.1933 (FI, G, S, UC); *Reboud*, Sahara Eirgla de Khafile au nord de Ngoussa (Mzab) 11.1.1857 (G, P); *Murat* 1426, Sahara occidental méridional, région du Tiris, sebkha d'Oum Dferat (holotype of *T. balansae* var. *microstyla* Maire, P). TUNISIA: *Pitard*, Fafsa, in aridis desert. .4.1909 (B, BM, E, G). EGYPT: *Davis* 8121, Gebel Abu Roash, sandy depression on crystalline limestone (rare) 22.1.1945 (E).

Observation: The author's interpretation of *T. amplexicaulis* is based strictly on the original description and that of Bunge (which was also based on the original material). The author was not able to see the type, which was lost in Berlin during World War II.

47. **T. dubia** Bge., Tentamen, 18 (1852) [Plate XLVII]

Type: IRAN: *Buhse* 1349, 2.c!, Persia pr. Yesd .4.1849 (holotype P; isotypes G, LE).

Shrubby tree or low shrub with brown to purple bark, entirely glabrous. Leaves sessile with narrow base, 1–4 mm long. Inflorescences usually simple. Racemes 2–4 cm long, 7–9 mm broad. Bracts usually longer than pedicels. Pedicels as long as calyx. Calyx pentamerous. Sepals 1.5–1.75 mm long, ovate, more or less erose-denticulate, connate at base, the outer ones acute and slightly keeled. Corolla

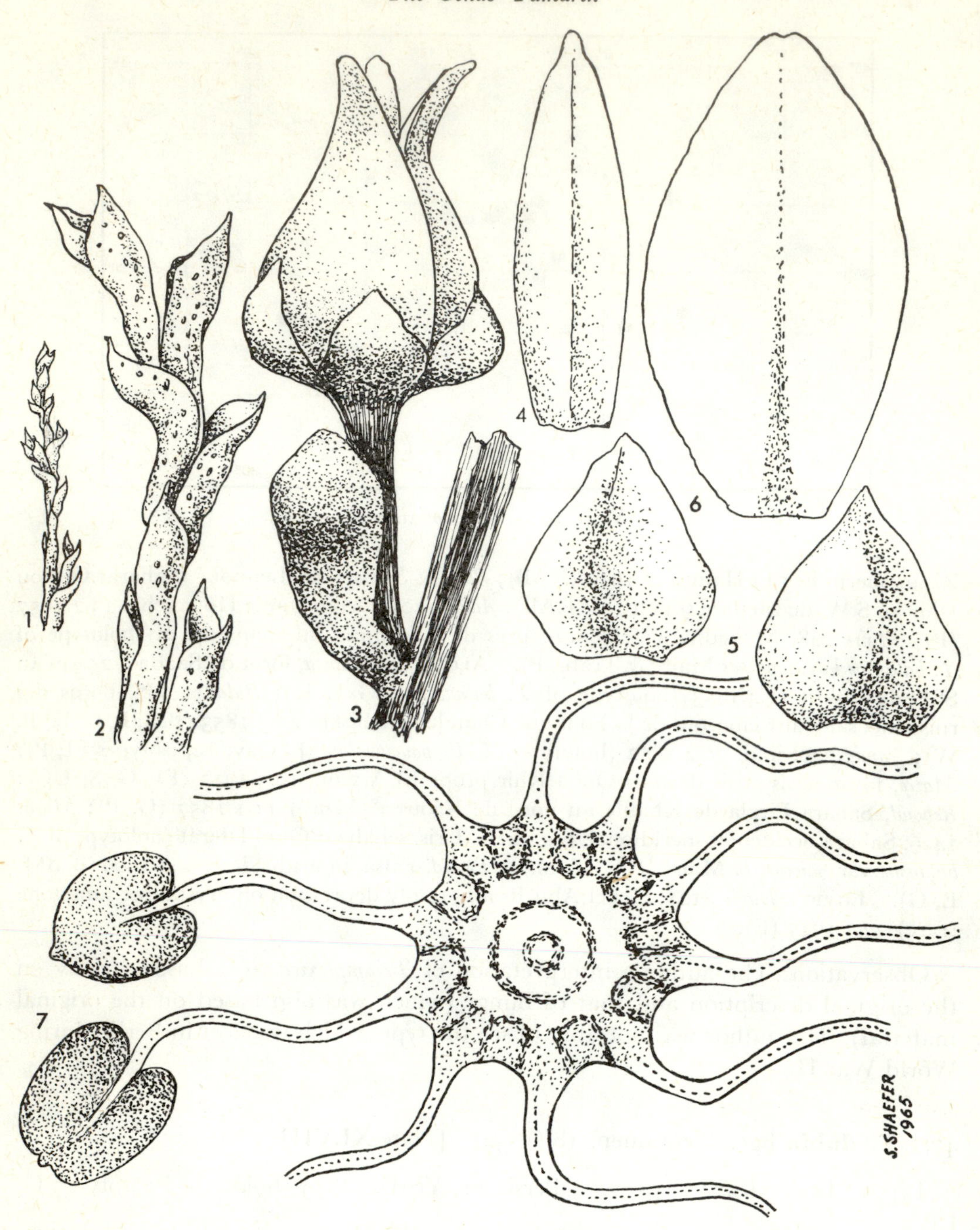

Plate XLVII *T. dubia*
1. Young twig (x 5); 2. id (x 20); 3. Fruit on elongated pedicel (x 20); 4. Bract (x 15); 5a. Outer sepal (x 15); 5b. Inner sepal (x 15); 6. Petal (x 15); 7. Androecium (x 20).

pentamerous, caducous. Petals broadly elliptic, 3–3.5 mm long. Androecium diplostemonous, of 5 antesepalous stamens and of 5 antepetalous stamens with slightly shorter filaments; insertion of filaments peridiscal; disk paralophic.

Flowering: March to April.

Habitat: Humid places in deserts.

Distribution: Iran (rare), Afghanistan (rare) (see Map 47).

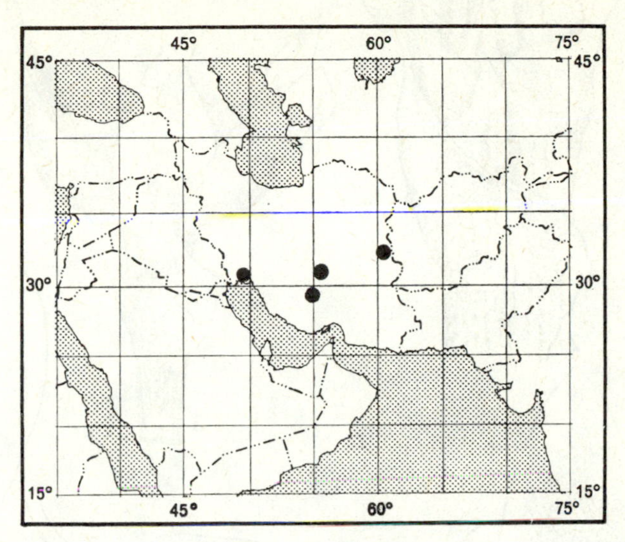

Map 47: *T. dubia*

Selected specimens: IRAN: *Bunge*, Saerdaki prope Ssertschah 10.3.1859 (FI, G, L, P); *Rechinger, Aellen* & *Esfandiari* 2886, Yazd Anar to Batram-Abad 23.4.1948 (W, Herb. Esf.). AFGHANISTAN: *Khoie* 4211, Jiza 10.4.1949 (C, W).

48. **T. ericoides** Rottl., Ges. Naturf. Freunde Berlin Neue Schr., 4:214 (1803) [Plate XLVIII]

T. mucronata Smith, in: Rees, Cyclop., 35(1):70, No. 5 (1817).
Myricaria vaginata Desv., Ann. Sci. Nat. Bot., I, 4:350 (1824).
T. tenacissima Buch.-Ham. ex Wall., Cat. 131, No. 3757 (1831), nom. nud.
Trichaurus ericoides (Rottl.) Arn., in: Wight & Walker-Arn., Prodr. Fl. Ind. Or., 1:40 (1834), p.p. excl. descr. (pars altera = *T. ladachensis*).
T. tenacissima Buch.-Ham. ex Walp., Repert. Bot. Syst., 2:115 (1843), pro syn.
Trichaurus vaginata (Desv.) Walp., *loc. cit.*

Type: INDIA: *Rottler*, '*Tamarix ericoides* Nob. May or April 25, 1801 ab amico Heyne' (holotype K; isotypes G, GL, HBG).

Shrub or bushy shrub with black bark, glabrous to slightly papillose. Leaves waginate in lower part, clasping in upper, long acuminate, 3–7 mm long, sometimes with minutely papillose margins, with inconspicuous salt glands. Inflorescences

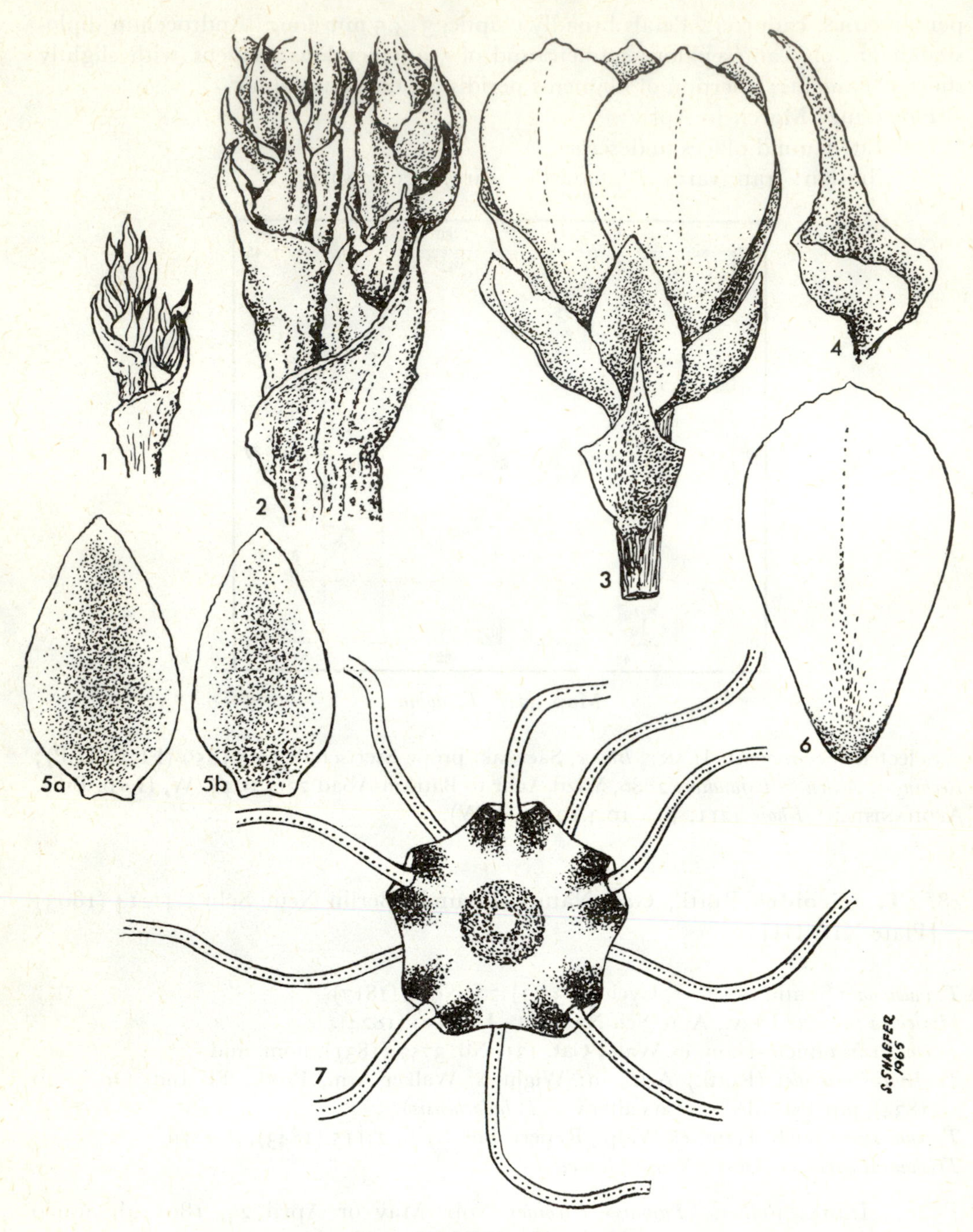

Plate XLVIII *T. ericoides*

1. Young twig (x 5); 2. id (x 10); 3. Flower (x 10); 4. Bract (x 20); 5a. Inner sepal (x 20); 5b. Outer sepal (x 20); 6. Petal (x 10); 7. Androecium (x 15).

of simple vernal or aestival racemes. Racemes 6–15 cm long, occasionally up to 20 cm, about 10 mm broad. Bracts semi-amplexicaul, triangular, acute, with somewhat denticulate margins, about as long as or longer than pedicels. Pedicel shorter than calyx, pentamerous, urceolate. Sepals 2–2.5 mm long, ovate, minutely and irregularly denticulate, especially at apex, the outer 2 acute, the inner obtuse. Corolla pentamerous, persistent. Petals 4–5 mm long, occasionally up to 7 mm, elliptic to usually obovate, densely and very minutely erose-denticulate. Androecium diplostemonous, of 5 antesepalous stamens inserted hypodiscally and of 5 shorter, antepetalous stamens inserted peridiscally; the latter are more or less embedded in fleshy lobe-like structures.

Flowering: September to February.

Habitat: River beds and banks.

Distribution: Endemic to India (see Map 48).

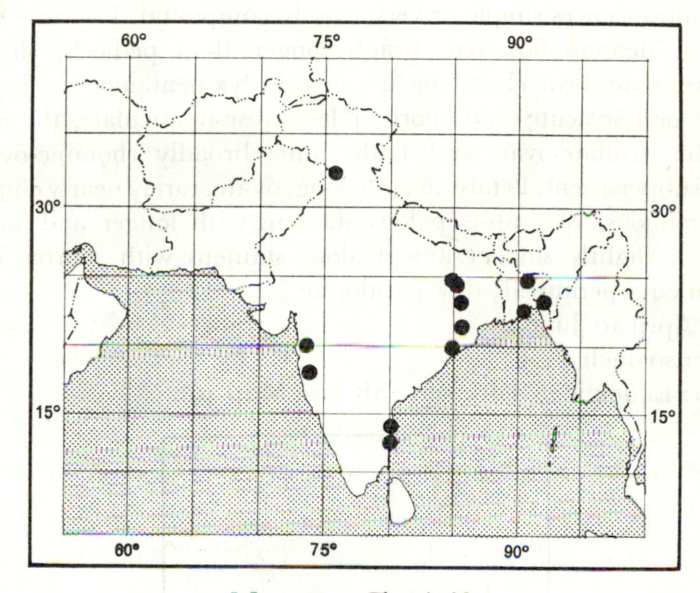

Map 48: *T. ericoides*

Selected specimens: INDIA: *Desvaux* (autograph), 'India orientalis Myricaria vaginata' (holotype of *M. vaginata* Desv., P); *Wright* 951, peninsula Ind. orientalis (with Arnott's autograph '*Trichaurus ericoides* Arn.', E, G, GL, P); *T. mucronata* Smith, Linnaean Herb No. 383–4 (LINN); *Wallich* 3757, e Banka (holotype of *T. tenacissima* Buch.-Ham., K; isotype BM); *Stocks*, Ind. or. Boucan (CGE, E, FI, G, K, L, OXF, P, U, W); *Campbell*, Teckamutta (E, G, GL); *Schlagintweit* 10682, Panjab Kohat to Kalabagh on the left side of the Indus 10–14.2.1857 (BM, E, US); id. No. 11963 (HBG); *Perrottet* 363, Nilgerries 1837–1838 (G, K, W); *Hook f.* & *Thomson*, Behar Reg. Trop. (BM, K, L, P, UPS, W).

Observations; (a) The holotype of *T. ericoides* is mounted on one sheet together with another small specimen which also belongs to the type collection: '*Tamarix ericoides* Nob. Seartu Jan. 2. 1800 in Indiae torrentibus'. (b) It is possible that an isotype of *T. ericoides* is in HBG, since its label bears the name '*T. ericoides* Nob.' in the same handwritting. (c) The type of *Trichaurus ericoides* (Rotel.) Arn. is in GL. Britten (1885) states that 'His [Arnott's] large collections subsequently became the property of the University of Glasgow'. (d) Arnott (*loc. cit.*) says that *Trichaurus* (*ericoides*) 'differs from Tamarix by the beaked seed'. The author has not observed this characteristic.

49. **T. komarovii** Gorschk., Not. Syst. Leningrad, 7:96 (1937) [Plate XLIX]

Type: Turkmen SSR: *Sedmuradov* 1144, Transcaspia Krasnovodsky distr., near railway station of Aidin 3.4.1910 (holotype LE).

Bushy tree or shrub with brown to reddish-brown bark, younger parts papillose to sparsely papillulose. Leaves amplexicaul, fleshy, remote, imbricate when young, 1 mm long. Inflorescences simple or very loosely compound. Racemes 1–5 cm long, 6–7 mm broad, densely flowered. Bracts longer than pedicels, slightly keeled, triangular-acuminate. Pedicels as long as calyx. Calyx pentamerous. Sepals 1.25 mm long, all more or less acute and more or less erose-denticulate, the outer 2 more acute, somewhat trullate-ovate keeled, the inner broadly rhombic-ovate. Corolla pentamerous, subpersistent. Petals 2.25 mm long, ovate, rarely nearly elliptic. Androecium diplostemonous, of 5 antesepalous stamens with longer and broader-based filaments and 5 slightly smaller antepetalous stamens with shorter filaments; insertion of filaments peridiscal; disk paralophic.

Flowering: April to June.

Habitat: On solonchak.

Distribution: Endemic to Turkmen SSR (see Map 49).

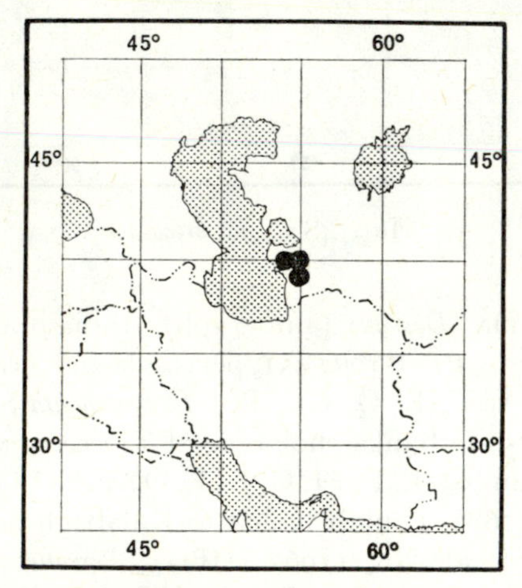

Map 49: *T. komarovii*

Selected specimens: Turkmen SSR: *Androssov* 22, Akhcha-Kuima 17.6.1910 (LE); *Russanov* 100, Djabeck near railway station, in solonchak 17.6.1930 (LE)

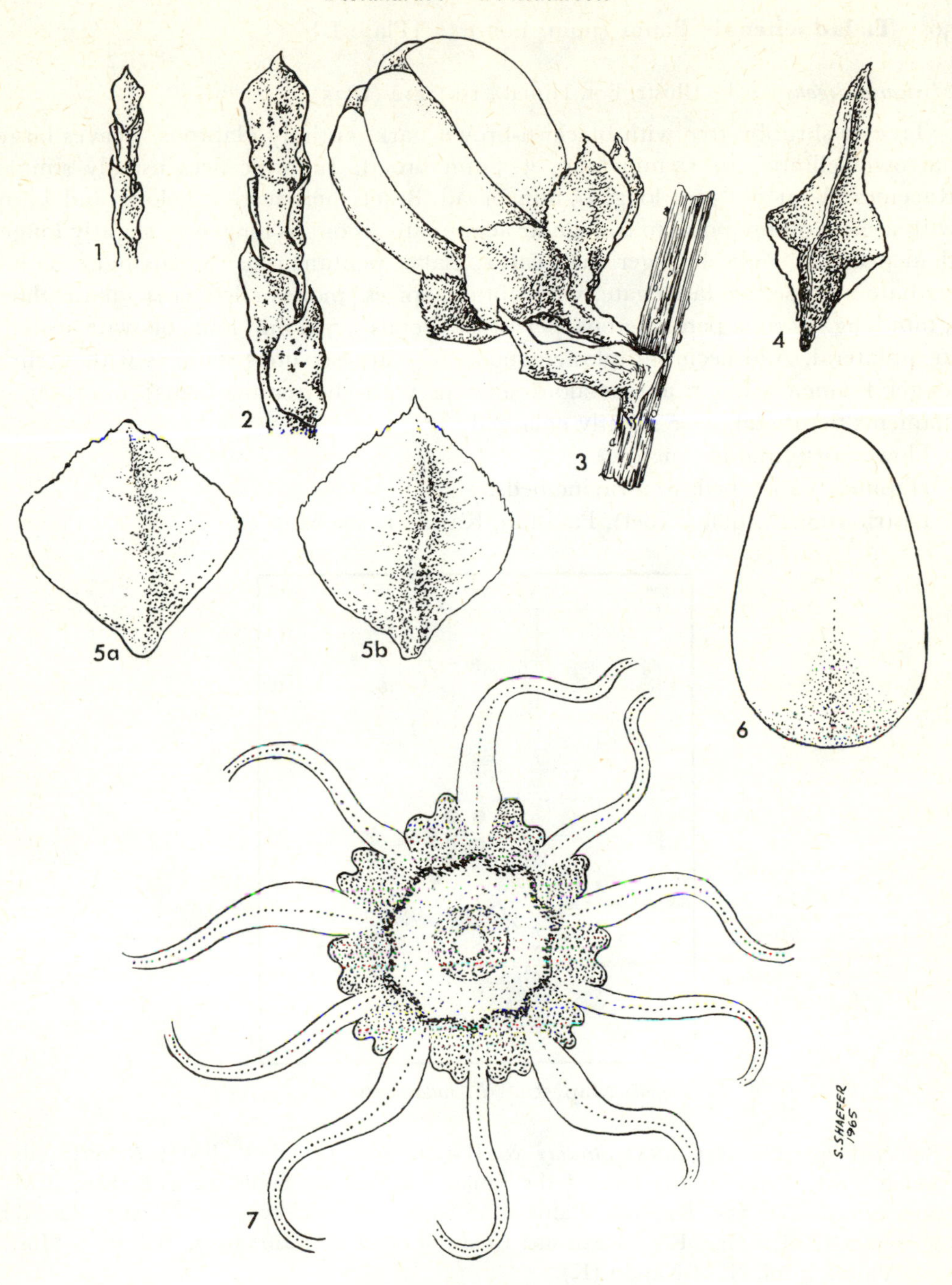

Plate XLIX *T. komarovii*

1. Young twig (x 5); 2. id (x 10); 3. Flower (x 10); 4. Bract (x 20); 5a. Inner sepal (x 20); 5b. Outer sepal (x 20); 6. Petal (x 20); 7. Androecium (x 30).

50. **T. ladachensis** Baum (nom. nov.)[34] [Plate L]

Myricaria elegans Royle, Illustr. Bot. Himal., 1(6):214 (1835).

Tree or shrubby tree with blackish-brown bark, entirely glabrous. Leaves large, narrowly elliptic, 10–15 mm long, 3–5 mm broad. Inflorescences usually simple. Racemes about 7–15 cm long, 15 mm broad. Bracts herbaceous, oblong and blunt with a short obtuse point to triangular acuminate, about as long as to slightly longer than pedicels. Pedicel longer than calyx. Calyx pentamerous. Sepals more or less connate at base, trullate-ovate, with obtuse apices, more or less erose-denticulate, 3 mm long. Corolla pentamerous, persistent. Petals 5.5–6 mm long, obovate, usually inequilateral. Androecium diplostemonous, of 5 antesepalous stamens with slightly longer filaments and 5 antepetalous stamens with shorter filaments; insertion of filaments peridiscal, base slightly enlarged.

Flowering: June to August.

Habitat: Valley beds and ravine beds.

Distribution: China (Tibet), Pakistan, Kashmir (see Map 50).

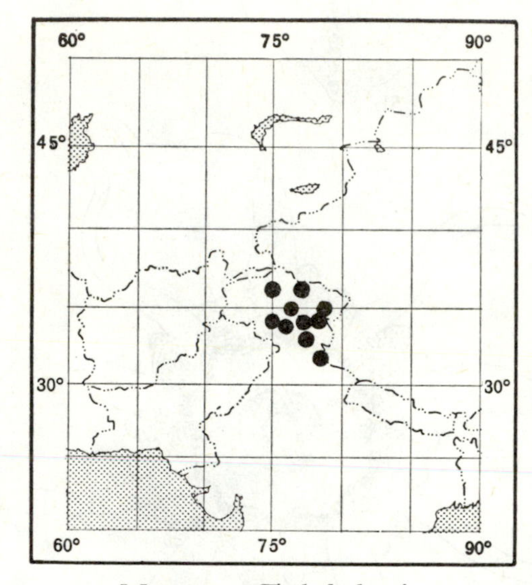

Map 50: *T. ladachensis*

Selected specimens: China: *Strachey* & *Winterbottom* 1, Tibet (BM); *Royle*?, Tibet Tsankar prov. Padir at the foot of the Sinku La Pass to Suble 20–21.6.1856 (BM). Pakistan: *Ludlow* 355, Kashmir, Baltistan .6.1928 (BM); *Hook. f.* & *Thomson*, Ladakh Reg. temp. (BM, FHO, K). Kashmir: *Grady*, *Webster* & *Nasir* 5957, Baltistan, Hushi River Valley, 3 mi. N. of Kandu (K).

Observations: (a) The author was not able to examine the holotype of *Myricaria elegans* Royle (Inglis, Lippa in Kunawur), which, according to Stansfield (1953), is in Liverpool. (b) This is the only *Tamarix* species with beaked seeds and flat leaves.

34 The epithet 'elegans' is not available because of *T. elegans* Spach, a synonym of *T. chinensis* Lour.

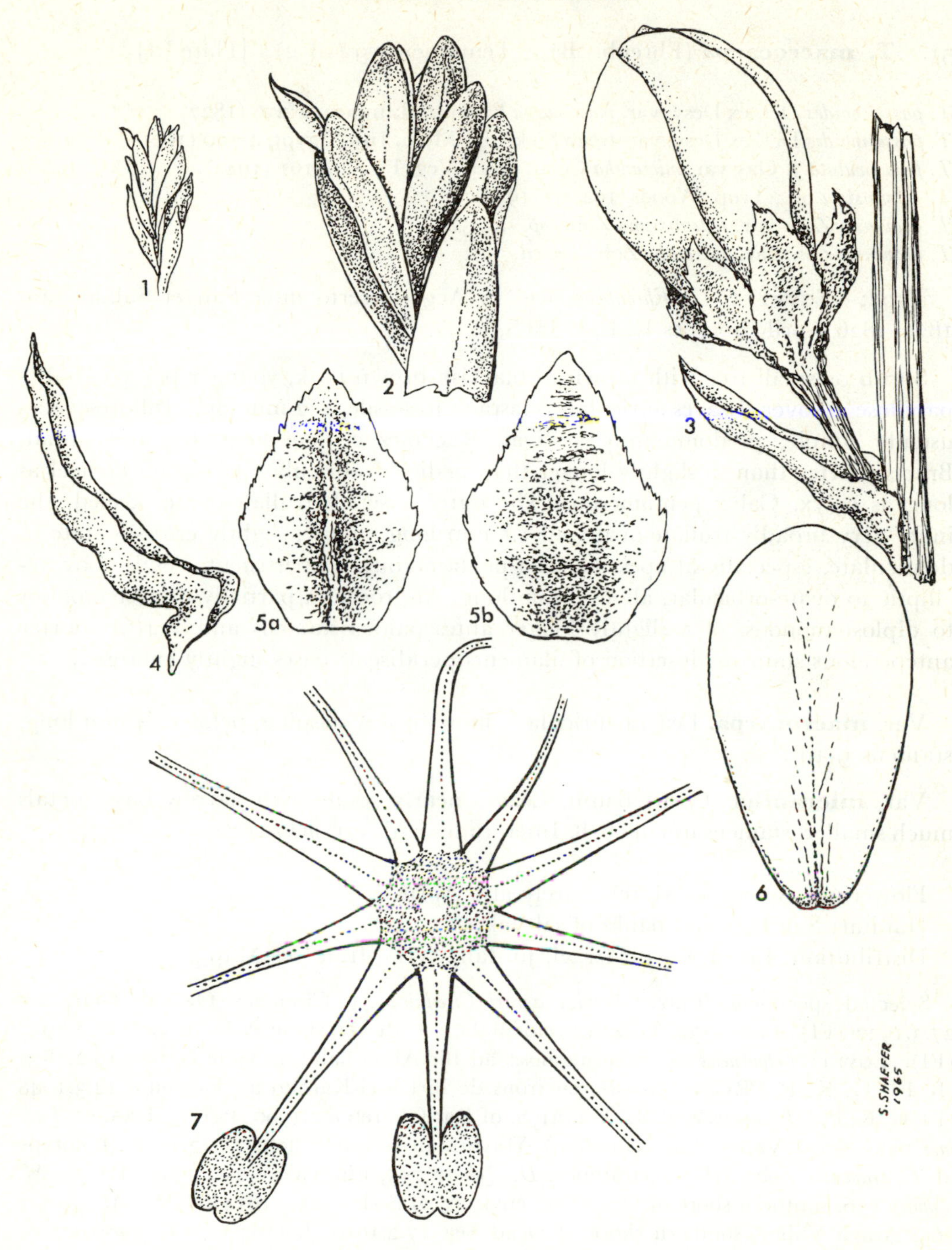

Plate L *T. ladachensis*
1. Young twig (x 1); 2. id (x 2); 3. Flower (x 7); 4. Bract (x 7);
5a. Outer sepal (x 10); 5b. Inner sepal (x 10); 6. Petal (x 10);
7. Androecium (x 10).

51. **T. macrocarpa** (Ehrenb.) Bge., Tentamen, 79 (1852) [Plate LI]

T. passerinoides Del. ex Desv. var. *macrocarpa* Ehrenb., Linnaea, 2:276 (1827).
T. passerinoides Del. ex Desv. var. *typica* Sickenb., Mem. Inst. Egypt, 4:190 (1901).
T. pauciovulata J. Gay var. *micrantha* Corti, Flora Veg. Fezzan, 191 (1942).
T. aravensis Zoh., Trop. Woods, 104:50 (1956).
T. aravensis Zoh. var. *patentissima* Zoh., *op. cit.*, 52.
T. aravensis Zoh. var. *micrantha* Zoh., *loc. cit.*

Type: EGYPT: *C. G. Ehrenberg*, leg. in Aeg. Deserto inter San et Salahie am 1820–1826 April (isotypes K, L, P, UPS, W).

Shrub or small tree with brown to blackish-brown bark, younger parts variably papillose all over. Leaves auriculate, clasping to sessile, 1–2 mm long. Inflorescences usually simple, predominantly aestival. Racemes 2–7 cm long, 6–7 mm broad. Bracts shorter than to slightly longer than pedicels, auriculate to sessile. Pedicel as long as calyx. Calyx pentamerous. The outer 2 sepals trullate-ovate, keeled, the inner very broadly trullate-ovate, all 1.5 mm long, abtuse, slightly erose-dentate to denticulate, especially at apex. Corolla pentamerous, subpersistent. Petals obovate-elliptic to ovate-orbicular, about 4 mm long. Androecium partially diplostemonous to diplostemonous, of 5 slightly longer antesepalous stamens and 1–4(5) shorter antepetalous stamens; insertion of filaments peridiscal; bases slightly enlarged.

Var. **macrocarpa.** Leaves auriculate, more or less clasping, petals c. 4 mm long, stamens 9(10).

Var. **micrantha** (Corti) Baum. Leaves nearly sessile with narrow base, petals much smaller, stamens usually 6–8. Intergrades with var. *macrocarpa*.

Flowering: August to March, rarely also April.
Habitat: Salt flats and banks of salt lakes.
Distribution: Libya, Egypt, Israel, Jordan, Syria, Iran (see Map 51).

Selected specimens: LIBYA: *Servizi agrari Cirenaica* 44, Cirenaica Oasi di Giarabub 27.3.1937 (FI); *Corti* 1532, Fezzan occidentale reg. di Gat dune N.E. di Gat 26.2.1931 (FI). EGYPT: *Letourneux* 31, in salsuginosis ad Bir Abou-Balah in valle Gessen 16.2.1877 (B, FI, G, K, P, PRC, S); *Kralik*, environs de Birket-el-Karoun au Faoyoum 14.3.1948 (FI, G, K, P); *Täckholm et al.* 84, km. 21 S. of Ras Ghareb 8.2.1960 (CAI). ISRAEL: *Tadmor* 748, Arava Valley, Ein (Ghadian) Yotvata in a water hole 25.12.1952 (holotype of *T. aravensis* Zoh., HUJ), paratypes: *D. Zohary* 755, Ein Yahav 23.3.1950 (HUJ), *M. Zohary* 756, southern shore of Dead Sea, environs of Sodom 23.3.1954 (HUJ); *M. Zohary* 765, Arava Valley, southern shore of Dead Sea 17.2.1950 (holotype of *T. aravensis* var. *patentissima* Zoh., HUJ); *D. Zohary* 629, Ein Yahav, banks of brackish spring 9.4.1950 (holotype of *T. aravensis* var. *micrantha* Zoh., HUJ). JORDAN: *Kasapligil* 125b, salt flats at Azraq Hunting Aero Survey 23.4.1955 (E). SYRIA: *Mouterde*, Palmyre salinés et autres terrains (Herb. Mount.). IRAN: *Pabot* 428, base des dunes de Kharkab près village Tohma N. Ahwaz sables, 12.4.1959 (Herb. Pabot, HUJ).

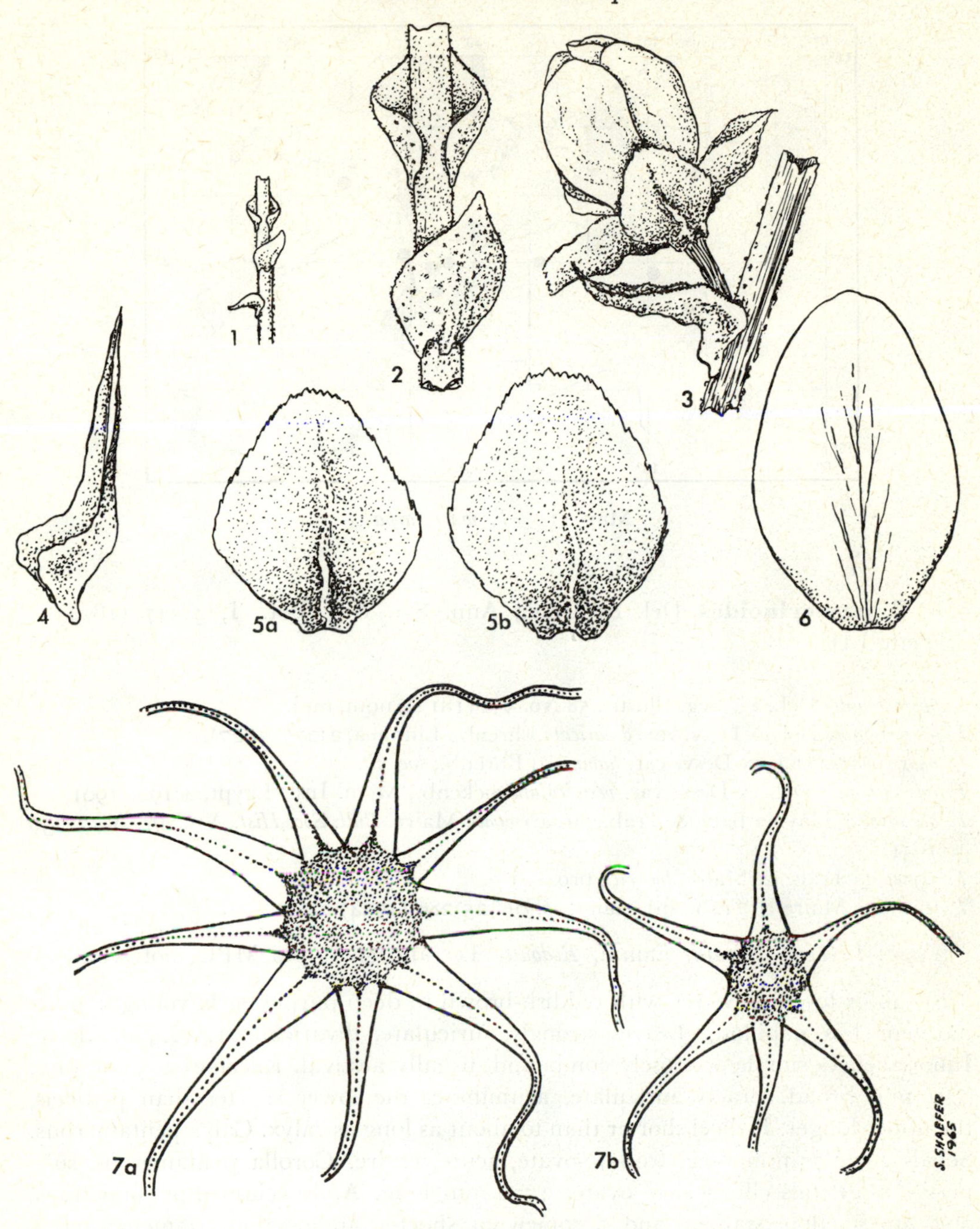

Plate LI *T. macrocarpa*
1. Young twig (x 5); 2. id (x 10); 3. Flower (x 10); 4. Bract (x 20);
5a. Outer sepal (x 20); 5b. Inner sepal (x 20); 6. Petal (x 20);
7a. Androecium with 4 antepetalous stamens (x 30);
7b. id, with only 1 antepetalous stamen (x 20).

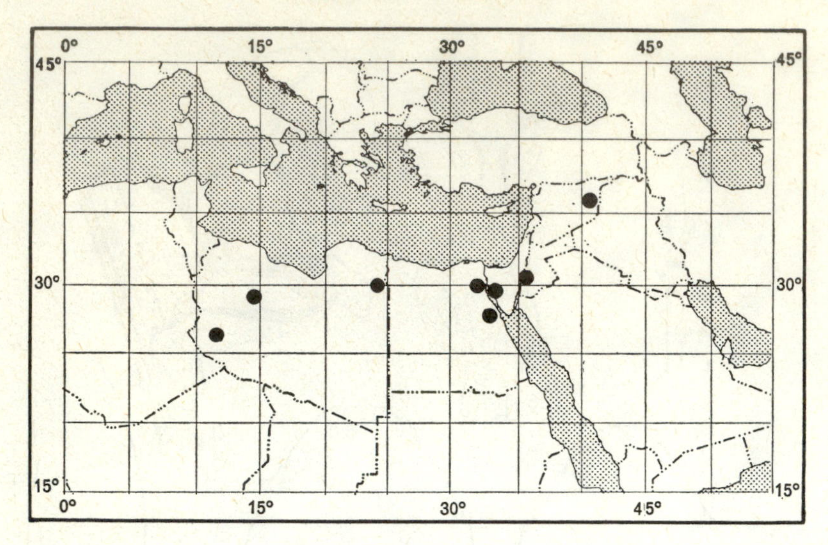

Map 51: *T. macrocarpa*

52. **T. passerinoides** Del. ex Desv., Ann. Sci. Nat. Bot., I, 4:347 (1824) [Plate LII]

T. passerinoides Del., Fl. Aeg. Illustr., 58 No. 352 (1813), nom. nud.
T. passerinoides Del. ex Desv. var. *divaricata* Ehrenb., Linnaea, 2:275 (1827).
T. passerinoides Del. ex Desv. var. *hammonis* Ehrenb., *loc. cit.*
T. passerinoides Del. ex Desv. var. *vermiculata* Sickenb., Mem. Inst. Egypt, 4:190 (1901).
T. balansae J. Gay ex Batt. & Trab. var. *oxysepala* Maire, *Bull. Soc. Hist. Nat. Afr. N.*, 22:36 (1931).
T. oxysepala Trab. ex Maire, *loc. cit.*, pro. syn.
T. tenuifolia Maire & Trab., in: Maire, *op. cit.*, 25:286 (1934).

Type: EGYPT: *Jomard*, Fajum, *Redouté*, Terrane (syntypes MPU, not seen).

Shrub or low bushy tree with reddish-brown to deep purple bark, younger parts more or less papillose. Leaves strongly auriculate, divaricate, 1.5–2.5 mm long. Inflorescences simple or loosely compound, usually aestival. Racemes 2–5 cm long, 8–10 mm broad. Bracts auriculate, acuminate, the lower shorter than pedicels, the upper longer. Pedicel shorter than to about as long as calyx. Calyx pentamerous. Sepals 2.5–2.75 mm long, trullate-ovate, acute, entire. Corolla pentamerous, subpersistent. Petals elliptical to ovate, 3.5–4 mm long. Androecium diplostemonous, of 5 antesepalous stamens and 5 somewhat shorter antepetalous stamens; insertion[35] of filaments practically peridiscal; filaments slightly enlarged towards base.

35 The insertion of the antesepalous stamens is slightly hypodiscal and should be examined carefully.

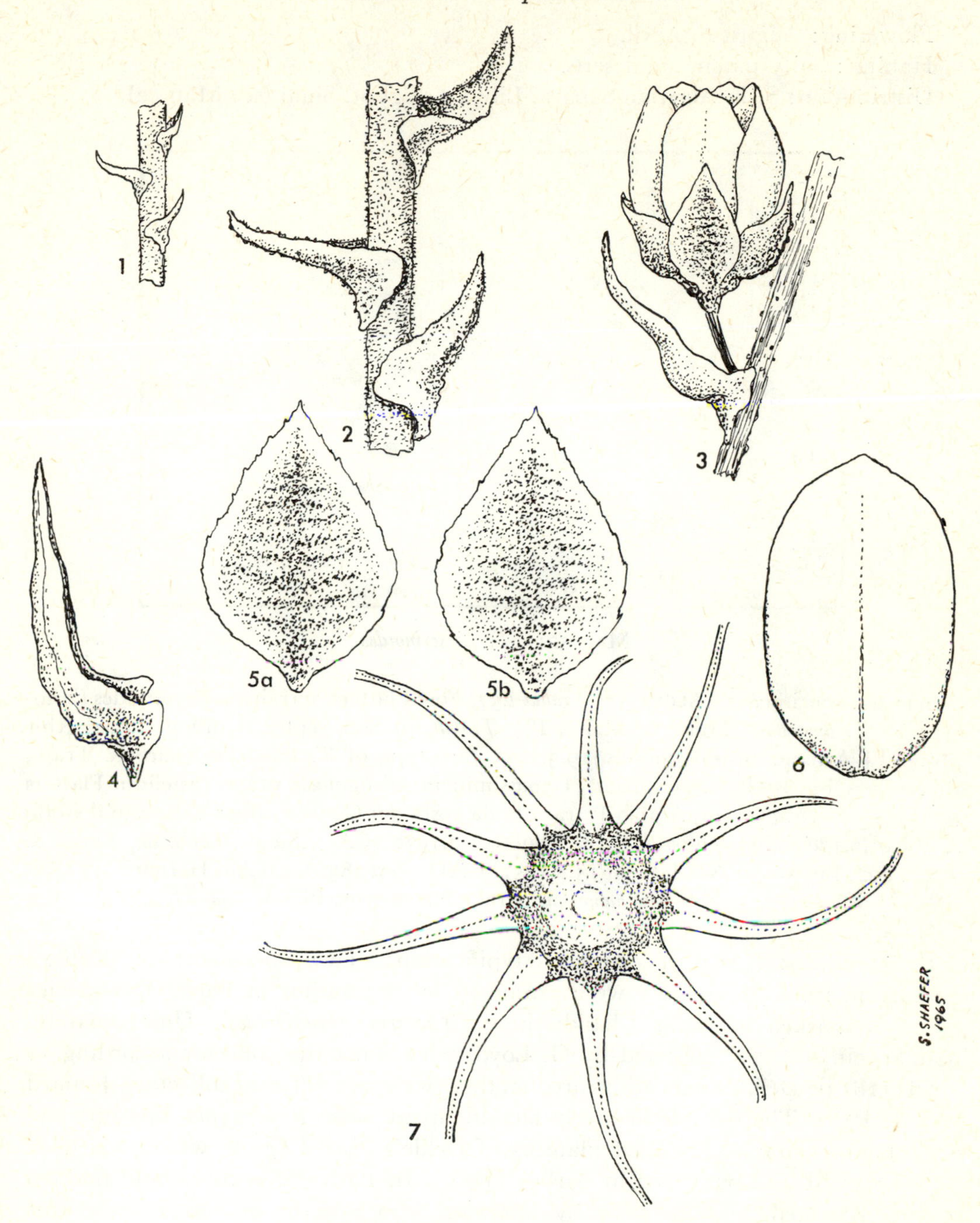

Plate LII *T. passerinoides*

1. Young twig (x 5); 2. id (x 10); 3. Flower (x 10); 4. Bract (x 20); 5a. Inner sepal (x 20); 5b. Outer sepal (x 20); 6. Petal (x 12); 7. Androecium (x 20).

Flowering: August to April.
Habitat: Salty patches in deserts.
Distribution: S. E. Algerian Sahara, Libya ?, Egypt, Sinai (see Map 52).

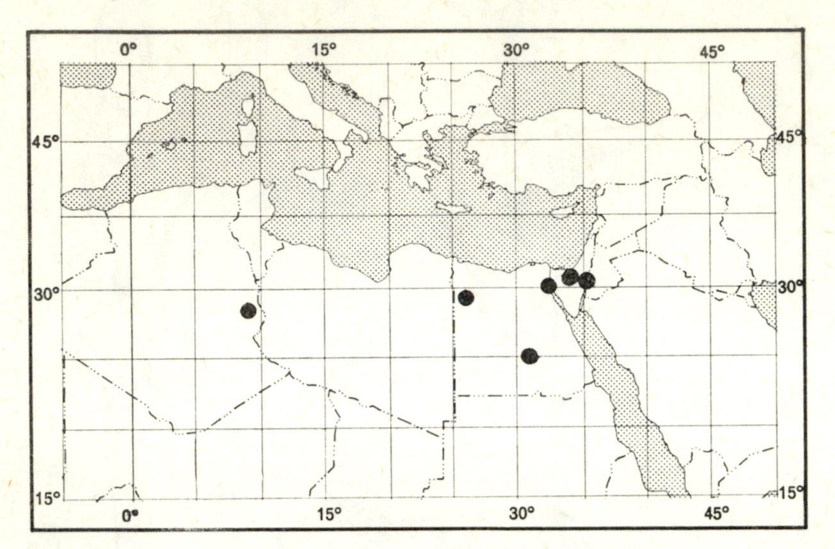

Map 52: *T. passerinoides*

Selected specimens: ALGERIA: *Trabut* 257, Fort Flatters, terrains salés humides (holotype of *T. oxysepala* Trab. ex Maire, P); *Trabut*, in Sah. septentr. ditione Oued Rhir prope El Arfiane in salsuginosis 19.3.1933 (holotype of *T. tenuifolia* Maire & Trab., P; isotypes FI, RAB); *Maire* 1621, Temassimin in salsuginosis prope castellum Flatters 3.5.1928. EGYPT: *Tigari*, Arabia Petraea Sulla costa del Golfo du Suez e di tutto il Golfo Arabico .3.1867 (FI); *Oliver*, Kharga Boulaq 8.2.1931 (CAI). Sinai; *Täckholm, Kassas & Jadras* 222, Bir Mazar North Sinai 15.8.1951 (CAI); *Post* 183, Sinai Ain Harnara 1.3.1882 (BM); Ehrenb., '*T. passerinoides hammonis*' (fragment of type, PRC).

Observations: (a) *T. passerinoides* typification: Two specimens of Delile's herbarium from Montpellier were examined by the author in Paris. The species folder is marked in Delile's handwriting '*Tamarix passerinoides*'. Unfortunately, both specimens were collected by G. Loyd, who is not the collector according to Delile (1813). Delile's indications are that there were two different collectors, Jomard and Redouté. The two plants are in fact identical with *T. tetragyna* Ehrenb. and do not match the unpublished Plate 63 of Delile's *Flore d'Egypte*, which was later photographed and published in Barbey (1882). In Paris the author could find no specimen named *T. passerinoides* by Desvaux, who was the first to describe this species. The author's conception of this species is based mainly on J. Gay's determinations and observations in Kew. (b) One specimen collected by L. Humbolt (No. 11) in the Comoro Islands in 1885 might be significant. So far, it is doubtful whether the species is native there.

53. **T. salina** Dyer, in: Hook. f., Fl. Brit. Ind., 1:248 (1874) [Plate LIII]

Type: PAKISTAN: *Fleming* 115, Kaffir Kate at Esakhail W. banks of Indus 4.3.1852 (holotype K).

Small bushy tree or shrub with brown to reddish-brown bark, younger parts usually densely papillose. Leaves amplexicaul or nearly so, the younger imbricate, 1–1.5 mm long. Inflorescences usually aestival and simple, sometimes loosely compound. Racemes 6 cm long, 7 mm broad. Bracts cordate-auriculate, shorter than pedicels. Pedicel as long as to slightly shorter than calyx. Calyx pentamerous. Sepals subentire, 1.5 mm long, the outer 2 ovate, somewhat connate at base, smaller than the inner, the inner much broader, very broadly ovate-trullate. Corolla pentamerous, caducous. Petals elliptic, 2.75 mm long. Androecium haplostemonous, rarely partially diplostemonous, of 5 antesepalous stamens, occasionally with 1 or 2 additional antepetalous stamens with somewhat shorter filaments; insertion of filaments peridiscal; disk synlophic.

Flowering: January to March.

Habitat: Banks of rivers and lakes.

Distribution: Endemic to Pakistan (see Map 53).

Selected specimens: PAKISTAN: *Koelz* 7555, India Sind Khimpir Lake 19.1.1934 (US); *Schlagintweit* 11322, Sindh Shikarpour 20.2.1857 (US).

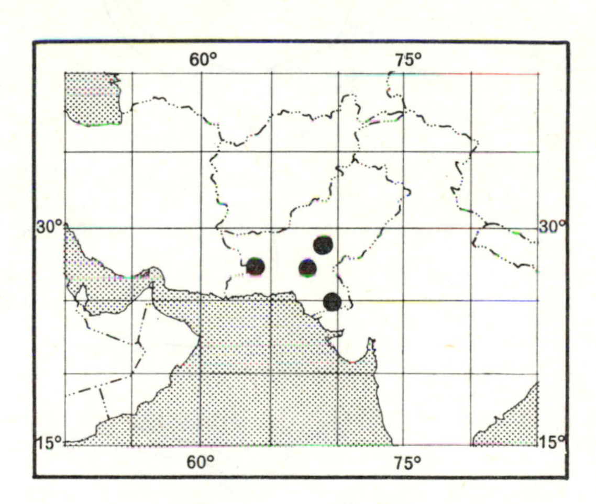

Map 53: *T. salina*

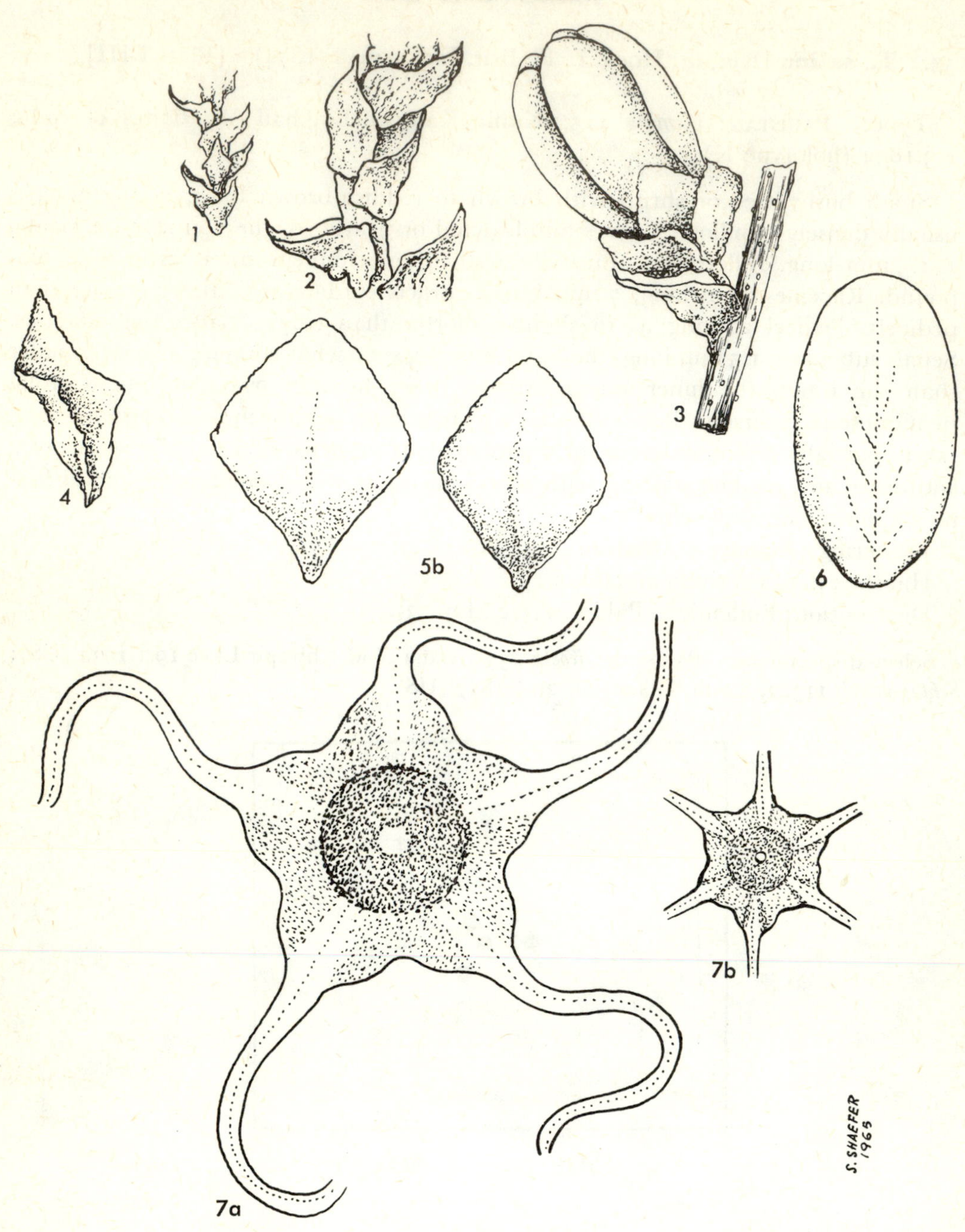

Plate LIII *T. salina*

1. Young twig (x 5); 2. id (x 10); 3. Flower (x 10); 4. Bract (x 20); 5a. Inner sepal (x 20); 5b. Outer sepal (x 20); 6. Petal (x 15); 7a. Haplostemonous androecium (x 25); 7b. Partially diplostemonous androecium (x 15).

54. **T. stricta** Boiss., Diagn. Pl. Or. Nov., II, 2:57 (1856) [Plate LIV]

Type: PAKISTAN: *Stocks* 274, *Tam. decandra* Scinde and Balutchistan 1850 (holotype G; isotypes CGE, G, K, P, W).

Tree or shrub with brown to grey bark, entirely glabrous. Leaves vaginate, 2–3 mm long, with a small acute and scarious point. Inflorescences as in *T. aphylla*. Racemes 2–6 cm long, 4–5 mm broad, with subsessile flowers. Bracts triangular, auriculate, acuminate, longer than pedicels. Pedicels much shorter than bracts. Calyx pentamerous. Sepals 1.75 mm long, obtuse to truncate, finely and densely denticulate, the 2 outer smaller and broadly ovate, the inner obovate-cuneate, truncate. Corolla pentamerous, persistent? Petals 2–2.25 mm long, elliptic-obovate. Androecium diplostemonous, of 5 antesepalous and 5 somewhat shorter antepetalous stamens; insertion of filaments peridiscal.

Flowering: August.

Habitat: Sandy deserts.

Distribution: Pakistan, S. Iran (see Map 54).

Selected specimens: PAKISTAN: *Schlagintweit*, Pandzab Peshaur (P, US); *Lander*, Baluchistan .2.1902 (BM). IRAN: *Sharif* 438E, Balouchestan Bampour 3.8.1949 (W, Esfand. Herb.).

Observation: This is the only species of Sect. *Polyadenia* with vaginate leaves.

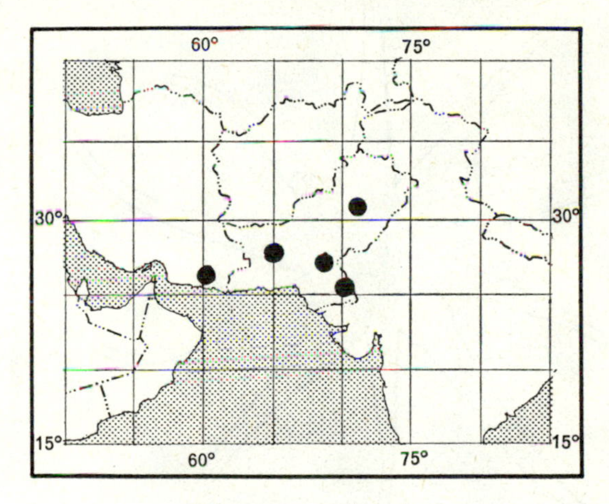

Map 54: *T. stricta*

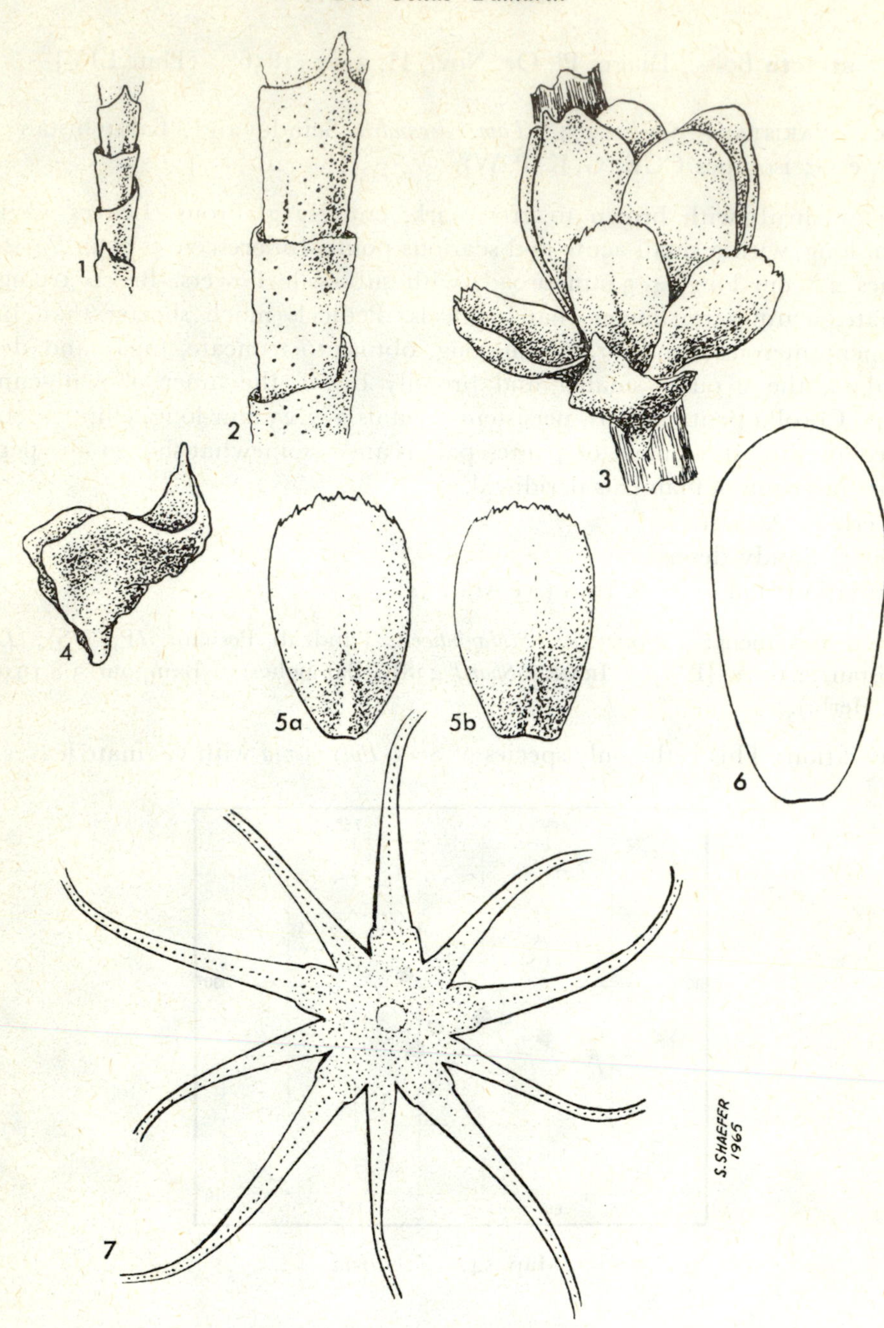

Plate LIV *T. stricta*

1. Young twig (x 5); 2. id (x 10); 3. Flower (x 10); 4. Bract (x 20); 5a. Inner sepal (x 20); 5b. Outer sepal (x 20); 6. Petal (x 20); 7. Androecium (x 30).

Species Non Satis Notae

T. arabica Pall., Nova Acta Acad. Petrop., 10:376 (1797), nom. nud.
T. gallica L. var. *narbonensis* Ehrenb. f. *effusa* Ehrenb., Linnaea, 2:267 (1827).
T. furas Buch.-Ham. ex Royle, Illustr. Bot. Himal., 2(6):213 (1835), nom. nud.
T. tetramera Fisch. & Mey. ex Karel., Bull. Soc. Nat. Moscou, 12:154 (1839), nom. nud.
T. gallica L. var. *vulgaris* Ledeb., Fl. Ross., 2:135 (1843).
T. polystachya Ledeb. var. *cerifera* Regel, Acta Horti Petrop., 5:582 (1877).
T. hohenackeri Bge. f. *frondosa* Lipsky, Zapisk. Kiewsk. Obshch., 12:10 (1892).
T. kashgarica Hort. ex Lemoine, *Gard. Chron.*, III, 13:414 (1893).
T. getula Batt., Bull. Soc. Bot. France, 54:254 (1907).
T. pallasii Desv. var. *logodechiana* Regel & Mlokoss., in: Kusn., Busch & Fomin, Fl. Cauc. Crit., 3(9):90 (1909).
T. pallasii Desv. var. *laxiuscula* Regel & Mlokoss., *op. cit.*, 93.
T. pallasii Desv. var. *caspica* Regel & Mlokoss., *op. cit.*, 95.
T. pallasii Desv. var. *daghestanica* Regel & Mlokoss., *loc. cit.*
T. pallasii Desv. var. *longifolia* Regel & Mlokoss., *op. cit.*, 96.
T. pallasii Desv. var. *kumensis* Regel & Mlokoss., *op. cit.*, 98.
T. hohenackeri Bge. var. *glandulosa* Regel & Mlokoss., *op. cit.*, 101.
T. hohenackeri Bge. var. *pontica* Regel & Mlokoss., *loc. cit.*
T. hohenackeri Bge. f. *iberica* Regel & Mlokoss., *loc. cit.*
T. leptopetala Bge. var. *karabachensis* Regel & Mlokoss., *op. cit.*, 103.
T. mongolica Ndz., in: Engler & Prantl, Nat. Pflanzenfam., ed. 2, 21:286 (1925).
T. font-queri Maire & Trab., in: Emb. & Maire, Bull. Soc. Sci. Nat. Phys. Maroc, 11:90–114 (1931).
T. brachystylis J. Gay var. *stenonema* Maire & Trab., in: Jah. & Maire, Cat. Pl. Maroc, 2:489 (1932), nom. nud.
T. africana Poir. var. *faurei* Sennen, Cat. Fl. Rif Or., 42 (1933).
T. pulchella Sennen & Maur., in: Sennen, *op. cit.*, 43 (1933), nom. nud.
T. africana Poir. var. *brivestii* Maire & Trab., Bull. Soc. Hist. Nat. Afr. N., 24:193–232 (1933).
T. brachystylis J. Gay var. *stenonema* Maire & Trab., *op. cit.*, 26:184 (1935).
T. gallica L. var. *wallii* Maire, *ibid.*, 27:215 (1936).
T. africana Poir. var. *microstachys* Maire, *ibid.*, 27:241–270 (1936).
T. africana Poir. var. *palasyana* (Sennen & Maur.) Maire, *loc. cit.*
T. austromongolica Nakai, *J. Japan. Bot.*, 14:291 (1938).
T. tenuissima Nakai, *loc. cit.*
T. balansae J. Gay var. *squarrosa* Maire, Bull. Soc. Hist. Nat. Afr. N., 30:327 (1939).
T. africana Poir. f. *leptostachya* Maire, *op. cit.*, 31:99 (1940).

Observations: (a) The type of *T. narbonensis effusa* Ehrenb. might be in Herb. Chamisso in LE. (b) One of the characteristics of *T. tenuissima*[36] is a stamen number of 10. The author has not seen any 10-stamened tamarisk from this area. (c) From the description, *T. austromongolica*[36] seems to be conspecific with *T. ramosissima* or with *T. chinensis*. (d) The type of *T. furas* should be in Liverpool (cf. observation on *T. ladachensis*).

36 The author requested from MAK, TH, TI, TNS, TOFO that they send him the types and other material of the genus on loan; unfortunately, he received no reply.

T. germanica L., Sp. Pl., 1:271 (1753) = *Myricaria germanica* (L.) Desv., Ann. Sci. Nat. Bot., I, 4:349 (1824).

T. decandra Pall., Fl. Ross., 2:73 (1788), nom. illegit. = *Myricaria longifolia* (Willd.) Ehrenb., Linnaea, 2:279 (1827).

T. decandra Salisb., Prodr., 173 (1796) = ?

T. songarica Pall., Nova Acta Acad. Petrop., 10:374 (1797) = *Hololachna songarica* (Pall.) Ehrenb., *op. cit.*, 273.

T. germanica (decandrae) Pall., *op. cit.*, 376 = *Myricaria germanica* (L.) Desv., *loc. cit.*

T. monogyna Stokes, Bot. Mat. Med., 2:176 (1812) = *Myricaria* sp.

T. herbacea Willd., Abh. Akad. Berlin Physik, 1812–1813:84 (1816) = *Myricaria herbacea* (Willd.) Ehrenb., *op. cit.*, 350.

T. davurica Willd., *op. cit.*, 85 = *Myricaria dahurica* (Willd.) Ehrenb., *op. cit.*, 278.

T. longifolia Willd., *loc. cit.* = *Myricaria longifolia* (Willd.) Ehrenb., *loc. cit.*

T. songarica Ewersm. ex Steud., Nom., ed. 2, 2:661 (1841), nom. illegit. = *Haloxylon ammodendron.*

CONCLUSIONS AND DISCUSSION

Taxonomic Remarks

This revision includes the names of 54 accepted species out of about 200 published binomials. The problems of nomenclature in this genus are not less intricate than those of its taxonomy. All efforts have been made to typify not only the accepted names but also the bulk of the synonyms and to examine authentic specimens, even of out-of-the-way and forgotten collections.

The taxonomy of the genus has been studied by many botanists since Linnaeus, in most cases very seriously, but they failed to include the whole genus and to shed light on some obscure groups. The reason for this failure was the relatively poor material at their disposal. The author, on the other hand, was very lucky to have been able to examine 8,000 herbarium sheets scattered over 42 herbaria. The author has not introduced any new theories of systematics, but has followed the same approach to the species concept as his predecessors and has used the same methods; he hopes, however, that he has refined them somewhat. With the help of a dissection microscope the author succeeded in obtaining a clear insight into the structure of the flower disk and its diagnostic reliability. This compound character guided him into the labyrinth of the whole genus and led him to find correlations with other diagnostic features.

Among the many previously described taxa the author found only two undescribed species. He will certainly be accused of excessive 'lumping' because he put quite a large number of binomials and a variety of names into the synonymies, even in cases where there is some justification for keeping these taxa as infraspecific units. The author adheres to the modern notions of species developed by Du Rietz (1930), Stebbins (1951), Lam (1959) and others, in which the species are regarded as populations and are thus described accordingly. Consequently, the descriptions given in the systematic part are not limited to the strict physical make-up of the holotypes.

The author can well understand the reasoning that has led others to establish their 'new' species, but he believes that these workers would have proceeded otherwise had they seen the whole body of the genus. In some cases, however, the author can be accused of having joined the 'splitters'. For instance, *T. nilotica*, *T. mannifera* and *T. arborea* were given by Ehrenberg as varieties, but the author has followed Bunge in giving them specific rank; Bunge, however, did not supply sufficient evidence for this.

The author is well aware that despite his efforts there are a few species which are unevenly balanced and that there are clusters of species so close to one another that they deserve further treatment in the future side by side with the clearly delimited and well separated species.
The author will certainly be accused also of not having provided an easily manageable key; for the forester and those unskilled in botany this will be troublesome. The principal key characteristic used is not the mode of insertion of the filaments, which although much more clear-cut is more difficult to observe than some of the other external features he used, such as the leaves. For this he can only apologize, since he found no alternative in this intricate genus, and, despite all his efforts, no other key methods could be devised. The keys have, however, the advantage of being reliable, and once the reader has acquired some skill in dissecting the flowers he will find a way of naming all his specimens, especially when he compares his findings with the descriptions and illustrations of the species.
A word may also be devoted to the Sections and Series which the author believes bring some order into the perplexing 'heap' of species. These superspecific units, especially the Series, are intended to reflect the relationships between the species.
The author does not claim to have advanced any new theories in this revision. On the contrary, many modern approaches to taxonomy such as cytogenetics, palynology and chemotaxonomy, whose help to classic taxonomy is beyond doubt, have been applied in only a few cases. Even within this conservative framework, moreover, much has been left for future study — for instance, the distribution of some Central Asian, Mongolian and Chinese species (especially of Chinese Turkestan and Tibet); the relationships between some difficult species such as *T. aucheriana* and *T. pycnocarpa*; and the distribution of the species of the South Saharan and French West African–Sudanian belt, such as *T. senegalensis* and *T. arabica*.

Evolutionary Trends

On the basis of the conventionally agreed upon morphological progressions, some evolutionary trends can be noted concerning both individual organs and relationships between the taxa. The evolutionary trends of the androecium in *Tamarix* have been discussed by Zohary and Baum (1965) and will be briefly reviewed here along with evolutionary lines.
The following five phyletic processes are quite obvious and are in some way linked together in *Tamarix*:
1. A trend towards the formation of smaller flowers and narrower racemes (all species of Section One). This trend is also seen within the series, e.g., in Series 9, *T. salina*, *T. dubia*, *T. stricta* and *T. macrocarpa* var. *micrantha* have narrow racemes, while the other species of this series have broad racemes. The trend is also manifested in the small-flowered Series 6, which presumably arose from the large-flowered

Series 5. Large flowers are characteristic of species which are considered more primitive and less developed than others in respect to other features such as the biological adaptation to pollination. Large-flowered species display no true disks, while small-flowered ones have strongly nectariferous disk lobes.

2. A transition from diplostemony to haplostemony leads gradually to complete abortion of the antepetalous stamens. *T. ericoides* with its diplostemonous androecium has two distinct whorls and is probably the most primitive species or the closest living relative of the assumed ancestral *Tamarix* type. This supposition is supported by anatomical evidence such as the presence of ten stamen traces and of a vascular supply to the bracts (Murty, 1954). From this '*Ericoides* type' (Fig. 10) complete abortion of the antepetalous stamens has taken place in Section One' where the stamens are primarily replaced by staminode-like lobes. This process is also clearly manifested in Section Two (all the paralophic and synlophic species; see also Fig. 10) and in Section Three (*T. salina*). In Section One this reductive trend is strengthened, involving also the gradual rudimentation of the disk lobes from the hololophic state through the paralophic to the synlophic configuration (Zohary & Baum, 1965, and Fig. 10). The reduction of the disk lobes and their connation with the filament bases may be interpreted as a trend leading from entomophily to anemophily.

3. Fusion or adnation of the antepetalous whorl of stamens to the antesepalous one. This line probably initiated in species with hypodiscal insertion of the antesepalous filaments (*T. ericoides*, *T. dioica*, *T. karakalensis*, *T. nilotica*, *T. ramosissima*, *T. smyrnensis*, *T. chinensis*, *T. rosea*), passing through stages of hypo-peridiscal insertion (see Figs. 10 and 11) to the peridiscal insertion exemplified by most of the species.

4. Persistence of the corolla is no doubt secondary to caducity. This characteristic is limited to the majority of the hypodiscal species (*T. ericoides*, *T. dioica*, *T. smyrnensis*, *T. chinensis*, *T. ramosissima*) and is surprisingly only found in the glyciphilous species (except for *T. ramosissima*).

5. Glycophily is no doubt a less advanced characteristic than halophily. Most of the hypodiscal species are glycophytes, while the bulk of the species are halophytes. This is in agreement with the view that non-salt-tolerant species are less specialized than salt-tolerant ones.

These five closely-linked evolutionary processes are conventionally accepted by taxonomists and are very helpful in discussing the relationships between the species. The following additional phyletic trends may also be postulated here:

In Section Two there seems to be a clear progression from pentamery to tetramery, most probably through meiomery (as exemplified in *T. gracilis* and in *T. hampeana*; see Plates 25 and 33, respectively). In Section Three the splitting of the antepetalous stamens has led to so-called triplostemony, as presented in Series 8. In Section Three, and especially in Series 9, it is difficult to fill in the gaps between species. Moreover, the species concerned here display a very restricted range of distribution, suggesting relict character. Dioecism, a process confined to Series 3, is found in only two species, *T. dioica* and *T. usneoides*.

Croizat (1960) states that, among others, 'Flacourtiaceae/Tamaricaceae represent a group of primeval angiospermous antiquity', and that 'there exists a marked similarity between the female flowers of *Populus* (e.g., *P. euphratica*) and those of *Daphniphyllum* and *Tamarix*. Daphniphyllaceae and Tamaricaceae may indeed bind as one Ulmaceae, Salicaceae, Flacourtiaceae at a very low level of angiosperms phylogeny and morphology'. So far no pre-Pleistocene fossil evidence has been found for *Tamarix*.

An examination of the present distribution of the species of *Tamarix*, together with the morphological data and the deduced phyletic trends, may supply a conclusive view not only on the origin, speciation centres and migration of the genus but also on its phylogeny. It is seen that, in accordance with the above assumption, the supposedly most primitive species are limited to India (*T. ericoides*, *T. dioica*) or the immediate vicinity (to be called here the Indo-Turanian Centre), which also harbours other primitive species (*T. chinensis*, *T. ramosissima*, *T. smyrnensis*, *T. rosea*). On the other hand, the presumably most advanced species, displaying, *inter alia*, a synlophic configuration of the disk, inhabit the extreme parts of the general distribution range of the genus, where none of the species assumed to be primitive occurs. Thus, for instance, *T. gallica*, *T. canariensis* and *T. africana* are found in N. W. Africa and S. W. Europe, and *T. angolensis* and *T. usneoides* in S. W. Africa.

It is therefore most probable that the earliest *Tamarix* species (*T. ericoides*) developed in the southeastern part of the Indo-Turanian Centre. As seen from the table and from the distribution charts of the individual species, this centre, still displaying such a variety of primitive forms, should no doubt be considered the most important speciation centre of the genus.

This speciation centre harbours nearly 50% of the species of *Tamarix*, and it also gave birth to many tetramerous species. One of its southeastern corners may have been the cradle of the genus. A second, probably secondary, centre of speciation seems to be located in the East Mediterranean (Egypt-Palestine-Turkey) area. Here the *Nilotica* and the *Smyrnensis* evolutionary series of Series 1 and 2 (see Fig. 12) developed. This second centre seems to be closely related in time and space to the first by the species of the *Rosea* evolutionary series and by some Arabian-Sudanese species of Series 9 (*Pleiandrae*). Series 3 (*Vaginantes*) seems to be a very ancient one because of its widely scattered and disjunctive species (Fig. 12 and see maps of distribution). The centre of speciation of this series may be located near the Indian sector of the first centre. The reasons for this assumption are: (a) *T. dioica*, the supposed leading species of the 'vaginate-leaves' series, occurs today in India and penetrates slightly into S. Iran; (b) *T. bengalensis* (see Map 21), the species closest to *T. dioica*, has a restricted-relict range of distribution in Bengal; (c) *T. stricta*, a member of Section Three, though it has vaginate leaves and is closely related to *T. ericoides* and to species of Series 3, is confined to Baluchistan and shows a relict nature in its range of distribution.

Other speciation nuclei seem to have developed in various locations along the migration routes of *Tamarix* (Fig. 12), which, as deduced from actual distribution, are more or less similar to 'standard plant dispersal tracks' such as those of *Suaeda*. The migrations could have taken place in the following directions: (a) from India to C. Asia, i.e., by shifting of the hypothetical primeval speciation centre northward to the actual Indo-Turanian nucleus; (b) migration from C. Asia westward, which gave rise to the E. Mediterranean speciation centre; (c) general westward migration, which is possibly one of the earliest tracks of dispersal, leaving several relict species, e.g., *T. pycnocarpa*, *T. dubia*, *T. stricta* and others more recent such as *T. arceuthoides–T. gallica* (W. Europe direction), *T. tetragyna–T. africana*, or *T. tetragyna–T. boveana* or *T. macrocarpa–T. amplexicaulis* (N. W. African direction), *T. arabica–T. senegalensis–T. canariensis* (Sudanese–N. W. African direction), and others (see Fig. 12); (d) Indo-Turanian–S. / S. W. African migration route, as exemplified in the Series *Vaginantes*.

APPENDIX

Formal Diagnosis of New Sections, Series and Species

Section 1. TAMARIX

Folia subsessilia basi angusta, subauriculata vel vaginata. Racemi 3–5 mm lati, praeter species diocias paucas cum racemis 5–7 mm latis. Bracteae pedicellis longiores. Flores pentameri vel in speciebus nonnullis interdum tetra-pentameri vel in racemis vernalibus solum tetrameri. Petala 1–2.25 mm longa. Androecium haplostemonum, staminibus(4–)5 antesepalinis et structuris variis discalibus compositum, stamina antepetalina carens.

Species typica: *T. gallica* L.

Series 1. GALLICAE Baum (ser. nov.)

Plantae omnino glabrae, papillas carentes. Folia plerumque sessilia, basi angusta, praeter *T. mascatensim* cum foliis valde amplexicaulibus.

Species typica: *T. gallica* L.
Species inclusae: *T. arceuthoides* Bge., *T. gallica* L., *T. korolkowii* Regel et Schmalh., *T. mascatensis* Bge., *T. palaestina* Bertol., *T. ramosissima* Ledeb., *T. smyrnensis* Bge.

T. bengalensis Baum (sp. nov.) [Plate 21]

Arbores, saepe fruticosae, *T. dioicae* similes sed monoicae, partibus junioribus glabris. Cortex fuscus vel canescens. Folia amplexicaulia, pseudo-vaginantia, 1.5–2.25 mm longa. Inflorescentiae simplices vel e racemis 4–11 cm longis, 4–5 mm latis laxe compositae. Bracteae simplices, trullatae, acuminatae, pedicellis longiores vel iis interdum subaequales. Pedicelli calyce breviores. Calyx pentamerus; sepala 1.25–1.5 mm longa, orbicularia usque obovata, margine confertim inciso-denticulata, duo exteriora interioribus magis obovatis, interdum mediocriter carinatis paulo breviora, magis orbicularia et carinata. Corolla pentamera, persistens; petala 2 mm longa, obovata usque late obovata. Stamina 5 antesepalina; filamentorum insertio hypoperidiscalis, discus hololophus.

Typus: *Hook. f.* & *Thomson*, Bengal or. reg. temp. (holotypus W; isotypi B, BM, CGE, G, K, OXF, P, S, U, W).

T. dalmatica Baum (sp. nov.) [Plate 31]

Arbores *T. africanae* vel *T. hampeanae* similes, glabrae. Cortex rubro-niger, nigro-fuscus vel nigrescens. Folia 2.5–4 mm longa, sessilia, basi angusta. Inflorescentiae vernales simplices; racemi 2–6 cm longi, 8–10 mm lati, floribus praesertim tetrameris apicalibus paucis pentameris. Bracteae calyces aequantes vel eos multo excendentes, triangulares, obtusae, superantes longiores cum mucrone brevi latoque, saepe diaphanae, margine scabriculo-papillulosae. Pedicelli calycibus triplo breviores. Calyx tetra (–penta) merus; sepala 3.5 mm longa, trullato-ovata, plus minusve carinata, carinis interdum minutissime papillosis, duo exteriora ovata interioribus trullati acutiora. Corolla tetra (–penta) mera, subpersistens; petala 2.5–4.5 (–5) mm longa, anguste elliptico-obovata, unguiculata. Stamina 4–5 antesepalina, interdum 1–3 antepetalina filamentis paulo brevioribus; filamentorum insertio peridiscalis; discus paralophus.

Typus: *Fiala*, an der Narenta bei Capljina Herzegovina .5.1892 (holotypus PRC; isotypus WU).

Series 6. Arbusculae Baum (ser. nov.)

Bracteae pedicellis paululum breviores vel iis longiores. Flores tetrameri, tetrandri. Petala 2–2.25 mm longa. Racemi simplices, 3–5 mm lati.

Species typica: *T. parviflora* DC.
Species inclusae: *T. androssowii* Litw., *T. kotschyi* Bge., *T. parviflora* DC.

Series 7. Fasciculatae Baum (ser. nov.)

Bracteae pedicellis multo breviores vel eos interdum aequantes. Flores tetrameri, tetrandri. Petala 2–2.25 mm longa. Racemi 3–6 mm lati, floribus supremis in fasciculos apicales umbelliformes contractis.

Species typica: *T. polystachya* Ledeb.
Species inclusae: *T. litwinowii* Gorschk., *T. polystachya* Ledeb.

Series 8. Arabicae Baum (ser. nov.)

Androecium florum nonnullorum quoque racemo partim triplostemonum (saltem 2 stamina antepetalina breviora inter antespalina longiora) usque ad triplostemonum.

Species typica: *T. pycnocarpa* DC.
Species inclusae: *T. aucheriana* (Decne.) Baum, *T. pycnocarpa* DC.

Illustrations

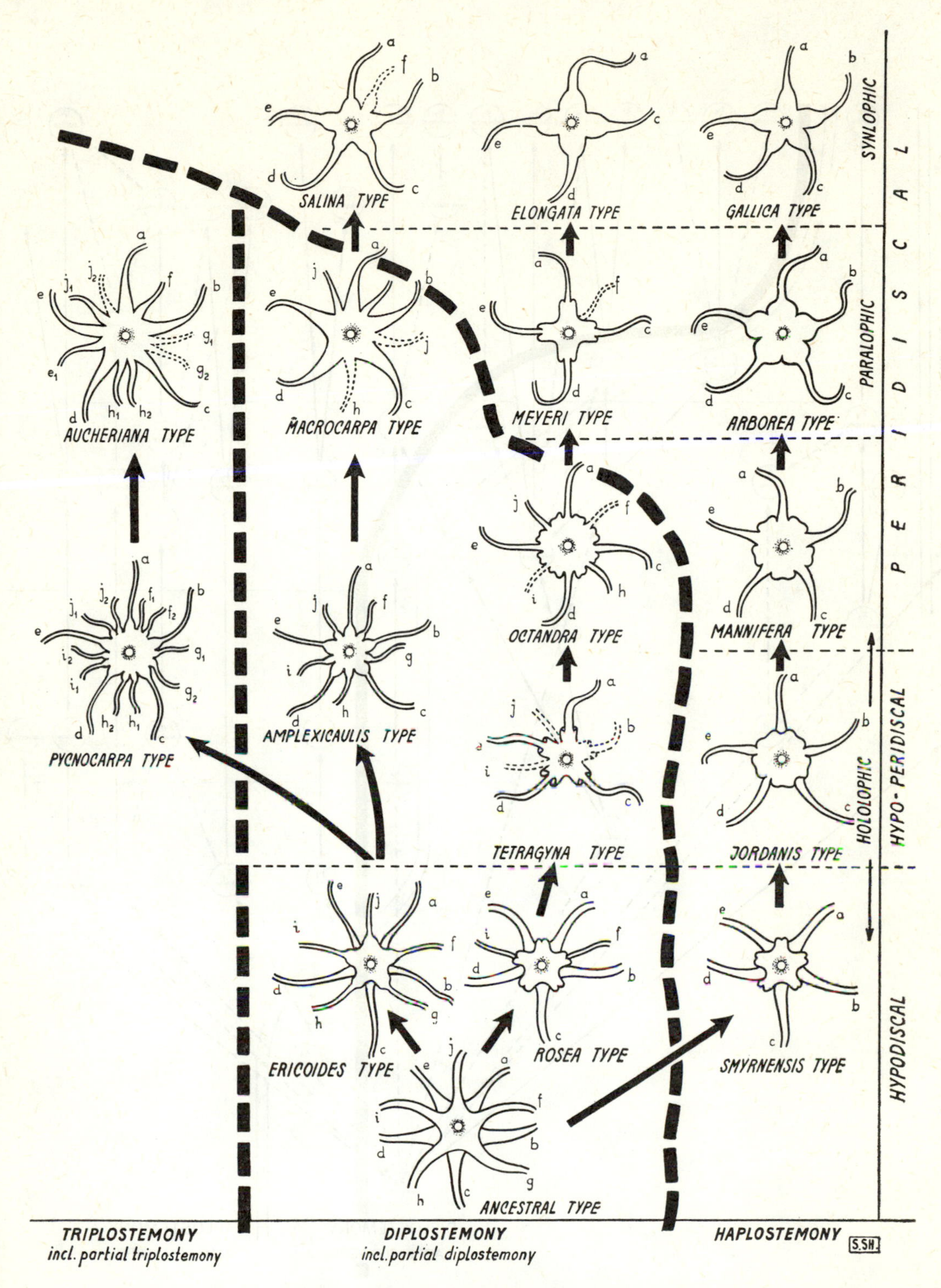

Evolutionary trends in the androecium of *Tamarix* (see explanation in text and in Zohary & Baum, 1966). The letters *a* to *e* denote the antesepalous stamens and their split-offs

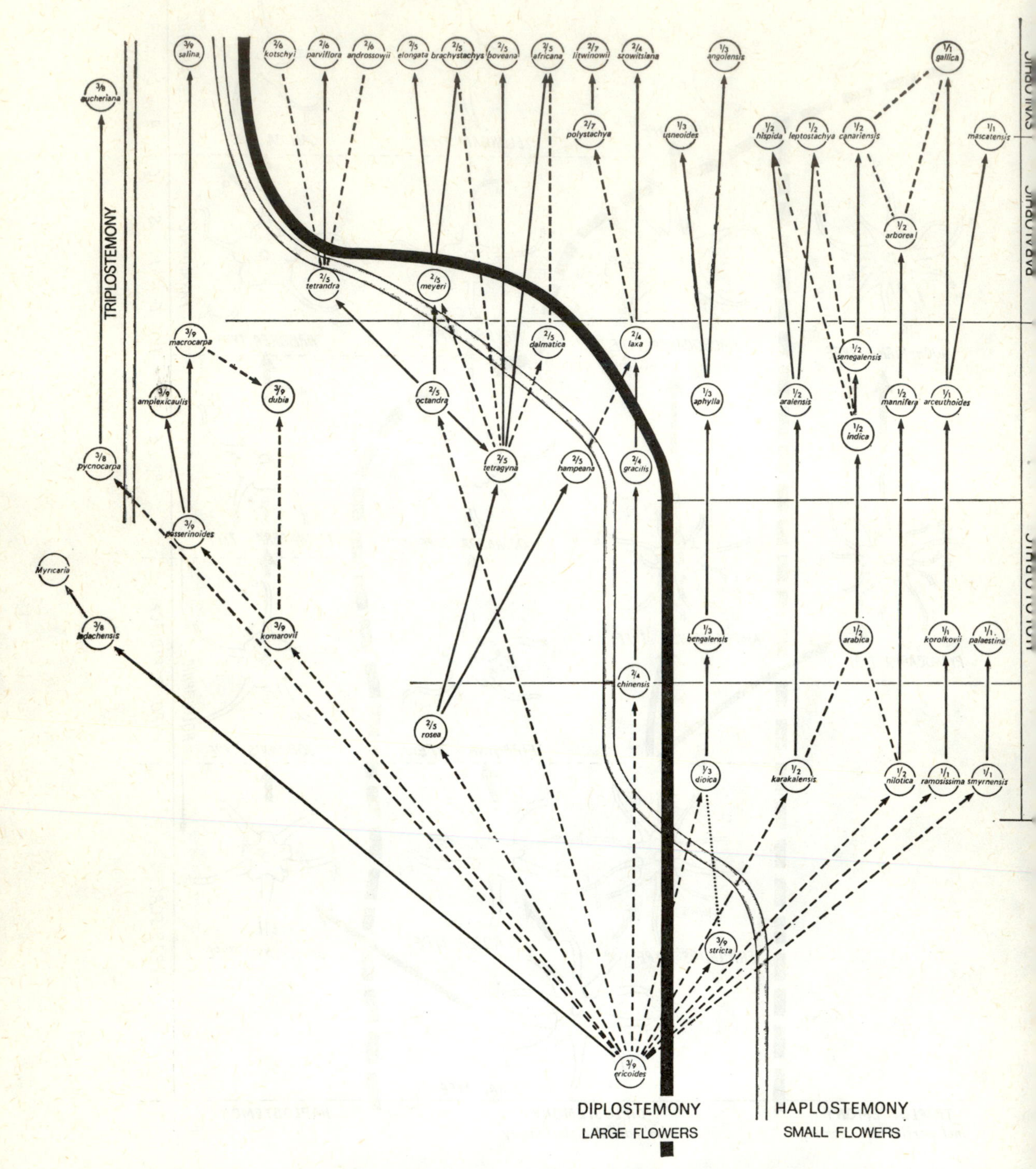

Assumed phylogenetic relationship between the species of *Tamarix*. In each case, such as 3/9 (*ericoides*), the section numbers is to the left and the series number to the right of the slanted line

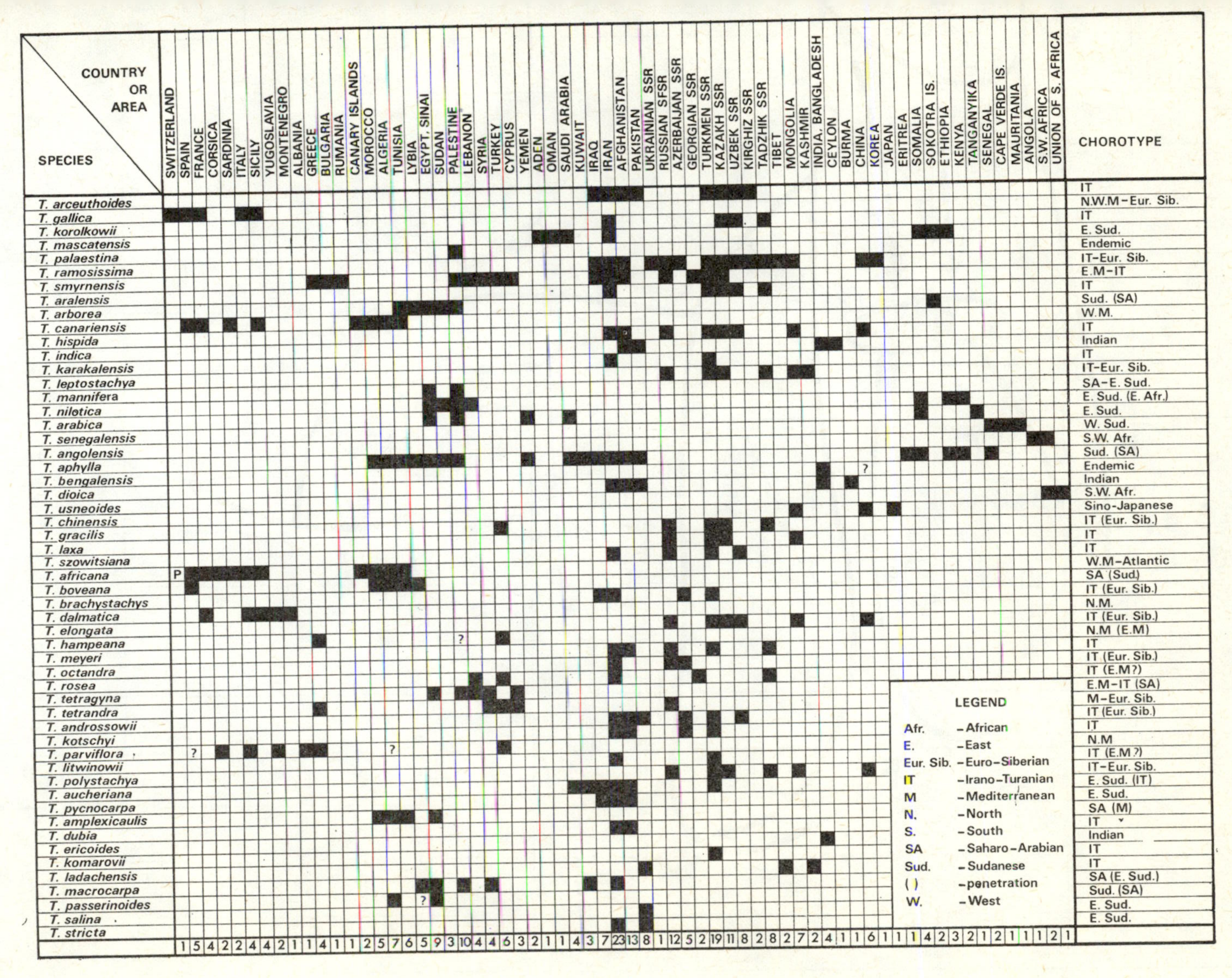

Summary of the distribution of the species of *Tamarix*

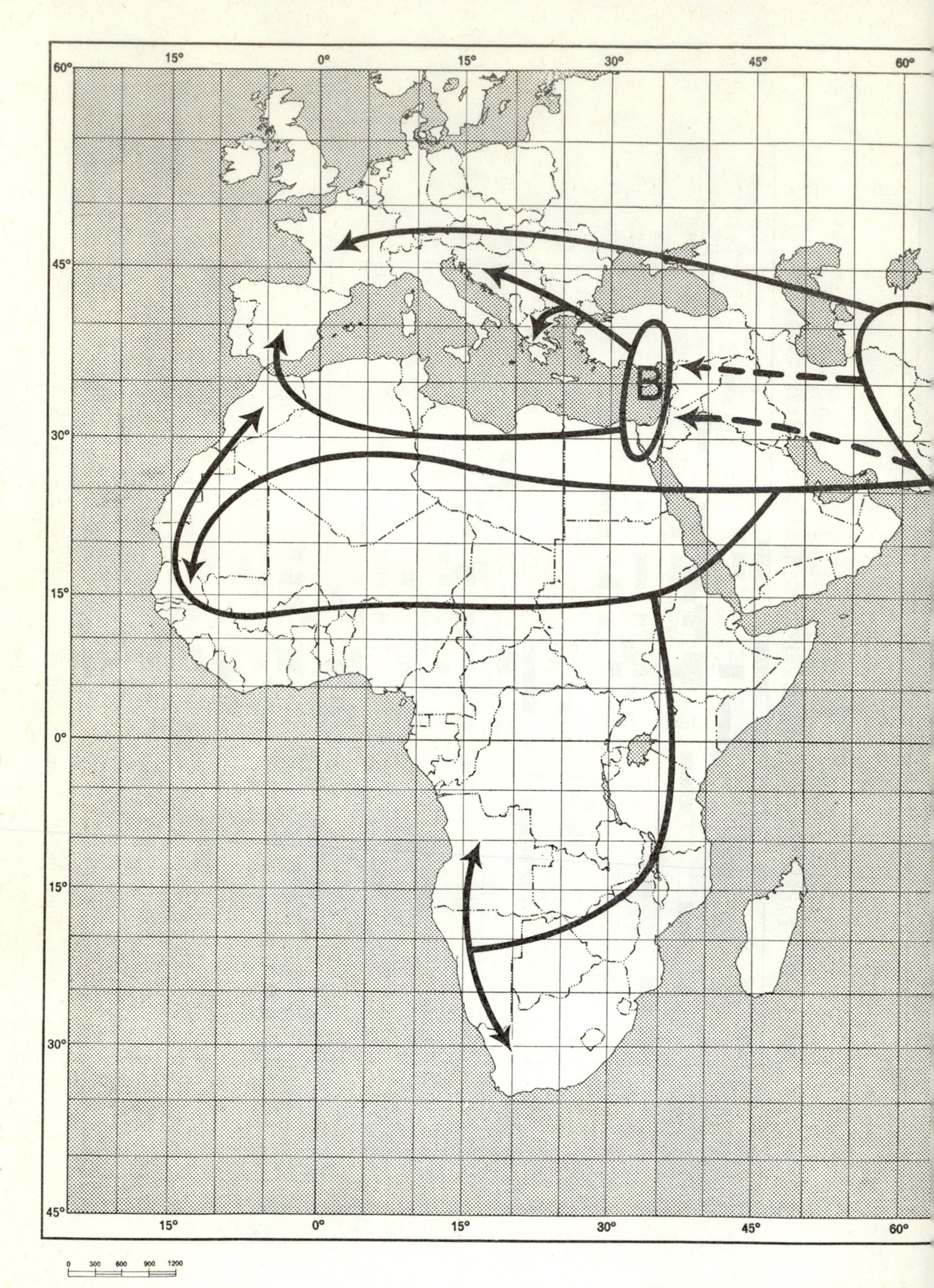

Diagrammatic representation of t
A denotes the Indo-Turanian centr

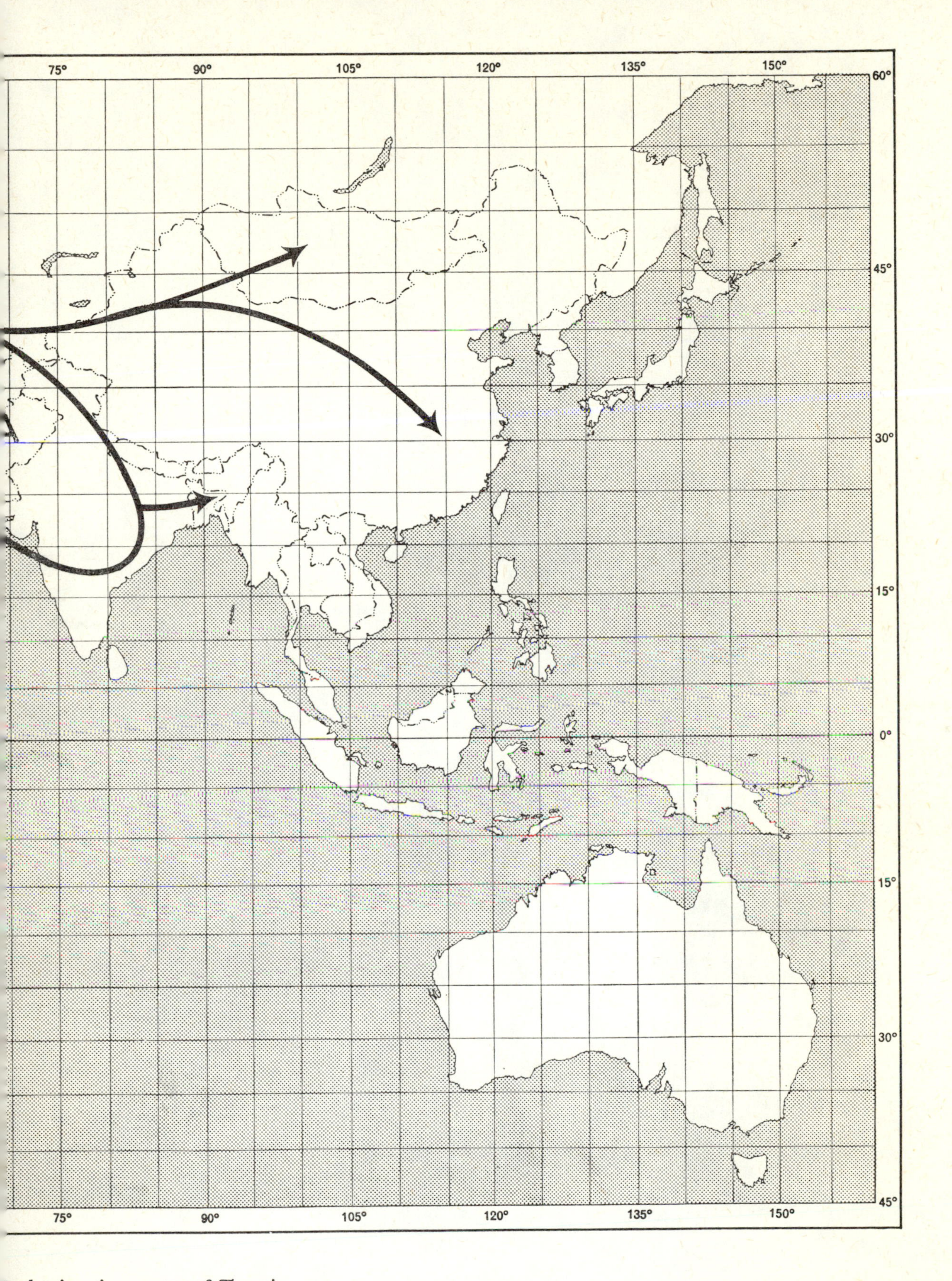

pal migration routes of *Tamarix*
otes the East-Mediterranean centre

BIBLIOGRAPHY

Aitchison J. E. T. (1888) 'The Botany of the Afghan Delimitation Commission', *Trans. Linn. Soc. Bot.*, 3: 41–42.

American Geographical Society (1947) *Map of the World.*, 1:30 000 000 Eq. scale.

Ardoino H. (1862) *Catalogue des plantes vasculaires qui croissent spontanément aux environs de Menton et de Monaco avec l'indication des principales espèces de Nice, Sospel, Vintimille, San Romo, . . .*, 13, Turin.

Arendt G. (1926) 'Beiträge zur Kenntnis der Gattung Tamarix', Inaugural Dissertation, Borna-Leipzig.

Arnott; see Wight & Walker-Arnott.

Ayyar T. V. R. (1934) 'Notes on Indian Thysanoptera with Descriptions of New Species', *Rec. Indian Mus.*, 36: 496.

Bailey L. H. (1937) *Standard Cyclopedia of Horticulture,* III, New York, pp. 3307-3309.

Balachowsky A. (1928) 'Contribution à l'étude des coccides de l'Afrique mineure', *Bull. Soc. Hist. Nat. Afr. N.*, 19:122.

— (1930) 'Addition à la faune Nord-Africaine avec description de trois espèces nouvelles', *ibid.*, 21:119–125.

— (1933) 'Contribution à l'étude des Coccides de France—(14e note) Nouvelles recherches sur la faune indigène de la Corse', *A. Soc. Ent. Fr.*, 102: 35–40.

Ball J. (1873) 'Descriptions of Some New Species, Subspecies and Varieties of Plants Collected in Morocco by J. D. Hooker, G. Maw and J. Ball', *J. Bot.*, 12: 301–302.

— (1877) 'Spicilegium Florae Maroccanae', *J. Linn. Soc. Bot.*, 16: 372.

Barbey C. & W. (1882) 'Planches 63 et 64 de la Flore d'Égypte par M. Delile in *Herborisations au Levant, Egypte, Syrie et Méditerranée*', Lausanne, pp. 175–176, Pl. VIII.

Battaglia E. (1941) 'Contributo all' embriologia dello Tamaricaceae', *Nuovo G. Bot. Ital.*, 48: 575–612.

— (1942) 'Alcune osservazioni del gametofito femmineo della Myricaria germanica Desv.', *ibid.*, 49: 464–466.

Battandier J. A. (1907) 'Revision des Tamarix algériens et description de deux espèces nouvelles', *Bull. Soc. Bot. Fr.*, 54: 232-257.

Battandier J. A. & L. Trabut (1889) *Flore de l'Algérie,* Dicotyledones, I (2), Algiers-Paris, pp. 321–323.

Baum B. (1964) 'On the Vernales-Aestivales Character in Tamarix and its Diagnostic Value', *Israel J. Bot.*, 13: 30–35.

Bentham G. & J. D. Hooker (1862) *Genera Plantarum,* I, London.

Bergevin E. de (1932) 'Description d'une nouvelle espèce de Psyllidae Aphalarinae (Mission du Hoggar)', *Bull. Soc. Hist. Nat. Afr. N.*, 23: 8–10.

— (1932) 'Liste des Hémiptères Oasi di Giarabub', *Annalli Mus. Civ. Storia Nat. Giacomo Doria,* 55: 32, 35.

Bernard F. (1932) 'Remarques, sur le comportement du Platygonatopus polychromus Marsh', *Bull. Soc. Ent. Fr.*, 37: 71–73.

Bertoloni A. (1853) *Miscellanea Botanica*, 14: 13–19, or *Atti Accad. Sci. Ist. Bologna Memorie*, 4: 423–424.

Billerbeck J. (1824) *Flora Classica*, Leipzig, p. 84.

Blanche I. & C. Gaillardot (1854) *Catalogue de l'Herbier de Syrie*, Paris, p. 10.

Blatter E., F. Hallberg & C. McCann, 'Contributions towards a Flora of Baluchistan, from Materials Supplied by Col. J. E. B. Hotson, I.H.R.O.', *J. Indian Bot.*, 1: 86.

Bodenheimer F. S. (1929); see Warburg, pp. 45–88.

Boissier E. (1849) *Diagnoses Plantarum Orientalium Novarum*, I, 10, Paris, pp. 8–10.

— (1856) *Diagnoses Plantarum Novarum Praesertim Orientalium Nonnullis Europaeis Boreali-Africanisque Additis*, II, 2, Leipzig.

— (1867) *Flora Orientalis*, I, Basel-Geneva, pp. 763–769.

Bongard G. H. & C. A. Meyer (1841) 'Verzeichniss der im Jahre 1838 am Saisang-Nor und am Irtysch Gesammelten Pflanzen—ein zweites Supplement zur Flora Altaica', *Mém. Acad. St. Pétersb.*, IV, 6: 190.

Bonstedt C. (1931) *Pareys Blumengartnerei, Beschreibung, Kultur und Verwendung der Gesamten Gartnerischen Schmuckpflanzen*, I, Berlin, p. 915.

Boulger G. S. (1897) 'Roxburgh Collection', in: *Dictionary of National Biography* (ed. L. Sidney), XLIX, pp. 368–370.

Bourdot H. & A. Galzin (1927) *Hyménomycètes de France—Hétérobasidies, Homobasidies, Gymnocarpes*, Paris, p. 304.

Boureau E. (1951) 'Sur l'anatomie comparée et les affinités d'échantillons fossiles de Tamaricaceae, découverts en Somalie française et en Mauritanie', *Bull. Mus. Hist. Nat., Paris*, 23(4):462–469.

Bowden, Wray M. (1945) 'A List of Chromosome Numbers in Higher Plants, II: Menispermaceae to Verbenaceae', *Am. J. Bot.*, 32: 195.

British Museum (1904) *History of the Collections Contained in the Natural History Department of the British Museum*, I, London, pp. 79–193.

British Museum, Natural History Department, *Catalogue of Library*, London.

Britten J. (1885) 'The Forster Herbarium', *J. Bot.*, 23: 362.

— (1885) *Arnott, George Arnott Walker*, in: *Dictionary of National Biography* (ed. L. Stephen), II, pp. 120–121.

Brown J. G. & T.G. Scandone (1953) 'A Botanical Disease of Tamarisk', *Plant Dis. Reporter*, 37: 524–525.

Brown W. R. (1919) 'The frash (Tamarix articulata)', *Agric. J. India*, 14: 758–761.

Brunner C. (1909) 'Beiträge zur vergleichenden Anatomie der Tamariceen', *Jb. Wiss. Anat. Hamb.*,: 89–162.

Brunswick H. (1920) 'Über das Vorkommen von Gipskrystallen bei den Tamaricaceae', *Akad. Wien. StizBer.*, 129 (2, 3): 115–136.

Buchanan Hamilton; see Royle, Wallich.

Buchenau Fr. (1868) *Bot. Ztg.*, 8: 5.

— (1878) 'Miscellen. Beachenswerthe Fälle von Fasciation, 3: Beiträge zu der von M. Masters (Vegetable Teratology, p. 20) gegebenen Liste fasciirter Pflanzen', *Abh. Naturw. Ver. Bremen*, 5: 645–648.

Buhse F. (1899) *Die Flora des Alburs und der Kaspichen Südküste*, Riga.

Bunge A. von (1833) 'Enumeratio Plantarum Quas in China Boreali Collegit, . . .', *Mém. Acad. St. Pétersb. Sav. Etr.*, 2: 75–148.

— (1851) 'Alexandri Lehmann. Reliquiae Botanicae. . .', *Mém. Acad. St. Pétersb.*, 7: 290–296.

— (1852) *Tentamen Generis Tamaricum Species Accuratius Definiendi*, Dorpat.

Burtt B. L. & Patricia Lewis (1954) 'On the Flora of Kuweit, III: Tamaricaceae', *Kew Bull.*, 9 : 388–391.

Caballero A. (1911) 'Enumeración de las Plantas Herborizadas en El Rif', *Mem. Soc. Esp. Hist. Nat.*, 8: 256.

Cambessedes J. & J. Decaisne (1843–1844) *Voyage dans l'Inde par Jacquemont pendant les années* 1828 *à* 1832, *Botanique, Plantae Rariores, quas in India orientali*, IV: Paris.

Campbell W. S. (1902) 'The Tamarisk', *Agr. Gaz. N. S. W.*, 13: 825–829.

Candolle A. P. de (1828) *Prodromus Systematis Naturalis Regni Vegetabilis*, III, Paris, pp. 95–98.

Candolle Alph. de (1880) 'Énumération Alphabétique d'Auteurs et Collecteurs avec indication des herbiers', in: *La Phytographie*, Paris, pp. 391–462.

Capra F. (1928–1930) 'Risultati zoologici della Missione inviata dalla R. Societa Italiana per l'esplorazione dell' oasi di Giarabub (1926–1927) — Due nuovi Coccinellidi di Cirenaica', *Annali Mus. Civ. Storia Nat. Giacomo Doria*, 53: 241–242.

Carleton M. A. (1914) 'Adaptation of the Tamarisk for Dry Lands', *Science*, 39: 692–694.

Carrière E. A. (1868) 'Tamarix plumosa', *Revue Hort.*, 40: 358.

Carrisso W. L. (1937) *Conspectus Florae Angolensis, elaborado pelo Instituto Botanico de Coimbra com a colaboraçao do Museu Britanico*, Vol. I, Fasc. I: 'Ranunculaceae-Malvaceae' by A. W. Exell & F. A. Mendonca, Lisbon, p. 117.

Chapman R. E. (1934) 'The Composition of the Salts on Leaves of Some Desert Plants', *Ann. Bot.*, 48: 777.

Chapman V. J. (1960) *Salt Marshes and Salt Deserts of the World*, London-New York.

Chaubard M. & M. Bory de St. Vincent (1838) *Nouvelle Flore du Péloponnèse et des Cyclades*, Paris-Strasbourg, p. 20.

Chiovenda E. (1929) *Flora Somala*, Rome, p. 93.

Chittenden F. J. (1951) *Royal Horticultural Dictionary of Gardening*, Oxford, pp. 2078–2079.

Chow H. F. (1934) 'The Familiar Trees of Hopei', *Peking Nat. Hist. Bull.*, 9: 326–331.

Chrétien P. Bize (1926) *Amat. Papillons*, 3: 4–11, 17–24, 33–38, 49–58.

Christensen C. (1922) 'Index to Pehr Forsskål : Flora Aegyptiaco-Arabica 1775, with a Revision of Herbarium Forsskålii Contained in the Botanical Museum of the University of Copenhagen', *Dansk Bot. Ark.*, 4: 33.

Christensen E. M. (1962) 'The Rate of Naturalization of Tamarix in Utah', *Am. Midl. Nat.*, 68: 51–57.

— (1963) 'Layering in Tamarix Induced by Algae', *ibid.*, 70: 250.

Clokie H. N. (1964) *An Account of the Herbaria of the Department of Botany in the University of Oxford*, Oxford.

Clute W. W. (1924) 'Flowering Habit of Tamarix', *Am. Bot.*, 30: 34.

Cooper C. F. (1963) 'Tamarix as a Potential Noxious Shrub', *J. Austr. Inst. Agric. Sci.*, 29: 178–181.

Corti R. (1942) *Flora e vegetazione del Fazzan e della regione di Gat*, Florence.

Cosson E. (1854) 'Rapport sur un voyage botanique en Algérie, d'Oran à Chott-el-Chergui entrepris en 1852', *Annl Sci. Nat., Bot.* IV, 1: 223, 239.

— (1855) 'Rapport d'un voyage botanique en Algérie, de Philippeville à Biskra et dans les Monts Aurès, entrepris en 1853', *ibid.*, 4: 283.

Couppis T. A. (1956) 'Reclamation of Sand Dunes with Particular Reference to Ayia Erimi Sand Drifts, Cyprus', *Emp. For. Rev.*, 35: 77–84.

Croizat L. (1960) *Principia Botanica*, Caracas.

Debeaux O. (1879) 'Contributions à la flore de la Chine: Florule de Tien-Tsin, prov. de Pe-tche-ly', *Acta Soc. Linn. Bordeaux* IV, 3: 26–105.

Decaisne J. (1835) 'Énumération des plantes recueillies par M. Bové dans les deux Arabies, la Palestine, la Syrie et l'Égypte. Florula Sinaica', *Annl Sci. Nat., Bot.*, II, 3: 260–261.

— (1843) see Cambessedes & Decaisne, IV, pp. 58–60: ('this part was written by Decaisne himself only').

Decker J. P. (1961) 'Salt Secretion by Tamarix pentandra Pall.', *Forest Sci.*, 7: 214–217.

Decker J. P., W.G. Gaylor & F.D. Cole (1962) 'Measuring Transpiration of Undisturbed Tamarisk Shrubs', *Pl. Physiol.*, 37: 393–397.

Decker J. P. & B. F. Wetzel (1957) 'A Method for Measuring Transpiration of Intact Plants under Controlled Light, Humidity, and Temperature', *Forest Sci.*, 3: 350–354.

Deleuze M. (1823) *Histoire et description du Muséum Royal d'Histoire Naturelle*, I, Paris, pp. 319–320.

Delile A. R. (1813) 'Florae Aegyptiacae Illustratio', in: *Description de l'Egypte*, II, Paris, p. 58.

Deogun P. N. (1939) 'Succession in Plantations of Light Crowned Species', *Indian Forester*, 65: 201–204.

Desfontaine R. L. (1798) *Flora Atlantica*, I, Paris, p. 269.

Desvaux M. (1824) 'Sur la nouvelle famille de plantes fondée sur le genre Tamarix', *Annl Sci. Nat., Bot.*, I, 4: 344–350.

Diels L. (1917) 'Beiträge zur Flora der Zentral-Sahara und ihrer Pflanzengeographie', *Bot. Jb.*, 54: *Bibl.* n. 120.

Dieuzeide R. (1931) 'Une Cecidomyie du Tamarix africana Poiret (Ambliardiella tamricum Kieffer) et ses parasites', *Bull. Soc. Hist. Nat. Afr. N.*, 22: 261–270.

Dippel L. (1893) *Handbuch der Laubholzkunde*, III, Berlin, pp. 7–11.

Doney C. F. (1945) 'Shrubs for Special Uses', *Pls and Gds*, 1: 18–51.

Drege J. F. (1843) *Zwei Pflanzengeographische Dokumente, nebst einer Einleitung von E. Meyer*, pp. 61, 63, 92, 94, 225.

— (1847) 'Vergleichungen der von Ecklon und Zeyher und von Drege gesammelten südafrikanischen Pflanzen', *Linnaea*, 19: 659.

Du Rietz G. E. (1930) 'The Fundamental Units of Biological Taxonomy', *Svensk Bot. Tidskr.*, 24: 333–428.

Dyer Thiselton W. T. (1874) 'Tamariscinea' (ed. J. D. Hooker), in: *Flora of British India*, I, London, pp. 248–250.

Ecklon C. F. & C. Zeyher (1837) *Enumeratio Plantarum Africae Australis Extratropicae*, Hamburg, p. 330.

Edgeworth M. P. (1862) 'Florula Mallica', *J. Linn. Soc. Bot.*, 6: 188.

Ehrenberg C. G. (1872) 'Über die Manna-Tamariske nebst allgemeinen Bemerkungen über die Tamariscineen', *Linnaea*, 2: 241–344.

Eichwald K. E. von (1831) *Plantarum novarum vel minus cognitarum, quas in itineri Caspico-Caucasico observavit*, Fasc. 1, Vilna–Leipzig, p. 12; Pl. 8.

Endlicher S. (1841) *Enchiridion Botanicum Exhibens Classes et Ordines Plantarum Accedit Nomenclator Generum et Officinalium vel Usalium Indicatio*, Leipzig–Vienna, p. 544.

Engler A. & O. Drude (1921) *Die Vegetation der Erde*, IX: 'Die Pflanzewelt Afrikas . . . 3 (2) Characterpflanzen Afrikas . . .', Leipzig, pp. 531, 850.

Erdtman G. (1952) *Pollen Morphology and Plant Taxonomy — Angiosperms*, Stockholm.

Exell A. W. & F. A. Mendonca; see Carrisso.

F., H. (1928) 'Dünenaufforstung in Palästina', *Forstwiss. Centralbl.*, 50: 135–136.

Fahn A. (1958) 'Xylem Structure and Annual Rhythm of Development in Trees and Shrubs of the Desert, I: T. aphylla, T. jordanis var. negevensis, T. gallica var. marismortui', *Trop. Woods*, 109: 81–94.

Falk J. P. (1786) *Beiträge zur topographischen Kenntniss des Russischen Reichs* (revised by J. G. Georgi), II, St. Petersburg, pp. 151–152.

Fedde F. (1906) 'Species Novae ex Schedae ad Herbarium Florae Rossicae a Museo Botanico Academiae Imperialis Scientarum Petropolitanae Editum', *Reprium Nov. Spec. Regni Veg.*, 3 : 390.

— (1907), *ibid.*, 4: 8.

— (1909), *ibid.*, 6: 104.

— (1911), *ibid.*, 9: 554.

— (1913), *ibid.*, 13: 27.

Fedtschenko O. & B. Fedtschenko (1911) 'Conspectus Florae Turkestanicae', *Beih. Bot. Zbl.*, 28: 8–13.

Fedtschenko B. A. (1922) 'De Generis Tamaricis Species Nova Annua', *Notul. Syst. Inst. Cryptog. Horti Bot. Petropol.*, 3: 44–48, 182–184.

Fletcher T. B. (1932) 'Life Histories of Indian Microlepidoptera (Ser. 2), Alucitidae (Pterophoridae), Tortricinae and Gelechiadae', *Imp. Counc. Agric. Res. (Calcutta) Sci. Monogr.*, 2: 1–58.

— (1933) 'Life Histories of Indian Microlepidoptera (Ser. 2), Cosmopterygidae to Neopseustidae', *Imp. Counc. Agric. Res. (Delhi) Sci. Monogr.*, 4: 1–85.

Fodor A. & R. Cohn (1929) 'Notiz über die chemische Zusammensetzung des Tamariskenmannas', see Warburg, p. 89.

Forskål P. (1775) *Flora Aegyptiaco-Arabica* (ed. C. Niebuhr), Copenhagen, pp. 206–207.

Franchet A. (1883) *Mission Capus. Plantes du Turkestan*, Paris.

French C. (1933) 'New Records of Plants Attacked by Native Insects, 10: The Elephant-Beetle of the Orange (Orthorrhinus cylindrostris Fabr.)', *Victorian Nat.*, 50: 190.

Freyn J. (1903) 'Plantae ex Asia Media', *Bull. Herb. Boissier* II, 3: 1058-1063.

Gaillardot C.; see Blanche & Gaillardot.

Gandoger M. (1918) 'Sertum Plantarum Novarum,' Part I, *Bull. Soc. Bot. Fr.*, 65: 27.

— (1910) *Novus Conspectus Florae Europae*, Paris–Leipzig, pp. 190–191.

Gary H. L. (1960) 'Utilization of Five-Stamen Tamarisk by Cattle', *Rocky Mtn Forest Range Exp. Stn, Research Notes*, No. 51.

— (1963) 'Root Distribution of Five-Stamen Tamarisk, Seep-Willow, and Arrow Weed', *Forest Sci.*, 9: 311–314.

Gay J. (1852–1853) 'Tamariscineae. Études sur le genre Tamarix', MS. No. 22 in Kew Library, London, unpublished (*c.* 1852–1853).

— (1856); see Blanche & Gaillardot.

Georgi J. G. (1800) *Geographische, physikalische und naturhistorische Beschreibung des russischen Reiches*, Konigsberg, IV (5), pp. 611–1461.

Gimingham C. H. (1955) 'A Note on Water Table, Sand Movement and Plant Distribution in a North African Oasis', *J. Ecol.*, 43: 22–24.

Glading B., R. W. Enderlin & H. A. Hjersman (1945) 'The Kettleman Hills Quail Project', *Calif. Fish Game*, 31: 139–156.

Gomes B. A. (1868) 'Elogio historico do Padre João de Loureiro', *Mems R. Acad. Sci. Lisb.* (N.S.), 4(1): 25–31.

Gorschkova S. G. (1936) 'Novyie Vidi Roda Tamarix L.', *Sov. Bot.*, No. 4: 117–118 (French summary).

— (1937) 'Conspectus Specierum Generis Tamarix L. in USSR Crescentium', *Notul. Syst. Inst. Cryptog. Horti Bot. Petropol.*, 7: 75–98.

— (1949) 'Tamaricaceae', in: *Flora URSS* (ed. V. L. Komarov), XV, Moscow–Leningrad, pp. 290–327.

Grénier M. & M. Godron (1848) '*Flore de France*', I, Paris–Besançon, pp. 600–601.

Guillemin J. -A., S. Perrottet & A. Richard (1830–1833) *Florae Senegambiae Tentamen*, Paris.

Gupta A. C. (1952) 'Morphological and Anatomical Studies on Tamarix odessarioica', Ph. D. Thesis, Agra University.

Gutmann H. (1947) 'The Genus Tamarix in Palestine', *Pal. Jour. Bot. Jerusalem*, 9: 46–54.

Guttenberg H. von (1926) *Handbuch der Pflanzenanatomie*, V: 'Die Bewegungsgewebe', Berlin.

Hallier H. (1921) 'Zur morphologischen Deutung der Diskusgebilde in der Dikotylenblüthe', *Meded. Bot. Mus. Utrecht*, 16 (41): 1–14.

Hampe G. E. L. (1842) 'Correspondenz' in No. 4, 28.1.1842, *Flora* (Regensburg), 25: 62.

Handel- Mazzetti (1912) 'Pteridophyta und Anthophyta aus Mesopotamien und Kurdistan sowie Syrien und Prinkipo', *Annln Naturh. Mus. Wien*, 26: 57.

— (1913) *ibid.*, 27:17.

Harold St. J. (1958) *Nomenclature of Plants*, New York.

Hartmann R. (1884) 'Die Nilländer', *Wissen der Gegenwart*, 24: 215.

Hartwich C. (1883) 'Uebersicht der technisch und pharmaceutisch verwendeten Gallen', *Arch. Pharm., Berl.*, 21: 820–872.

Harvey W. H. (1894) 'Tamariscineae Desv.', in: *Flora Capensis* (eds. W. H. Harvey & O. W. Sonder), I, Ashford, Kent, pp. 119–120.

Hawkins P. J. (1958) 'What Trees to Plant on the Downs', *Qd. Agric. J.*, 84: 368–374.

Hefley H. M. (1937) 'The Relations of Some Native Insects to Introduced Food Plants', *J. Anim. Ecol.*, 6: 138–144.

Heldreich Th. (1877) 'Die Pflanzen der attischen Ebene', in: *Griechischen Jahreszeiten A. Mommsen*, V, Schleswig, pp. 471–597.

Hemsley A. L. S. (1888) 'Enumeration of all the Plants Known from China Proper, Formosa, Hainan, the Corea, the Luchu Archipelago and the Island of Hongkong', *J. Linn. Soc. Bot.*, 23: 346.

Hemsley W. B. (1875) 'An Outline of the Flora of Sussex,' App. to *J. Bot.*

Henderson G. (1873) 'Lahore to Yarkand . . . under T. D. Forsyth', *J. Bot.*, 11: 217–218.

Henriques J. A. (1900) 'Contribuiçao para a flora africana', *Bolm Soc. Broteriana*, 17: 83.

Herriot R. (1942) 'The Athel Tree, an Evergreen Tamarisk', *S. Aust. Dep. Agr. J.*, 46: 58–59.

Heyer G. (1876) 'Die Forsten von Lower Sind', *Allgemeine Forstund Jagd-Zeitung*, 13–16.

Hiekisck Carl (1884) 'Prschewalskys dritte Reise nach Centralasien', *Ausland*, 221–226, 245–249, 264–268.

Hiern W. P. (1896) *Catalogue of the African Plants Collected by Dr. Friedrich Welwitsch in 1851–61*, I; p. 55.

Hohenacker R. F. (1838) 'Enumeratio Plantarum quas in Itinere per Provinciam Talysch Collegit R. Fr. Hohenacker', *Bull. Mosk. Soc. Nat.*, 6: 363.

Hole R. S. (1919) 'A New Species of Tamarix', *Indian Forester*, 45: 247–249.

Hooker J. D.; see Dyer.

Hooker J. D., B. D. Jackson, and their successors (1885) *Index Kewensis: An Enumeration of the Genera and Species of Flowering Plants*, I–II, Oxford, 959.

Hooper D. (1909) 'Tamarisk Manna', *J. Asiat. Soc. Beng.*, 5: 31–36.

Horton J. S. (1957) 'Inflorescence Development in Tamarix pentandra Pallas', *Swestern Naturalist*, 2: 135–139.

Horton J. S., Mounts F. C. & J. M. Kraft (1960) 'Seed Germination and Seedling Establishment of Phreatophyte Species, *Rocky Mountain Forest and Range Experiment Station, Forest Service* (*USDA*) Station Paper No. 48.

Hunt D. R. (1963) 'Typification of Thuja aphylla L.', *Kew Bull.*, 12: 481–482.

Jahandiez E. & R. Maire (1932) *Catalogue des Plantes du Maroc* (*Spermatophytes et Ptéridophytes*), II, Algiers, pp. 487–489.

Johri B. M. & D. Kak (1954) 'The Embryology of Tamarix L.', *Phytomorphology*, 4: 230–247.

Joshi A. C. & L. B. Kajale (1936) 'A Note on the Structure and Development of the Embryo-Sac, Ovule, and Fruit of Tamarix dioica Roxb.', *Am. Bot.*, 50: 421.

Juel H. O. (1918) 'Bemerkungen über Hasselquist's Herbarium', *Svenska Linnèsållsk. Arsskr.*, I: 95–125.

Just L. (1874–1890) *Botanischer Jahresbericht — Systematisch geordnetes Repertorium der Botanischen Literatur aller Länder*, I–XVI, Berlin.

Karelin G. (1839) 'Enumeratio Plantarum quas in Turcomania et Persia Boreali', *Bull. Mosk. Soc. Nat.*, 12: 154–155.

Karelin G. & J. Kirilow (1841) 'Enumeratio Plantarum Anno 1840 in Regionibus Altaicis et Confinibus Collectarum', *ibid.*, 14: 423–424.

— (1842) 'Enumeratio Plantarum in Desertis Songoriae Orientalis et in Jugo Summarum Alpium Alatau Anno 1841 Collectarum', *Cais. Soc. Nat. Scrutat. Moscow*, 15: 353–354.

Karsten H. (1880–1883) *Deutsche Flora — Pharmaceutisch-medicinische Botanik, Berlin*, p. 641.

Kassas M. & M. Iman (1954) 'Habitat and Plant Communities in the Egyptian Desert, III: The Wadi Bed Ecosystem', *J. Ecol.*, 42: 424–441.

Kew (1894) *Hand-List of Trees and Shrubs Grown in Arboretum*, London, p. 35.

— (1899) 'Catalogue of the Library of the Royal Botanic Gardens', *Kew Bull. Add. Ser.*, 3.

— (1919) 'Catalogue', *op. cit.*, *Kew Bull.*, *Suppl.*

Kirpicznikov M. (1962) 'Index Auctorum ad Taxa Adductorum', *Notul. Syst. Inst. Cryptog. Horti Bot. Petropol.*, Indexes and Bibl. to Vols. I–XX, pp. 249–271.

Klebahn H. (1884) 'Die Rindenporen', *Jenäische Z. Naturw.*, 10: 537–592 (includes anatomy of the lenticels of *Tamarix gallica*).

Koch C. (1841) 'Catalogus Plantarum quas in Itinere per Caucasum, Armeniamque Annis 1836 et 1837', *Linnaea*, 15: 706.

Koch K. H. E. (1869) *Dendrologie*, 1: 452–458.

Komarov V. L. (1896) 'Materiali po Flor Turkestankavo Nagaria', *Trav. Soc. Imp. Nat. St. Pétersb.*, 26: 142.

— (1949) *Flora URSS*, Moscow–Leningrad; see Gorschkova, 1949.

Korshinsky S. (1882) 'Umriss der Flora der Umgebung von Astrachan', *Arbeit. Naturf.-Ges. Univ. Kazan*, 10: 1–63.

Krausel R. (1939) 'Ergebnisse der Forschungstage Prof. E. Stromers in der Wüste Ägyptens, *Abh. Bayer. Akad. Wiss.*, 47: 97.

Krupenikov I. (1947) 'Tamariksi ego soleustoichivost' ('Tamarix and its Salt Resistance'), *Priroda*, 40 (7): 65–66.

Krüssmann G. (1951) *Die Laubgehölze. Alphabetisches Verzeichniss nebst Beschreibung und Bewertung der in Deutschland winterharten Laubgehölze. Eine Dendrologie für die Praxis*, (2nd ed.), Berlin, p. 366.

Kudriashev S. (1932) 'On Plants of the Central Part of the Gissarskal Mountain Range which Produce Tannin Substances' (in Russian, with German summary), *Acta Univ. Asiae Mediae Bot.*, 3: 31–38.

Kuntze O. (1887) 'Plantae Orientali-Rossicae', *Acta Horti Petrop.*, 10 (1): 175.

— (1891–1893) *Revisio Generum Plantarum Vascularum*, I-III, Würzburg.

Kurz S. (1877) *Forest Flora of British Burma*, I: 'Ranunculaceae to Cornaceae', Calcutta.

Kusnezew N., N. Busch & A. Fomin (1909) *Flora Caucasica Critica*, 3 (9): 85–111.

Lam H. J. (1959) 'Taxonomy. General Principles and Angiosperms', in: *Vistas in Botany* (ed. W. B. Turrill), London–New York–Paris–Los Angeles, pp. 3–75.

Lambert A. B. (1811) 'Some Accounts of the Herbarium of Professor Pallas', *Trans. Linn. Soc.*, 10: 256–265.

Lavalle Alph. (1877) *Arboretum Segrezianum*, Paris.

Lecomte M. H. (1907) *Flore générale de l'Indochine*, I, Paris, pp. 278–280.

Ledebour (1833); see Eichwald.

Ledebour C. F., C. A. Meyer & A. Bunge (1829) *Flora Atlantica*, I, Berlin, pp. 421–426.

— (1831), *ibid.*, III, pp. 224–225.

— (1831) *Icones Plantarum Novarum*, III, Riga, Pls. 253–256.

— (1843) *Flora Rossica*, II, pp. 132–136 (prob. Sept).

Lemoine (1893), *Gdnrs' Chron., III.*, 13:414.

Leontiev V. L. (1952) 'Plants Suitable for the Stabilization of Banks and Dams of the Main Turkman Canal', *Bot. Zh. SSSR*, 37: 434–441 (in Russian).

Lewin M. & A. Reibenbach (1957) 'The Chemical Composition and Fibre Properties of T. articulata from the Negev', *Bull. Res. Coun. Israel*, 6A: 256–270.

Link H. F. (1821) *Enumeratio Plantarum Horti Regii Botanici Berolinensis Altera*, I, Berlin, p. 291.

Linné C. von (1753) *Species Plantarum*, I, pp. 270–271.

— (1755) *Amoenitates Academicae*, p. 32, n. 96 (Abr. D. Juslenius. W.G. 5).

Lipski W. (1892) 'De la mer Caspienne jusqu'à la mer Noire', *Zap. Kiev Obshch. Estest.*, 12: 352.

Litwak M. (1957) 'The Influence of Tamarix aphylla on Soil Composition in the Northern Negev of Israel', *Bull. Res. Coun. Israel*, 6D: 38–45.

Litwinow D. (1905) 'T. androssowi Litw.; T. laxa var. parviflora Litw.; T. karelini var. hirta Litw.', *Sched. Herb. Fl. Ross.*, 5: 41, 79.

— (1906, 1907, 1909, 1911, 1913); see Fedde.

— (1907) 'Plantae Turcomaniae (Transkaspiae)', *Trav. Mus. Bot. Acad. Sci. Russ.*, 3: 112.

— (1910) 'Florae Turkestanicae Fragmenta', *ibid.*, 7: 72.

Loureiro J. de (1790) *Flora Cochinchinensis*, I, Berlin, p. 182.

Lyon H. L. (1924) 'The Athel in Hawaii', *Hawaii Plrs' Rec.*, 28: 508–510.

Maheshwari P. (1950) *An Introduction to the Embryology of Angiosperms*, (5th ed.), New York–Toronto–London.

Maire R. (1929) 'Champignons nords-africains nouveaux ou peu connus, IV', *Bull. Soc. Hist. Nat. Afrique N.*, 20: 279.

— (1931) 'Contributions à l'étude de la flora de l'Afrique du Nord', *ibid.*, 22 : 30.

— (1933) *Ibid.*, 24 : 194–232.

— (1933) *Bull. Soc. Sci. Nat. Phys. Maroc*, 13: 263.

— (1934) *Bull. Soc. Hist. Nat. Afr. N.*, 25: 286.

— (1935) *Ibid.*, 26: 184.

(1936) *Ibid.*, 27: 203–233, 241–270.
— (1937) *Ibid.*, 28: 332.
— (1938) *Ibid.*, 29: 403.
— (1939) *Ibid.*, 30: 327.
— (1940) *Ibid.*, 31: 99.
Marloth R. (1887) 'Zur Bedeutung der Salz abscheidenden Drusen der Tamariscineen', *Dt. Bot. Ges.*, 5: 319–324.
Marschall von Bieberstein F.A.F. (1808) *Flora Taurico-Caucasica*, I, Leipzig, pp. 246–247.
Mauritzon J. (1936) 'Zur Embryologie Einiger Parietales — Familien,' *Svensk Bot. Tidskr.*, 30: 86–93.
Maury P. (1888) 'Anatomie comparée de quelques espèces caractéristiques du Sahara algérien', *Ass. Franç. Avanç. Sci. Toulouse*, 2: 604–632.
Mayr E., E. C. Linsley & R. L. Usinger (1953) *Methods and Principles of Systematic Zoology*, New York.
McClintock E. (1951) 'Studies in California Ornamental Plants, 3: Tamarisk', *J. Calif. Hort. Soc.*, 12: 76–83.
(1957) 'Memoria de los trabajos realizados durante la campaña 1955–56', *Anu. Inst. Nac. Invest. Agron. (Madrid)*, 6: 223–298.
Merrill E. D. (1935) 'A Commentary on Loureiro's *Flora Cochichinensis*', *Trans. Am. Phil. Soc.*, II, 24, Part 2.
Messeri A. (1938) 'Studio anatomico-ecologico del legno secondario de alcune piante del Fezzan', *Nuovo G. Bot. Ital.*, 45: 267–356.
Metcalf Z. P. (1954) 'The Construction of Keys', *Syst. Zool.*, 3: 38–45.
Meyer C. A. (1831) *Verzeichniss der Pflanzen*, Leipzig, p. 165.
Meyer E.; see Drege (1843).
Meyrick E. (1936) *Exot. Microlepidopt.*: 580.
— (1878) *Michelia*, 1: 206.
Moldenke H. N. & A. L. Moldenke, (1952) *Plants of the Bible*, Waltham, Mass.
Möller J. (1876) 'Verzeichniss von histologisch analysirten Holzarten'; see Just, IV, p. 394.
Morton J. F. (1926) 'Our Largest Psychid, Oiketicus dendrokomos n. sp. (Lepidoptera, Psychidae)', *Trans. Am. Ent. Soc.*, 52: 1–6.
Mundkur B. B. & S. Ahmad (1946) 'Revisions of and Additions to Indian Fungi, II', *Mycol. Pap. Imp. Mycol. Inst.*, 18: 1–11.
Murty Y. S. (1954) 'Studies in the Order Parietales, IV: Vascular Anatomy of the Flower of Tamaricaceae', *J. Indian Bot. Soc.*, 33: 226–238.
Nair P. K. K. (1962) 'Pollen Grains of Indian Plants, II', *Gdnrs' Bull. Lucknow*, 60: 6–9.
Nakai T. (1938) 'Species of the Genus Tamarix either Indigenous to or Cultivated in Inner Mongolia, North China, Manchuria, Korea, and Japan Proper', *J. Jap. Bot.*, 14: 289–293.
Nevsky V. (1928) 'The Plant-Lice of Middle-Asia, II: Sub-Tribe Aphinida; Sect. Xerophilaphidini', *Acta Univ. Asiae Mediae Zool.*, 3: 3–32.
Niedenzu F. (1895/1896) *De Genera Tamarice—Index Lectionum in Lyceo Regio Hosiano Brunsbergensi par heim*, Braunsberg.
— (1895) 'Tamaricaceae', in: *Pflanzenfamilien* (ed. Engler; 1st ed.), Vol. III, Part 6, Berlin, p. 296.
— (1925), *ibid.* (2nd ed.), II, pp. 282, 289.

Notaris De; see Ardoino.

Nyman C. F. (1879) *Conspectus Florae Europaeae*, Örebro, p. 253.

Oliver D. (1868) *Flora of Tropical Africa*, I: 'Ranunculaceae to Connaraceae', Ashford, Kent, p. 151.

Oliver F. W. (1947) 'Dust Storms in Egypt as Noted in Maryut: A Supplement', *Geogrl J.*, 108: 221–226.

Pallas P. S. (1795) *Tableau physique et topographibue de la Tauride*, St. Petersburg, p. 49.

— (1796) *Physikalisch-topographisches Gemälde von Taurien*, St. Petersburg, pp. 92–124.

— (1796) *Verzeichniss der in Taurien wildwachsenden Pflanzengattungen*, VII, St. Petersburg-Leipzig, pp. 426–438.

— (1797) 'Catalogue des espèces de végétaux spontanés, observés en Tauride', *Nova Acta Acad. Petrop.*, 10: 309.

— (1797) 'T. songarica', *ibid.*, p. 374; Pl. 10; Fig. 4.

— (1797) 'Plantae Novae ex Herbario et Schedis Defuncti Botanici Iohanis Sievers', *ibid.*, p. 376.

— (1788), in: *Flora Rossica*, II, St. Petersburg, p. 72.

Paroli V. (1940) 'Contributo allo studio embriologico delle Tamaricaceae', *Annuar. Bot. Roma*, 22: 1–18.

Pau C. (1906) 'Synopsis Formarum Novarum Hispanicorum cum Synonimis Nonnullis Accedentibus', *Bull. Acad. Geogr. Bot.*, 16: 75.

— (1918) 'Une ligera visita botanica a Tous', *Butl. Inst. Cat. Hist.*, *Nat.*, 18: 160.

— (1922) 'Nueva contribución al estudio de la flora de Grenada', *Memóries Mus. Ciènc. Nat. Barcelona*, *Bot.*, 1: 43.

— (1925) 'Contribución a la flora Española, plantas de Almeria', *ibid.*, 3: 18.

— (1924) 'Plantas del norte de Yerbala (Manuecos)', *Mem. Soc. Esp. Hist. Nat.*, 12: 293.

Pau C. & E. Huguet-del-Villar (1927) 'Novae Species Tamaricis in Hispania Centrali', *Broteria Bot.*, 23: 100, 104, 106, 107, 109.

Paunero E. (1950) 'Catalogo de plantas recogidas par D. Arturo Caballero en Guadelup (Coceras), 1948–1949', *An. Inst. Bot. A. J. Cavanillo*, 10/1: 24.

— (1950) 'Species Novae de A. Caballero', *ibid.*, 75.

P'ei C. (1948) 'Notes on Tamaricaceae of China', *Bot. Bull. Acad. Sin.*, *Shanghai*, 2: 18.

— (1948) 'Flowering plants of North-Western China, I', *ibid.*, p. 96–106.

Perrottet S. (1833); see Guillemin, etc.

Peyerimhoff P. de (1929) 'Les Nanophyes (Col. Curculionidae) du Tamarix aphylla L.', *Bull. Soc. Ent. Fr.*, 1929: 179–185.

— (1931) 'Mission scientifique du Hoggar—Coléoptères', *Mem. Soc. Hist. Nat. Afr. N.*, 2: 173.

Pfeiffer Hans H. (1953) 'Rapid Chromosome Methods', *Taxon*, 2: 86–87.

Pfeiffer L. (1874) *Nomenclator Botanicus*, II, Part 2, Cassel, pp. 1347–1348.

Poiret J. L. M. (1789) *Voyage en Barbarie*, II, Paris, p. 139.

Pritzel G. A. (1871) *Thesaurus Literaturae Botanicae* (new ed.), Milan.

Pujiula J. (1942) 'Contribución al conocimiento anatomico-fisiologico de algunas disposiciones en el reino vegetal', *Bull. Inst. Catal. Hist. Nat.*, 37: 87-97.

Puri V. (1939) 'Studies in the Order Parietales—A Contribution to the Morphology of Tamarix chinensis Lour.', *Beih. Bot. Zbl.*, 59A: 335–349.

Rabenhost L. (1873) *Fungi Europaei Exsiccati—Centuria XVII*', Dresden.

Radde G. (1901) *Die Sammlungen des kaukasischen Museums*, II, Tiflis, p. 68.

Rayss T. (1943) 'Contribution à l'étude des Deutéromycètes de Palestine', *Palestine J. Bot. Jerusalem Ser.*, 3:22–51.
Rees A. (1817) *Cyclopedia or Universal Dictionary of Arts, Sciences and Literature*, XXXV, London, (s. v. Tamarix).
Reese G. (1957) 'Über die Polyploidiespektren in der nordsaharischen Wüstenflora', *Flora*, 144: 598–634.
Regel R. & J. Mlokossewicz (1909); see Kusnezew et al.
Regel E. (1877) 'Descriptiones Plantarum Novarum et Minus Cognitarum', *Acta Horti Petrop.*, 5: 582.
Reich L. (1891) 'Les Tamarix et leurs applications, leur valeur au point de vue de reboisement', *Bull. Soc. Natn. Acclim. Fr.*, 38: 362–368.
Retzius A. J. (1791) *Observationes Botanicae*, VI, Leipzig, p. 27.
Richard A. (1833); see Guillemin et al.
Richardson A. M. (1954) 'Propagating the Athel Tree', *Qd Agric. J.*, 79: 335–337.
Rickett H. W. & F. A. Stafleu (1959) 'Nomina Generica Conservanda et Rejicienda Spermatophytorum', *Taxon*, 8: 213–243, 256–274, 282–314.
— (1960), *ibid.*, 9: 67–86, 111–124, 153–161.
— (1961), *ibid.*, 10: 70–91, 111–121, 132–149, 170–194.
Rivas G. S. (1945) 'Contribución al estudio del Schoenetum nigricantis de Vasconia', *Boln R. Soc. Esp. Hist. Nat.*, 43: 261–273.
Rivas G. S. & I. A. Amor (1945) 'Suelo y sucesion en el Schoenetum nigricantis de Quero Villacanas (Prov. de Toledo)', *An. Inst. Esp. Edafol. Ecol. y Fisiol. Veget.*, 4: 148–184.
Roth A. W. (1821) *Novae Plantarum Species Praesertim Indiae Orientalis ex Collectione Dr. Benjamin Heynii*, Halberstadt, pp. 184–185.
Rottler J. P. (1803) 'Botanische Bemerkungen auf der Hin- und Rückreise von Trankenbar nach Madras . . . mit Anmerkungen von C. L. Willdenow', *Ges. Naturf. Freunde Berlin Neue Schr.*, 4: 214, Pl. 4.
Roxburgh W. (1814) *Hortus Bengalensis*, Serampore, p. 22.
— (1820–1824) *Flora Indica*, I-II, Calcutta–London.
Royal Society of London (1800–1873) *Catalogue of Scientific Papers*, I-VIII, London.
Royle J. F. (1835) *Illustrations of the Botany and Other Branches of the Natural History of the Himalayan Mountains and of the Flora of Cashmere*, Vol. II, Part 6, London, p. 214.
Rusanov F. N. (1949) *Sredniyeaziatskie Tamariksi* ('Tamarisks of Central Asia'), Tashkent.
— (1950) 'Tamarisk Trees within the Shelter-Belts', *Les. Khoz.*, 3: 40.
Saint-Laurent J. de (1932) 'Études sur les caractères anatomiques du pois et du liber secondaire dans les éssences du Sahara et particulièrement du Hoggar', *Bull. Stn Rech. Fo. N. Afr.*, 2: 1–48.
Sale G. N. (1948) 'Note on Sand Dune Fixation in Palestine', *Emp. Fo. Rev.*, 27: 60–61.
Saunders E. R. (1934) 'Comments on Floral Anatomy and its Morphological Interpretation', *New Phytol.*, 33: 127–169.
— (1937–1939) *Floral Morphology*, I-II, Cambridge (especially II, pp. 247–248).
Savage S. (1945) *Catalogue of the Linnaean Herbarium*, London.
Schinz H. (1894) 'Beiträge zur Kenntnis der Afrikanischen Flora (Neue Folge)', *Bull. Herb. Boissier I*, 2: 183–185.
Schnarf K. (1929) 'Embryologie der Angiospermen', in: *Handbuch der Pflanzenanatomie*, Vol. X, Part 2, Berlin.
Schopf J. M. (1960) 'Emphasis on Holotype (?)', *Science*, 131: 1043.

Schumann C. (1900) *Symbolae Physicae*, Berlin, Pls. I, II, T. mannifera; Pl. XXV, T. passerinoides.

Schwarten L. & H. W. Rickett (1958) 'Abbreviations of Titles of Serials Cited by Botanists'. *Bull. Torrey Bot. Club.*, 85: 277–300.

— (1961) *ibid.*, 88: 1–10.

Sennen E. C. (1928) 'Une seconde semaine d'herborisation sur le littoral de Tarragone entre le Francoli et l'Èbre', *Annls Soc. Linn., Lyon*, 73: 9–10.

— (1928) 'Plantes d'Espagne, par le F. Sennen—Diagnoses et commentaires', *Boln Soc. Iber. Ciénc. Nat.*, 27: 66–67.

— (1931) 'Campagne botanique au Maroc', *Bull. Soc. Bot.*, 73: 193.

— (1932) 'Brèves diagnoses des formes nouvelles parues dans nos exsiccata "Plantes d'Espagne — F. Sennen" ', *Bull. Inst. Catal. Hist. Nat.*, 32: 90.

Sennen E. C. & Mauricio (1933) *Catalogo de la flora del Rif Oriental y principalmente de las cabilos limitrofes con Melilla*, Melilla, pp. 42–43, 147.

Sharma Y. M. L. (1939) 'Gametogenesis and Embryology of T. ericoides Rottl.', *Ann. Bot. II.*, 3: 861.

Sickenberger E. (1901) 'Contributions à la flore de l'Égypte', *Mém. Inst. Égypte*, 188–190.

Siebold Ph. F. & J. G. Zuccarini (1840) *Flora Japonica*, Sect. 1, Leipzig, p. 132, Pl. 71.

Simpson G. G. (1961) *Principles of Animal Taxonomy*, New York.

Slepyan E. I. (1962) 'Gally, pochkovye teratozy i ikh vozbuditeli na Tamarix L.' ('Galls and bud teratisms and their pathogens in Tamarix L.') *Byull. Mosk. Obshch. Ispyt. Prir.*, 67: 61–65.

Smith G. E. P. (1941) 'Creosoted Tamarisk Fence Posts and Adaptability of Tamarisk as a Fine Cabinet Wood', *Univ. Ariz. Agric. Exp. Sta. Tech. Bull.*, 92: 219–254.

Smueli E. (1948) 'The Water Balance of Some Plants of the Dead Sea Salines', *Pal. J. Bot. Jerusalem*, 4: 117–143.

Spach E. (1836) *Histoire naturelle des végétaux—Phanérogames*, V, Paris, pp. 481–483.

Stansfield (1953) *N. Western Nat.*, 24: 250–265.

Stearn W. T. (1957) 'Introduction' to *Carl Linnaeus' Species Plantarum* (facsimile of the 1st ed.), London.

Stebbins G. L. (1951) *Variation and Evolution in Plants*, New York.

Steckhan H. (1943) 'Beobachtungen, über Klima und Pflanzenleben der Krim im Jahre 1941-1942', *Biologe*, 12: 11–20.

Steenis-Kruseman M. J. & W. T. Stearn (1954) 'Dates of Publication', *Flora Malesiana*, 4: clxiii–ccxix.

Steudel E. G. (1840–1841) *Nomenclator Botanicus* (2nd ed.), Stuttgart-Tübingen, pp. 661–662.

Strasburger E. (1923) *Das botanische Prakticum*, Jena, p. 783.

Sukhowkov K. T. (1929) 'Tannin Content in Some Plants of the Lower Volga Region', *Jour. Exp. Landw. Südosten Eur. Russlands*, 7: 89–97. (in Russian with German summary).

Sydow H. & S. Ahmad (1939) 'Fungi Panjabenses', *Annls Mycol.*, 37: 439–447.

Tchihatcheff P. (1860) *Asie Mineure*, Part 3, Paris, pp. 254–256.

Teesdale J. (1897) 'Manna', *Nature*, 55: 349.

Teesdale M. J. (1897) 'The Manna of the Israelites', *Sci. Gossip, New Ser.*, 3: 229–232.

Teng S. C. (1947) 'The Forest Regions of Kansei and Their Ecological Aspects', *Bot. Bull. Acad. Sin., Shanghai*, 1: 187–200.

Théry A. (1931) 'Notes d'entomologie marocaine et Nord Africaine, 9', *Bull. Soc. Sci. Nat. Phys. Maroc*, 11: 142.

Thomas H. H. (1921) 'Some Observations on Plants of the Libyan Desert', *J. Ecol.*, 9: 75–89.
Thümen F. (1877) 'Fungi Nonnulli Novi Austriaci', *Österr. Bot. Z.*, 12.
Thunberg C. P. (1784) *Flora Japonica*, Leipzig, p. 126.
Timothy B. (1897) 'The origin of Manna', *Nature* 55: 440.
Tournefort J. P. (1719) *Institutiones res Herbariae* (3rd ed.), Paris.
Trabut L. (1926) 'Tamarix articulata Vahl'. *Bull. Stn Rech. For. N. Afr.*, 1: 336–249.
Trautvetter E. R. (1866) 'Enumeratio Plantarum Songoricarum a Dr. Alex Schrenk Annis 1840–1843 Collectarum', *Byull. Mosk. Obshch. Ispyt. Prir. Biol.*, 39: 310–311.
— (1873) 'Plantae a Capit. Maloma Annis 1870 et 1871 in Turcomania Collectae', *Acta Horti Petrop.*, 1: 273.
— (1874) 'Enumeratio Plantarum Anno 1871 a Dr. G. Radde in Armenia Rossica et Turciae Districtu Kars Lectarum', *ibid.*, 2: 533.
Turrill (1932) 'Tamarix hampeana Boiss. & Heldr. var. aegaea Turrill', *Hooker's Icon. Pl.*, 32: Pl. 3153.
Urban I. (1917) 'Geschichte des Königlichen Botanischen Museums zu Berlin-Dahlem (1815–1913) nebst Aufzählung seiner Sammlungen', *Beih. Bot. Zb.*, 34: 1–457.
Vahl M. (1791) *Symbolae botanicae*, II, Copenhagen, p. 48, Pl. 32.
Vierhapper F. (1907) 'Beiträge zur Kenntniss der Flora Südarabiens und der Inseln Sokotra, Samba und Abd-el-Kuri', *Denkschr. Akad. Wien Math. Naturw.*, 71: 390–392.
Vilbouchevitch J. (1890) 'Les Tamarix et leurs applications, leur valeur au point de vue de reboisement', *Bull. Soc. Natn Acclim. Fr.*, 37: 349-856, 906–915.
— (1892) 'Le tamarix articule', *ibid.*, 39: 186–189.
Villar M. (1948) 'Diagnose d'une variété nouvelle du Tamarix africana: Tamarix africana Poir. var. rungsii H. V.', *Bull. Soc. Sci. Nat. Phys. Maroc*, 28: 36–37 (pub. 1950).
Vogl A. (1875) 'Ueber Tamarisken-Gallen', *Lotos*, 25: 133–136.
Volkens G. (1887) 'Zu Marloth's Aufsatz "Ueber die Bedeutung der Salz abscheidenden Drüsen der Tamariscineen" ', *Ber. Dt. Bot. Ges.*, 5: 434–436.
— (1887) *Die Flora der Aegyptisch-Arabischen Wüste auf Grundlage Anatomisch-Physiologischer Forschungen dargestellt*, Berlin, pp. 107–108.
Wachtl Fr. A. (1886) 'Ueber Gallmücken', *Wien Ent. Ztg.*, 5: 308–310.
Waisel Y. (1960) 'Ecological Studies on Tamarix aphylla (L.) Karst', *Phyton*, 15: 7–17, 19–28. 1960.
— (1961) *Pl. Soil*, 13: 356–364.
Wallich N. (1831) *Numerical List of Dried Specimens of Plants in the East India Company's Museum*, London, pp. 131–132.
Walpers G. G. (1843) *Repertorium Botanices Systematicae*, II, Leipzig, pp. 114–117.
Walter H. & E. Walter (1953) 'Einige allgemeine Ergebnisse unserer Forschungsreise nach Südwestafrika 1952–53', *Ber. Dt. Bot. Ges.*, 66: 228–236.
Warburg O. (1929) 'Bemerkungen über die Tamarisken des Sinai-Gebirges sowie der Sinaihalbinsel', in: *Ergebnisse der Sinai-Expedition 1927 der Hebräischen Universität, Jerusalem*, (eds. F. S. Bodenheimer & O. Theodor), Leipzig, pp. 133–140.
Ware G. H. & W. M .T. Penfound (1949) 'The Vegetation of Lower Levels of the South Canadian River in Central Oklahoma', *Ecology*, 30: 478–484.
Warming E. & M. C. Potter (1934) *Handbook of Systematic Botany* (2nd ed.), London.
Webb P. B. (1841) 'Tamarix gallica of Linnaeus', *Hooker's J. Bot.*, 3: 422–431, Tab. 25.
Webb P. B. & S. Berthélot (1840) *Histoire naturelle des Iles Canaries*, Vol. III, Part 2, 'Phytographia canariensis', Sect. 1, Paris, p. 172.

Wight R. & G. A. Walker-Arnott (1834) *Prodromus Florae Peninsulae Indiae Orientalis*, I, London, p. 40.

Wilgus F. & K. C. Hamilton (1962) 'Germination of Salt-Cedar Seeds', *Weeds*, 10: 332–333.

Wilkinson D. S. (1936) 'Microgasterinae: Notes and New Species (Hym. Brac.)', *Proc. R. Ent. Soc. London Ser. B, Tax.*, 5: 81–88.

Willdenow K. L. (1816) 'Beschreibung der Gattung Tamarix', *Abh. Akad. Berlin Physik.*, 1812–1813: pp. 76–85.

Wilson R. E. (1944) 'Tree Planting and Soil Erosion Control in the Southwest', *J. For.*, 42: 668–673.

Woodbridge M. (1931) 'Protecting of Orchards by use of Windbreaks', *Calif. Citrogr.*, 15: 170–173.

Zabban B. di (1938) 'Osservazioni sulla embriologia di Myricaria germanica Desv.', *Ann. Bot. Roma*, 21: 307–321.

Zeyher C.; see Ecklon.

Zhemehuznikov E. A. (1946) 'On Salt Resistance in Trees and Shrubs', *Dokl. Akad. Nauk SSSR*, 51: 67–71.

Zohary M. (1956) 'The Genus Tamarix in Israel', *Trop. Woods*, 104: 24–60.

Zohary M. & A. Fahn (1952) 'Ecological Studies on East Mediterranean Dune Plants', *Bull. Res. Coun. Israel*, 1: 38–53.

Zohary M. & B. Baum (1965) 'On the Androecium of Tamarix Flower and its Evolutionary Trends', *Israel Jour. Bot.*, 14:

Zwaluwenburg R. H. (1929) 'Tamarix jassid.', *Proc. Hawaii Ent. Soc.*, 7: 224.

Note

Since writing this work, which was originally submitted to the Israel Academy of Sciences and Humanities more than ten years ago, Dr Baum has published the following papers pertinent to the taxonomy of the genus *Tamarix:*

Horowitz A. & B. Baum (1967) 'The Arboreal Pollen Flora of Israel', *Pollen et Spores*, 9: 71–73.

Baum B. R. (1967) 'Introduced and Naturalized Tamarisks in the United States and Canada', *Baileya*, 15: 19–25.

— (1967) 'A New Species of Tamarix from South-Eastern Iran', *Österr. Bot. Zeitschr.*, 114: 379–382.

Baum B. R., I. J. Bassett & C. W. Crompton (1970) 'Pollen Morphology and its Relationships to Taxonomy and Distribution of Tamarix', *Series Vaginantes, Österr. Bot. Zeitschr.*, 118: 182–188.

— (1971) 'Pollen Morphology of Tamarix Species and its Relationship to the Taxonomy of the Genus', *Pollen et Spores*, 13: 495–521.

INDEX

כתבי האקדמיה הלאומית הישראלית למדעים
החטיבה למדעי־הטבע

הסוג אשל

מאת

ב׳ באום

ירושלים תשל״ח